An Introduction to Animal Behaviour

Using its powerful beak, a lorikeet gently preens its mate's feathers; young cheetahs rest together in the shade; fireflies beam light flashes to each other across a darkened landscape and a mongoose deftly bites its prey to death. The study of animal behaviour is about all these things and more. It involves static postures and active movements, all the noises and smells and changes of colour and shape that characterize animal life.

Taking the organization of behaviour within the individual animal as its core, this is a clear, concise and readable foray into the fascinating world of animal behaviour, in terms of Tinbergen's questions of causation, evolution, development and function. It provides lucid accounts of all levels of behaviour, from the nerve cell to the population. The broad biological approach of this new edition makes it an excellent choice for all students of animal behaviour and psychology and their teachers.

ANIMAL

An introduction to

BEHAVIOUR

Fifth Edition

**Aubrey Manning
and Marian Stamp Dawkins**

CAMBRIDGE
UNIVERSITY PRESS

PUBLISHED BY THE PRESS SYNDICATE OF THE UNIVERSITY OF CAMBRIDGE
The Pitt Building, Trumpington Street, Cambridge CB2 IRP, United Kingdom

CAMBRIDGE UNIVERSITY PRESS
The Edinburgh Building, Cambridge CB2 2RU, United Kingdom
40 West 20th Street, New York, NY 10011-4211, USA
10 Stamford Road, Oakleigh, Melbourne 3166, Australia

First, second and third editions first published by Edward Arnold
and © Edward Arnold 1967, 1972, 1979
Fourth edition first published by Cambridge University Press 1992
Fifth edition 1998

Printed in the United Kingdom at the University Press, Cambridge

Typeset in 10/14pt Monotype Plantin Light [SE]

A catalogue record for this book is available from the British Library

Library of Congress Cataloguing in Publication data
Manning, Aubrey.
    An introduction to animal behaviour / Aubrey Manning and Marian
  Stamp Dawkins. – 5th ed.
      p. cm.
  Includes bibliographical references (p.   ) and index.
  ISBN 0 521 57024 7 (hardcover). – ISBN 0 521 57891 4 (pbk.)
  1. Animal behavior.   I. Dawkins, Marian Stamp.   II. Title.
QL751.M22   1998
591.5–dc21   97-46775   CIP

ISBN 0 521 57024 7 hardback
ISBN 0 521 57891 4 paperback

To Niko Tinbergen FRS, Nobel Laureate, 1907–1988

Contents

Contents

Preface

One chapter of Tinbergen's The Study of Instinct (1951) – the first textbook of modern ethology – is called 'An attempt at a synthesis'. This encapsulates what we are trying to achieve in this new edition. The various fields in animal behaviour continue to advance rapidly, new courses and new texts dealing with them are now familiar. As a consequence contact between fields of study becomes tenuous at times. We believe it is of continuing importance to provide an introductory text which does not specialize but keeps some balance and introduces students to the whole range of levels of investigation. Thus we cover aspects of neurobiology, physiology, psychology, and behavioural ecology to augment our unifying themes which derive from Tinbergen's 'four questions for ethology'.

We retain the basic structure of previous editions which is founded on our conviction that the best way to introduce animal behaviour is not along functional lines, but by studying its organization within the individual. Within this framework there has been some considerable updating, rewriting and reorganization of material between chapters, all carried out with the aim of reinforcing this framework.

Dr Victoria Braithwaite gave valuable help with the re-writing in Chapter 5. We would also like to express our warm thanks to a number of people who have allowed us to reproduce their splendid photographs – they are individually acknowledged in the captions. We are grateful to Jane Bulleid for her skilful editing and most especially to Dr Tracey Sanderson of the Press for her interest, encouragement and patience.

Aubrey Manning
Marian Stamp Dawkins

1 Introduction

With great delicacy, a rainbow lorikeet uses its powerful bill for gently preening the feathers around its mate's eye; a flock of starlings sweeps in to roost; a mongoose swiftly and deftly bites its prey to death; young cheetahs rest quietly together, very close to sleep (Fig. 1.1). The study of animal behaviour is about all these things and many others. It is about the chase of the hunter and the flight of the hunted. It is about the spinning of webs, the digging of burrows, and the building of nests. It is about incubating eggs and suckling young. It is about the migration of a hundred thousand animals and the flick of a tail of one. It is about remaining motionless and concealed as well as about leaping and flying. Behaviour involves the static postures and active movements, all

Figure 1.1 Young cheetahs recline sleepily. Inactivity is as much a characteristic of behaviour as activity. (Photograph by M.S. Dawkins.)

the noises and smells and the changes of colour and shape that characterize animal life.

Animal behaviour is a very popular subject, not just with biologists but with the general public – so much so that it has even come to occupy a lot of prime time on television, the surest measure of real popularity! Since our earliest origins, human beings have always been fascinated by our fellow creatures. Apart from this intrinsic interest and the fact, which we hope to demonstrate, that the subject presents us with questions as challenging as any in science, the study of animal behaviour is also of great practical importance. The conservation of wild animals in their natural habitats and the welfare of those other species we have domesticated for our use, are both topics which command a lot of public attention.

Experimental studies with animals are now controlled by law in many countries, and rightly so. In the past, we have gained valuable information about behaviour from experiments, for example, deafening young birds or isolating a young monkey from its mother, which might now be deemed unacceptable. Nevertheless we shall describe some such results as part of this book. Not to do so would be perverse, since they have advanced our understanding and by this understanding we are better able to design observations and experiments which are not invasive or cruel in any way. Animal behaviour workers are very conscious of their responsibilities in this regard and go to some lengths to minimize any kind of disturbance to the lives of their subjects. The Association for the Study of Animal Behaviour in the United Kingdom and the Animal Behavior Society of the USA recently collaborated to publish a collection of papers, *Ethics in Research on Animal Behaviour*, which considers such problems, not only for laboratory studies but also for field work.* The way in which we design our experiments or observations and legislate about work with animals has moral, social and economic implications. The actions and policies which are required will be effective only if they are based on good information on how animals live and how they respond to their environment. This is the essence of good research with animals and it can yield results of real practical importance.

For example, the captive breeding of cheetahs is likely to be one important tactic for the long-term survival of this wonderful animal. Zoos have kept them for many years but their breeding success was pitifully low until we

Edited by M.S. Dawkins & M. Gosling and published by Academic Press, London. Undated.

obtained the results of field work in Africa (see Caro 1994). This showed that keeping males and females together – the standard zoo practice – was doomed to failure. Female cheetahs live and hunt alone, completely separately from males, except for the brief days of their oestrous. Once this aspect of the natural situation was reproduced in captivity, cheetahs proved quite easy to breed and several zoos – those in Fig. 1.1 are at London's Whipsnade – now have good numbers which could be the basis for re-introductions to the wild should this prove necessary.

We opened this chapter with a glimpse at the range of very diverse phenomena we are dealing with. Where in all this diversity do we start? What do people who study animal behaviour actually do and what do they want to find out?

There are two main approaches, the physiological and the 'whole animal'. Behavioural physiology is the study of how the body works, that is how the nerves, muscles and sense organs are coordinated to produce complex behaviour such as singing in a cricket or a bird. The 'whole animal' approach investigates the behaviour of the intact animal and the factors that affect it, for instance what it is in the environment of the cricket or bird that prompts them to sing at a particular time or why they sing at all. 'Whole animal' questions of this latter type can be studied both by looking at wild animals in their natural environments and also by observing captive or domestic animals living under more controlled conditions: it depends on the exact question involved. Physiological investigations often require bringing animals into a laboratory environment because they will involve 'probing beneath the skin', as it were, for example to get at the mechanisms that give rise to the behaviour of singing, or those which organize an animal's reponses to visual stimuli from predators.

In practice there is considerable overlap between the approaches. It is now possible to collect urine samples from animals in the wild – only tiny amounts are needed – and then back in the laboratory get an accurate measure of what hormones are circulating in the bloodstream. A few cells from the root of a recently shed hair or feather, or even those shed from the bowel with the droppings or shed from the inside of the mouth of a herbivore and left with its saliva on vegetation, can be cultured for DNA fingerprinting enabling relatedness within a social group to be worked out. The best understanding of behaviour nearly always comes when a combination of approaches is employed for, used well, physiological and 'whole animal' approaches complement each other – behavioural studies sparking off the search for physiological mechanisms and in turn putting physiological studies into a functional perspective. One of our aims in this book is to give some idea of the extent to which this is now possible.

Within the 'whole animal' approach, a distinction is often made between psychologists and ethologists, both of whom could be described as interested in the behaviour of intact, functioning animals. Psychologists working with animals have traditionally been mainly interested in learning and have tended to work in laboratories on the learning abilities of a restricted range of species, often rats and pigeons. Ethologists have been more concerned with the naturally occurring, unlearnt behaviour of animals, often in their wild habitats. Although this distinction still exists to some extent, there is now a fruitful coming together of the two. Ethologists have become interested in the role of learning in the lives of wild animals and psychologists are beginning to ask evolutionary questions about the learning abilities of a much broader range of species and to study their responses to more natural stimuli. Another of our aims is to show how much psychologists and ethologists are increasingly learning from each other, to mutual advantage.

But nobody – physiologist, ethologist or psychologist – can rely solely on one source of information; they must approach problems at the level which is appropriate for them. Some physiologists like to emphasize that their methods are the more fundamental, and it is true that increasingly it becomes possible to explain certain aspects of behaviour in terms of the functioning of the basic units of the nervous system, the neurons. However, this is effectively a task without end and since the main function of the nervous system is to produce behaviour and we must also investigate the end product in its own right. Even if we knew how every nerve cell operated in the performance of some pattern of behaviour, this would not remove the need for us to study it at a behavioural level also. Behaviour has its own organization and its own units which we must use for its study. Trying to describe the nest-building behaviour of a bird in terms of the actions of individual nerve cells would be like trying to read a page of a book with a high-powered microscope. Not only would it be incredibly laborious to discern the boundaries and make out the identity of each printed letter, we might miss out completely on the grouping of letters first into words, then sentences, then paragraphs and so on.

As we illustrated with our opening paragraph, behaviour includes all those processes by which an animal senses the external world and the internal state of its body and responds accordingly. Many such processes will take place 'inside' the nervous system and not be directly observable. What we see may involve violent activity or complete inactivity, but all equally rank as behaviour. This immediately presents us with the problems of observation and measurement. In the physical sciences, and often elsewhere in biology, we have universally recognized units – molecules, milliamps, pH units, metres, etc. – for measuring and classifying our observations. When we watch animals we have no such framework. Behaviour is continuous for as long as life persists; anything and everything may count. Put this way, the task would seem to be impossible and in fact we cannot ever study behaviour unless we abstract and simplify. We have to make decisions about what it is important to record and what can safely be ignored.

For example, if we are studying the courtship behaviour of sticklebacks or pheasants we will probably decide that there is no need to record the number of times that the animals breathe. Breathing is part of that continuous activity which constitutes behaviour but it will often be judged irrelevant for behavioural studies of courtship. However, if we are watching the courtship of newts, the rules are changed. Male newts court females on the bottom of ponds or streams and their courtship is punctuated by trips to the surface to breathe. Halliday and Sweatman (1976) have found that males can postpone breathing up to a point if courtship is proceeding smoothly, but if there are delays or the female moves off they rush to catch up on their breathing. Thus records of breathing are an important part of any study of newt courtship patterns, unlike those of pheasants. All this is simply to emphasize that the behavioural measures we use must be chosen to suit both the animal and the type of problem under investigation.

Questions about animal behaviour

We can see, then, that animal behaviour presents us with problems of description and selection. There are other difficulties that arise from the very beauty and fascination of the subject which have made it so popular. Natural history

TV programmes and books now offer films and photographs of such a high standard that we have become familiar with details of animal behaviour which used to be virtually impossible to observe. We can watch penguins fleeing from leopard seals by leaping up onto an Antarctic ice floe; we can watch, in close-up, the fantastic courtship displays of male birds of paradise; we can follow from start to frustrating finish the hunting chase of a cheetah who finally pulls down a gazelle only to be immediately driven off its prey by roving hyaenas. All this is splendid, but there can be a considerable temptation simply to gaze and wonder, rather than to think about analysing what remarkable things are happening.

We certainly must not lose our sense of wonder, but in order to use constructively the enthusiasm for further study which it gives us, we have to identify clearly what kinds of questions about behaviour need to be answered. In 1963 Niko Tinbergen, one of the founders of modern ethology, wrote a paper, as valuable today as when it was written, which he entitled, *On the aims and methods of ethology*. There he suggested that there are four main types of question which we need to ask of an animal's behaviour. They are really four different ways of asking the question 'Why?' and we can get into a very unhelpful muddle unless we clearly understand the nature of the distinctions between them.

Firstly, Tinbergen asserts, we ask why the animal performs a piece of behaviour in terms of its **function**, i.e. how it helps to improve the survival of the animal or its success at reproducing itself. Secondly, we ask about the **evolution** of the behaviour, i.e. how it has changed over the course of evolutionary time, in just the same way that we might study how a whale's flipper or a bat's wing changed from the ancestral limb into what we see today. Thirdly, we have to deal with its **causation**, what factors, both internal and external, lead to the performance of that particular piece of behaviour at that particular moment. Lastly there is the question of the behaviour's **development**, how the behaviour of a young animal changes as it matures and what factors, again both internal and external, affect this process and its end point.

Function and evolution are obviously interconnected and it is by selection acting on the processes of development that behaviour becomes *adapted* to the environment in which it has to operate. **Adaptation** is a subject we shall examine more closely in Chapter 6; here we may note that it arises by selection between individuals over many generations following genetically based changes, or by modifying behaviour to achieve the best response during an individual's own lifetime. It will often be important to keep in mind the distinction between these two pathways to adaptiveness.

Tinbergen's ideas have been very influential amongst behaviour workers, so much so that 'The Four Questions' have entered into much of our thinking, certainly they have for the authors of this book. We want to integrate some of the diverse ways of studying behaviour, for full understanding can come only when we have answers to each of the questions. Tinbergen directed his paper at ethologists but we should note that the same four questions can be asked no matter at what level of organization we are working. They are as relevant to studies of physiological mechanisms as to ecological ones.*

* We must distinguish between the meanings of 'function' and 'adaptation' as used here in the ethological sense and these same words as used by physiologists. For them function often means the way that an organ works, e.g. 'liver function is badly affected by too much alcohol'. They do not mean the contribution one's liver makes to survival. Again, adaptation in physiological terms often refers to sensory adaptation whereby we cease responding to continuous or repeated stimuli and has nothing to do with good fit to the environment.

Of course, when one begins to examine particular cases it rapidly becomes obvious that function, evolution, causation and development are not isolated topics: they overlap extensively and studies directed at one topic will nearly always contribute to the others. Studies at different levels and on different animals will lead to emphasis being laid on this question or on that, but all will benefit from taking as broad an approach as possible. We can best illustrate this by taking a look at two very contrasted examples of well-studied behaviour patterns – the escape response of the cockroach and the courtship displays of male sage grouse.

The escaping cockroach

Anyone who has handled a cockroach will know that it behaves like greased lightning when an attempt is made to catch it. Toads, which are natural predators of cockroaches, seem to have almost as much difficulty as we do. Probably the first question that springs to mind is that of **causation**: we want to know how the cockroach manages to avoid the lightning strike of the toad's tongue. We want to explore both the mechanisms inside the cockroach and the stimuli impinging on it from the outside that enable it to detect that it is about to be attacked and to dash away so effectively.

The first clues about causal mechanisms come from watching the behaviour of freely moving cockroaches and toads, or rather, analysing film of them because everything happens so quickly that the naked eye cannot follow it. Slowed-down film shows that just before the toad's tongue flips out of its mouth and strikes, the cockroach will turn rapidly and run. It seems, then, to have some way of detecting in advance that a toad is going to strike and from which direction before it actually does so. It always turns in the appropriate direction, away from the toad and about 16 milliseconds (ms) before the toad's tongue becomes visible. At this point, the toad is apparently already committed to a particular direction and the tongue may strike in the wrong place.

The most important cue by which the cockroach detects when the toad is about to strike seems to be tiny gusts of wind produced by the toad's movements. Camhi *et al.* (1978) showed that these slight air movements are picked up by the cockroach through many tiny wind-sensitive hairs on its cerci, paired appendages which protrude like little tails from the end of its abdomen (see Fig. 1.2). These cercal hairs are exquisitely sensitive and the cockroach will make a false run if a tiny gust of air is blown at a cercus. The smallest gust of air that generates escape behaviour is 12 millimetres per second with an acceleration of 600 mm/s. The hairs have to be physically moved, because if they are immobilized with glue, the cockroach is much less successful at escaping.

In behavioural experiments in which puffs of air were blown at cockroaches, it was shown that they started to show their escape behaviour just 44 ms after a puff started. Then, by measuring the wind generated by a toad when it strikes at prey, it was found that the critical (12 mm/s) gust of wind occurred on average 41 ms before the tongue emerged from the mouth – very close to the response latency of 44 ms shown by the cockroach to an artificial puff of wind. So it is the minute gust of wind generated by a toad preparing to strike that sets off the escape turning behaviour. By waiting until the very last moment before turning, the cockroach evades the tongue that has already started to move and thus escapes because the toad cannot, at that late stage, change the direction of its strike.

So far, we can see that a great deal can be learned about the mechanism of the behaviour by studying the behaviour of whole animals – intact toads and intact cockroaches. Film records of strike and escape, experiments stimulating the sense organs of the cockroach and measurements of the wind stimulus produced by the toad give us a basic idea of the mechanism underlying escape behaviour at this level.

The next stage, and this is often the case with studies of causation, is to go 'inside the skin' and look at the same behaviour but at the physiological level. Close examination shows that there are about 220 hairs on each of a

Figure 1.2 The cockroach, *Periplaneta americana.* The cerci at the hind end are indicated by the arrow. (Photograph by Paul Embden.)

cockroach's cerci and that each hair is hinged so that it can be moved most easily in just two directions at 180° to one another. They move less easily in directions at 90°. Different hairs have different biases, which means that wind from certain directions moves some hairs and wind from other directions moves others. The basis for the cockroach being able to discriminate wind direction so accurately thus appears to lie in the mechanical construction of the hairs. By recording from sensory nerve cells at the base of each wind-sensitive hair, Camhi *et al.* found that the nervous activity in these cells reflected the directionality of their hairs: the nerve cells responded much more to wind from some directions than from others.

Camhi *et al.* also found that there is a cluster of nerve cells, the terminal ganglion, at the hind end of the cockroach which contains a group of large cells known as Giant Interneurons (GIs). The GIs run up the nerve cord to the head, on the way passing through the thorax and linking to the motor nerves of the legs that do the rapid escape running. The GIs receive a pattern of information from the sensory nerves in the cerci which conveys the direction of the wind. They pass this on to the nerves which control the leg muscles: these nerves, in turn, command the legs to turn the cockroach and make it run in the opposite direction (Fig. 1.3).

Figure 1.3 Hind end of a cockroach showing the cerci with filiform hairs, which are the wind receptors. (From Camhi 1984.)

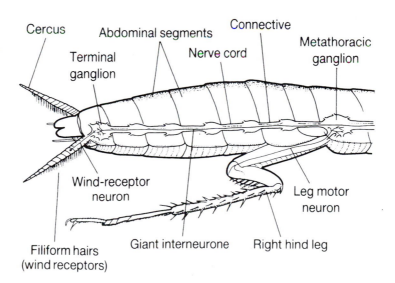

When the insect is stimulated with a wind puff, the GIs become very active (as recorded by microelectrodes inserted into them) and so do the motor neurons to the muscles of the hind leg. It is possible to inactivate some of the GIs by injecting them with an enzyme, leaving other GIs intact. This has the effect of making the cockroach turn in the wrong direction in response to a puff of a wind. So it appears that some sort of comparison between activity in different giant interneurons goes on in the normal cockroach, telling it which of its legs to move.

It is clear that for this piece of behaviour we have made very good progress, at both behavioural and physiological levels, towards understanding causation. For relatively few behaviours, particularly in vertebrates, do we understand the mechanisms in as much detail as this, right through from the detection of a stimulus to the command to the limbs. We can also give thoroughly satisfactory answers to the question of function, in the sense that we can readily understand how the escape behaviour aids the cockroach's survival. As we shall see, not all behaviour is so easy to interpret. However, note that the cockroach study does enable us to identify aspects of function in considerable detail. Whereas we might have assumed that it would be to the cockroach's advantage to run the instant it detected a toad – it might see its head turn for example – we can now understand the function of that 44 ms delay. It actually improves the insect's chances because it means that the toad's strike is now committed and if the cockroach can move a sufficient distance, the tongue is bound to miss – the toad cannot adjust. Many ethological studies of animals in their natural habitats have revealed just such detailed matching of behaviour to function.

What can we say about the related question, the evolution of this piece of behaviour? What were its origins and what sort of escape behaviour did ancestral cockroaches show? Since we have already seen that the function of the very rapid response is to evade predators, there must have been a kind of arms race between them and their prey, beginning with slower cockroaches and slower toads and other predators. Then, as now, both sides won some of the time, but some cockroaches would be a bit faster, perhaps because they had more sensitive cercal hairs which could detect smaller air currents. They would survive better and leave more offspring and sometimes their increased speed would have a genetic basis. When this was the case then their offspring would inherit their parents' extra speed. This would put selection pressure on the predators to become faster at grabbing cockroaches, which would in turn favour even faster cockroaches and so on.

This is a very general story suggesting how the skills of cockroaches became honed to a high degree over millions of generations. In this case it is difficult to be more precise because, for the most part, behaviour does not leave a fossil record and we do not have details of how the escape skills of modern cockroaches compare with those of their ancestors. However, we can deduce something of the course of escape behaviour's evolution by looking at the 'family tree' of such behaviour in living relatives.

Just as with bodily characteristics, so with behavioural ones we can get an idea of evolutionary history by comparing traits among groups of related species. If a wide range of species share a trait then we are fairly safe in assuming that it evolved before the different species diverged from a common ancestor. If a behavioural trait is shared by quite distant relatives then it pushes the time of its origin further back in time. On this sort of criterion we can be confident that the common ancestors of the whole ancient group of orthopteroid insects – the stick insects, crickets, grasshoppers and cockroaches – had escape responses organized very much along the lines just described. All of their descendants have cercus-like structures at their hind ends; all have giant fibres in their ventral nerve cords, often arranged in strikingly similar ways. Not all of them escape by running as does the cockroach; some jump, some may freeze into immobility to evade a predator's eyes, but the basic reponse and its mediation must have evolved very soon after the first insects appeared.

The last question we must address for the escape response concerns its development. As with questions about mechanism, questions about development can be asked at many different levels. Does a newly hatched cockroach escape as effectively from a toad as an adult or does it improve with practice? How do the connections between the sensory nerves and the giant interneurones form so that escape behaviour can be accurate?

In a way, there is a simple and direct answer to all developmental questions: it is always true that an animal's genetic makeup and the environment in which it grows up both contribute to the final behaviour it exhibits. Neither genes nor environment on their own would produce a fully functional, behaving animal. But the way in which genetic factors and environmental ones interact may be very complex, and this contributes to the fascination of developmental studies. For example, a cockroach that has just hatched and has had no previous experience of wind turns away from a puff of air just as accurately as an adult cockroach (Dagan & Volman 1982). From the very first puff, then, the escape behaviour is fully formed and very accurate even though the hatchling has only four sensory hairs on its cerci instead of an adult's 440. Nevertheless, this 'innate' behaviour can still be modified later in life. If the adult cockroach loses one of its cerci, its escape behaviour becomes at first very inaccurate but then gradually improves. Immediately after the loss, the cockroach erroneously turns towards a wind source instead of away from it. However, after 30 days, with or without practice, it improves markedly and turns consistently away from the wind source, despite still having only one cercus and no extra sensory hairs. Its nervous system has gradually changed its way of operating so that, by some mechanism not yet fully understood, the giant interneurons on both sides of the body come to respond to sensory input from the one intact cercus and the cockroach is able to adapt to its changed sensory picture of the world and escape in the correct direction. So, even when a behaviour is fully functional at hatching, modification and improvement are still possible. The genes and the environment continue to interact throughout life.

The courtship of the sage grouse

Let us see how far we can get with the 'four questions' approach applied to a very different type of behaviour, the very remarkable and highly complex pattern of courtship displays performed by sage grouse (*Centrocercus urophasianus*). These are large game birds of the American plains (see Fig. 1.4) which have been extensively studied by Wiley (1973a). For most of the breeding season it is the females who nest, incubate eggs and rear the young, but during the early part of the spring, male grouse gather on communal display grounds or *leks* and the females visit these and select a male to mate with.

Obviously courtship behaviour of the sage grouse takes us into realms far beyond the escape behaviour of the cockroach. To start with, we have to encompass the fact that several birds are interacting with each other and that what we call courtship is not a simple response over in a second or two – the cockroach turns and runs – sage grouse males may display for hours on end, even continuing through the night if there is enough moonlight! The females may not choose mates quickly, but visit the lek for several days in succession. Thus the four questions have really to be applied to the whole situation and it may be rather artificial and inadequate to apply them to individual elements of it.

Nevertheless we have to start somewhere and cannot be completely open-ended. We have to isolate something we identify as courtship and distinguish it from, say, feeding or parental behaviour. It helps to pick out behav-

Figure 1.4 Male sage grouse in full 'courtship strut' display on a lek or display ground. A female watches. (Drawing by Nigel Mann.)

iours which are to be seen only during courtship and not at other times. As far as the male sage grouse is concerned, the extraordinary display called the 'courtship strut', which is illustrated in Fig. 1.4, serves this purpose well for it is not seen outside the breeding season on the leks.

When we ask the four questions about the sage grouse's courtship, both the extent and the nature of the answers we can give are very different from those for the cockroach's escape response. In fact they will take us far beyond the scope of this introductory chapter but they will give some idea of the problems and the intellectual challenge of behavioural studies.

Function provided few problems with the cockroach but is much more difficult here. What is the function of courtship behaviour? We can begin with some deceptively simple answers; it brings the sexes together, it stimulates the female and brings her into a receptive condition, for example. The trouble is that the sage grouse's performance, from both male and female, seems terribly wasteful. The males go on for hours, the females take ages to make a choice, some males are never chosen and all the time there may be predators around who may find courting birds are not as alert to danger as normal. Why don't the birds just pair off as they arrive at the lek; why are the females so choosy; why do males who are not being chosen go on courting? Again we can ask why the males strut the way they do, what is there about this display which makes it functional; after all, other game birds have different displays. Peacocks have huge flamboyant trains, cockerels (domestic fowl) and pheasants have neck ruffs which they spread, and so on. Why haven't they all converged on the same 'best design'? Each of these 'why' questions relates to function, but we cannot arrive at answers without considering the evolution of courtship as well.

Comparisons with related species, just as with the cockroach, enable us to suggest quite plausibly that the strut display evolved by the exaggeration and amplification of feather erection and postures which occur anyway when game birds are highly aroused. Perhaps the particular manner in which this occurred in grouse, as compared with peacocks, meant that natural selection favoured exaggerating air sacs in the former and tails in the latter. The protracted nature of the displays and the 'choosiness' of the females may help them to pick out the most dominant and persistent males as mates. We can be sure that this is absolutely crucial for them because they are going to get no further assistance from their mate; at least they must ensure that he is fit and likely to father strong progeny. Such explanations do not mean that the functions we ascribed earlier to courtship are wrong. It does bring the sexes together and so on, but evolutionary considerations certainly enable us to extend the concept of function and understand the remarkable and apparently maladaptive aspects of courtship.

We are unable to approach the question of causation of courtship in anything like the detail we could muster for the cockroach. No physiological work has been done on sage grouse, but we can be fairly confident that both males and females begin to congregate on the lek when their secretion of sex hormone rises in response to the lengthening days of early spring. Apart from this internal factor there will also be external factors that initiate courtship. Probably the sight and calling of other males is a powerful stimulus. Wiley's observation that males will continue courting all night if there is a full moon certainly suggests that the males need to be able to see each other. The arrival of females also arouses the males to great heights of new activity, as do the 'quacking' calls females make as they circle the lek before alighting.

The last question is development, which asks how courtship behaviour first emerges in young birds and if it changes as they age. The question is not a simple one and actually has several components. For example, do young males court as persistently as older ones? Is the form of their strut displays the same? Do females choose to mate with males of all age groups, and so on?

From Wiley's (1973a, b) studies it would seem that males begin to visit the leks during the first spring after they hatch. All birds have to mature rapidly in terms of body size and strength and, in addition to this, yearling sage grouse already have the full adult plumage. However, they appear to be incapable of displacing older males and setting up their territories for strutting close to the centre of the lek. This is the part that females head for and as a result there is no evidence that yearlings ever succeed in getting females to mate with them. Wiley believes that mortality is high among the males and young ones that survive gradually move their territories towards the centre of the lek as the old males die off. It may take them several years to do so and by choosing older males the females, once again, will be ensuring that their mates are real survivors and accordingly fit.

Going back to the details of the strutting behaviour itself, there is one as yet unexplained difference between the performance of yearling and adult males. Wiley (1973a) describes how young males strut with a quicker tempo, although they are just as stereotyped as mature males. For the moment we know nothing more than this – it may or may not be a significant fact. It would be interesting to know how strutting begins in yearling birds performing the behaviour for the very first time; perhaps we could see early traces of the display, or elements of it, in much younger birds soon after fledging. The causation would not be the same, but the development of courtship might involve integrating various elements of other behaviour patterns to form a display which only comes fully together with sexual maturity. We have no evidence of this from sage grouse but not dissimilar developmental patterns are known for other birds, for example Groothuis's (1993) work with the threat and courtship displays of black-headed gulls (Fig. 1.5).

So when we apply Tinbergen's four questions to the courtship of the sage grouse, not only are there many possible answers and few definite ones but, as we warned at the outset, we inevitably raise a whole host of important issues relating to the way behaviour is organized. We shall be tackling some of them in more detail as we go through this book. The big differences between

Figure 1.5 The manner in which postures and movements of young black-headed gull chicks (*Larus ridibundus*) gradually become extended and stereotyped, developing into three very characteristic display movements of the adult gulls. We begin on the left with the more or less neutral posture of the young chick. As the chick matures, left to right, more discernible and regular postures begin to appear. These are (i) 'choking' (a display performed at the nest site; see Fig. 2.7, p. 48 for the homologous display in the kittiwake); (ii) the 'oblique' threat posture and (iii) the 'forward' threat posture. Note how the same posture in the chicks sitting with slightly extended wings, is a developmental stage common to both choking and oblique. (Modified from Groothuis 1993.)

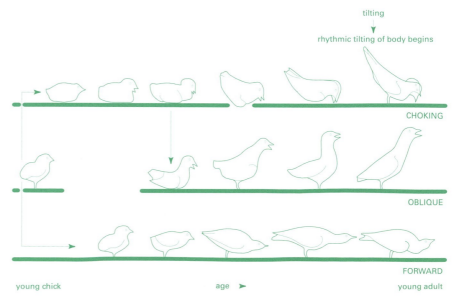

our understanding of the cockroach and of the sage grouse result mainly from differences in complexity. It does not denigrate insects, which comprise the huge majority of all animals, to conclude that their behaviour is organized along simpler lines than that of birds. There are good adaptive reasons for this. Yet complex behaviour is bound to share many features in common with simple patterns. They are, after all, both products of nervous sytems which, be they insect or avian, are constructed and operate along very similar lines.

We mentioned earlier the problem of choosing appropriate units for behavioural studies and the sage grouse's courtship provides a good example of one approach to such choices. It is quite impossible to describe, even for a moment, every twitch of every muscle or every small movement of a part of the body of an animal. What we have to do is to look for *patterns* in what we observe. Instead of describing the behaviour of an animal in terms of every small movement we can see (head down x, left front foot does y, tail does z, then after 1.5 seconds, head does $x1$, left front foot $y1$ . . . etc.), we hope to detect a recognizable pattern. We might notice that the head, the left foot and the tail always move in a coordinated sequence. There would then be no need to describe over and over again the exact way in which they moved because, like a repeated motif on a wallpaper, there are sequences of movement sufficiently similar from one occasion to the next to be given names like 'head-up display', 'facing away' or 'licking'.

The one dramatic element of male sage grouse's courtship behaviour offers us the opportunity to isolate a clear behaviour pattern in this way. It is the display we have already referred to, called by Wiley (1973b) 'the courtship strut' (Fig. 1.4), in which the bird inflates a huge air sac under his neck and breast feathers and then suddenly deflates it with sharp snapping sounds that can be heard over a kilometre away.

A fairly full description (although not down to the muscle twitch level) is given by Wiley, thus:

> The display begins from the strutting posture (tail fanned and cocked vertically, head raised, neck plumage erect) and consists in coordinated movements of the wings and the large oesophageal sac, together with associated movements of the legs, head and trunk. The oesophageal sac is inflated in the course of being twice lifted and dropped. Concurrently, the wings are extended forward and retracted twice. The culmination of the display follows immediately: a rapid compression and ballooning of the sac

accompanied by a third excursion of the wings. Complex acoustic signals accompany this culminating action. These include two sharp snaps produced by the inflated air sac and an intervening whistle that rises and falls in pitch. The final 0.2 s second part sounds roughly like POINK. Preceding this sound are three low-pitched coos, probably produced by the syrinx, and two swishing sounds generated by the wings rubbing against the sides of the chest.

Now each male sage grouse will go through this whole sequence every 6–12 seconds when they are clustered together on their display grounds or leks and, as we have described, sometimes continue for hours at a time. The similarity of each sequence from male to male and from one performance to the next means that the strut display can easily be recognized as a 'behaviour pattern' or unit of behaviour characteristic of this species and different observers will know exactly what this term means. Using slowed-down films of sage grouse displaying, Wiley showed that for 45 consecutive struts by one male, the mean interval from the first wing swish to the peak of the first 'snap' sound was 1.55 seconds and that its variation was less than 1%.

Extreme stereotypy of this type is typical of, although certainly not confined to, postures and displays which ethologists have called 'fixed action patterns'. They constitute part of the behavioural makeup of a species just as characteristic as the shape of its bill or the colour of its tail feathers. Their usefulness as a concept to us in the present connection is that they provide a handle for the quantification and further study of a male sage grouse's courtship. For example, we could compare the number of struts made by different males and see whether females were more likely to choose males that displayed the most.

We have already referred to some of the operations of the cockroach's nervous system and related them to the whole animal's behaviour as we can observe it. It is nowhere near as simple to do this for the sage grouse, but clearly having some reasonable and consistent units of behaviour for measurement is a start. We can begin by looking at ways in which units of behaviour may relate to the operation of the nervous system. Accordingly, we turn now to a brief description of the way in which the basic units of nervous systems work. Then we will try to make some comparisons between how the nervous system operates to produce very simple behaviour patterns, such as reflexes, and more complex behavioural units, such as the courtship strut. We will find that many of the principles that operate at the neuronal level are also to be found when we look at reflexes and more complex behaviour.

Units of the nervous system

The basic structural and functional unit of the nervous system is the nerve cell or neuron. Neurons are connected to each other in many complicated ways (Fig. 1.6), some individual cells being in contact with hundreds or even thousands of others. These connections are quite vital to the working of the nervous system since it is through them that nerve cells transmit information to each other. Each nerve cell has branching processes called dendrites that receive information from other cells; it has a cell body or soma and, connected to this, a long tubular axon that may extend over a considerable distance and transmits information to other cells (Fig. 1.7).

For the most part, information is transmitted through the nervous system by means of electrical impulses. However, at the junctions between cells, called synapses, there is usually chemical transmission. The synaptic terminal of the stimulating cell secretes small packages of a highly specific chemical – known as neurotransmitter – across the tiny gap that separates it from

Figure 1.6 Reconstruction of the cell body of a single motor neuron from the spinal cord of a cat. Parts of two dendrites can be seen (one branched) and the axon is indicated by the arrow. The cell and its dendrites are covered by synaptic knobs (boutons terminales) showing where the fine axons from hundreds of other neurons in the cord make contact with it. (After Walsh 1964.)

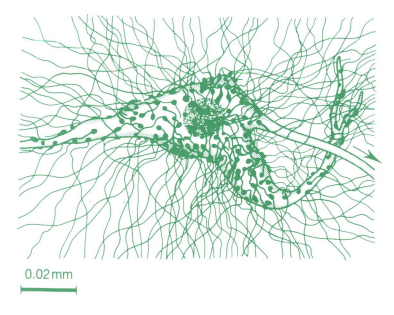

0.02 mm

the next cell. This neurotransmitter diffuses rapidly across the gap and when it arrives at the other side of the synapse, it gives rise to new electrical activity in the receiver cell; the excitation is once again transmitted electrically along the next axon until it reaches to the end of that. (Some synapses do not employ this chemical package method and are completely electrical – two cells being close enough together for direct electrical propagation.)

Each nerve cell is electrically active all the time and, when not transmitting information, maintains a potential difference across its cell membrane (known as the membrane or resting potential) which means that the inside of the cell is slightly electrically negative relative to the outside, i.e. it is polarized. 'Depolarization' is the term applied to changes in the membrane potential of a cell away from its normal value of 60–80 mV and these changes are the way cells transmit information.

Being stimulated by another cell through its dendrites causes a change in the membrane potential away from its resting value. This may be a very brief depolarization, but sufficient to pass on information if it gets propagated

Figure 1.7 Simplified drawings of (a) an invertebrate (arthropod) and (b) a vertebrate (mammalian) neuron. Real neurons have much more extensive branching than shown here.

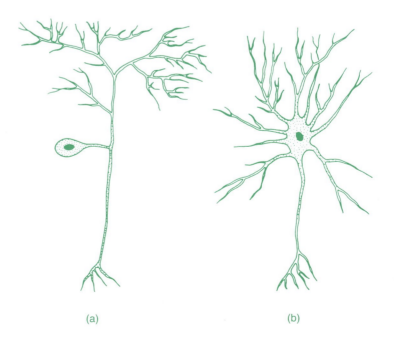

(a)   (b)

along the cell axon. Now, all cells are poor conductors – an electrical impulse simply does not travel very far down the cell (certainly not as far as if the cell were made of copper wire). Sometimes, when one cell is stimulated by another, all that happens is that there is a tiny electrical signal in the receiver cell which travels a few millimetres down a dendrite and then fades out because of the poor conductivity of the cell. The small induced signals are postsynaptic potentials (PSPs) and, although they do not last very long or travel very far, they nevertheless have a very important role in determining the kinds of operations that neurons can perform.

However, if PSPs were the only way in which nerve cells responded, long-distance transmission of information would be impossible. Accordingly, many neurons supplement PSPs with another type of signal altogether – a massive electrical change known as an action potential or nerve impulse (Fig. 1.8a). The nerve cell is capable of responding to a tiny PSP by generating an explosive action potential that can travel along the entire length of a long axon without a decrease in its amplitude. Sometimes one PSP alone is not enough to trigger a nerve impulse, but several PSPs coming rapidly over a short period of time, or from several different sources, will give rise to a nerve impulse. In this way, the nerve cell shows summation: it adds together the effects of several sub-threshold PSPs to give rise to the all-or-nothing response of the nerve impulse (Fig. 1.8b). At other times, the presynaptic neuron will deliver a barrage of several large action potentials to the receiver cell and each action potential results in more transmitter release than did the one before. Thus each PSP is larger than the one before, a phenomenon referred to as synaptic facilitation.

Interestingly, cells do not just excite one another. Some neurons also inhibit others. Their transmitters hyperpolarize the post synaptic membrane, making it *less* likely that receiver neurons will respond. As we will see through-out this book, the concept of inhibition is of very great importance for the understanding of all animal behaviour.

Having looked at some of the basic attributes of nerve cells – excitation, inhibition, sub-threshold responses, summation and facilitation – we can now turn to behaviour itself and see how these same phenomena are reflected in behavioural units which represent a more complex level of operation in the nervous system, and must involve the coordinated activity of several neurons at the very least.

(a)

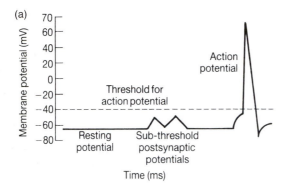

Figure 1.8 (a) Three examples of the electrical activity that can be recorded from neurons: the resting potential when the cell is at rest; postsynaptic potentials that do not lead to a 'nerve spike' or action potential; and the massive action potential itself.

(b) Some operations performed by neurons.

(b)

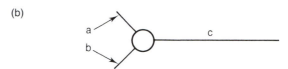

(i) A postsynaptic neuron, c, is shown diagrammatically with two presynaptic neurons, a and b.

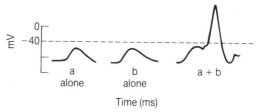

(ii) Spatial summation: the responses of presynaptic neurons a and b are summed by the postsynaptic neuron c.

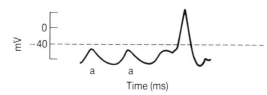

(iii) Temporal summation: the successive responses of one presynaptic neuron are summed.

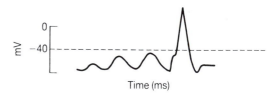

(iv) Facilitation: c responds more strongly with each successive input from a.

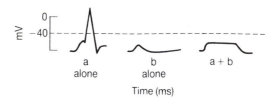

(v) Inhibition: c responds less strongly if both a and b are active.

Reflexes and more complex behaviour

Reflexes are often considered as among the simplest units of behaviour. The reflex which causes us to close our eyes when something flashes towards us or to withdraw a foot when we step on a sharp object, is functionally very important. Yet it seems so very different in scale both from the action of individual nerve cells on the one hand, and on the other from the kinds of behaviour with which most of this book will be concerned: courtship strutting, building a nest, running through a maze to get food and so on. Reflexes are somewhere between the two – not too dissimilar from the escape response of the cockroach but much simpler than a courtship display. It is certainly reasonable to regard them as behaviour patterns, albeit simple ones. We can learn some of the basic features of behavioural mechanisms from studying properties which reflexes share with more complex patterns and which can be clearly related to the properties of individual nerve cells, on the operation of which *all* behaviour depends.

We are, in any case, dealing with a continuum here and it is not possible to draw a firm line between reflexes and complex behaviour. Clearly the latter can incorporate many reflexes; the swallowing reflex is the culmination of elaborate food-seeking behaviour and the reflexes controlling balance and walking are involved in almost everything animals do. Nevertheless, there is much to be learnt from comparing the properties of nerve cells, reflexes and more complex behaviour and such comparison is a useful way of emphasizing the unity of the study of behaviour.

In 1906, Charles Sherrington published *The Integrative Action of the Nervous System*. Sherrington, more than any other single person, can be regarded as the founder of modern neurophysiology. In his book he considered the way in which reflexes operate and how the central nervous system integrates them into adaptive behaviour, combining information gathered from different sources, arranging sequences of action and allocating priorities.

In the first few chapters of his book, Sherrington discusses some of the properties of reflexes and contrasts them with those of the same movements when elicited by direct stimulation of the nerves to the muscles concerned. We shall, in turn, use part of his classification to compare the properties of reflexes and more complex behaviour.

Latency

Reflexes and complex behaviour both show latency in response – there is a delay between giving a stimulus and seeing its effect. The latency between a dog encountering a painful stimulus with its leg and showing the flexion reflex in which it withdraws its leg usually lies between 60 and 200 milliseconds. Only a small fraction of this delay is due to the time taken for nerve impulses to be conducted along axons; most of it is due to the delay at the synapses (a term we owe to Sherrington) between one neuron and the next. It is hardly surprising to find delays between stimulus and response in complex behaviour, for in the chain between receptors and effectors there are often dozens of synapses to cross. We have already noted the crucial 44 ms latency between the arrival of the tiny air movement which signals the movement of the toad's tongue and the first turning away of the cockroach, a delay which is actually beneficial.

Although it is often difficult to measure latencies for complex behaviour (unlike the cockroach escape situation, it is often impossible to fix the time of

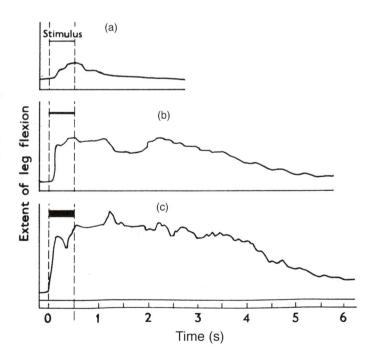

Figure 1.9 The extent and persistence of the dog's flexion reflex with three strengths of stimulus, each of 0.5 s duration. The area enclosed between the line and the *x*-axis gives a measure of the 'amount' of the response. Even with a weak stimulus (a) the after-discharge represents 75% of the total 'amount'; with the strongest stimulus (c) it represents over 90%. Note that the latency of the response decreases as the stimulus increases in strength. (Modified from Sherrington 1906.)

stimulus onset precisely), nevertheless, results are sometimes vivid. Wells (1958) describes how, when a tiny shrimp is presented to a newly hatched cuttlefish, there is no detectable response for perhaps as long as 2 minutes. Then the nearest eye of the cuttlefish turns to fixate on the shrimp. There is a further delay, but usually only a few seconds, before the cuttlefish turns towards the shrimp so that both of its eyes are brought to bear. Another brief delay follows and then it launches its attack and seizes the shrimp with its tentacles.

With reflexes, it is found that the stronger the stimulus, the shorter the latency (Fig. 1.9) and some evidence suggests that the same is true for certain complex behaviour. Hinde (1960) measured the latency between presenting various frightening stimuli to chaffinches and how soon they gave their first alarm calls. Just as with reflexes, the stimulus known on other grounds to be the strongest, produced the shortest latency.

Summation

We have seen that individual neurons sometimes respond only after they have received several postsynaptic potentials. This shows that they are able to summate excitation coming at different times (temporal summation) or from different places (spatial summation). Sherrington gives several clear examples of summation at the level of reflexes. The scratch reflex of the dog is elicited by an irritating stimulus anywhere on a saddle-shaped area of its back. The hind leg on the same side is brought forward and rhythmically scratches at the spot. Weak stimuli – say, a series of 5 or 10 touches given in rapid succession – may not evoke any response, but after 20 or 30 scratching appears: the stimuli have been summed in time. Figure 1.10 shows the spatial summation of stimuli from two areas of skin 8 cm apart. Neither is strong enough alone to provoke scratching, but they are effective when given together.

Dethier (1953) studied the stimuli which cause blowflies to extend their proboscis before they drink. The flies can detect sugars and other food substances with sensory hairs on their fore-tarsi. They search for food by running over a surface and extending the proboscis when the front legs encounter anything suitable. As measured by this proboscis extension, the flies can detect sugars at very low concentrations. Dethier found that when only one leg is dipped in a solution, the lowest concentration of sucrose to which 50% of the flies respond is 0.0037 molar. However, if both legs are stimulated together, twice the number of sensory fibres are sending signals into the brain; their effects summate and now 50% of flies respond to only 0.0018 M.

With more complex behaviour, summation frequently occurs between stimuli of quite different types perceived by different sense organs. We all know how the sight and smell of food summate when we are hungry. Beach (1942) showed that male rats respond sexually to a combination of olfactory, visual and tactile stimuli from a receptive female. Young males do not respond unless two such sources are available – it does not matter which two. Mature males, with previous sexual experience, will respond to one type of stimulus alone.

'Warm-up' or facilitation

Sherrington found that some reflexes do not appear at full strength at first but, even with no change to the stimulus, their intensity increases over a few seconds. Neurons, as we saw, show synaptic facilitation, each successive PSP being larger than the one before. At a behavioural level, Hinde (1954) found

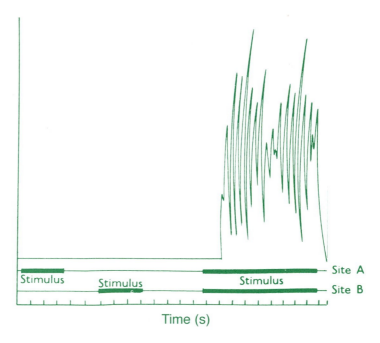

Figure 1.10 Spatial summation leading to the appearance of the scratch reflex in the dog. The tracing represents the movements of the dog's leg when scratching. Site A and Site B are two points on the shoulder skin. Weak stimuli, given singly first at A and then at B, do not evoke the reflex. When both points are stimulated simultaneously the reflex appears with a latency of about 1 s. (Modified from Sherrington 1906.)

Stimulus   Stimulus   Stimulus   Site A
                                    Site B

Time (s)

that chaffinches show a similar type of 'warm-up' effect when shown an owl. The birds' response to the owl is to give a mobbing call. Counting the number of calls given by a chaffinch in successive 10-second periods after the owl is shown to it indicates that it begins by calling at a relatively low rate and that the maximum calling rate is not reached for about 2.5 minutes, after which time it gradually declines (Fig. 1.11).

Sherrington was able to show that 'warm-up' in some reflexes is due to summation of stimuli which come to evoke a response from more and more nerve fibres, producing a stronger contraction. He called this phenomenon 'motor recruitment'. Some analogous process probably occurs with complex behaviour, but what we commonly see is not only a change in the intensity of response but in the nature of the behaviour as well. Sherrington (1917) provides an excellent example from what he calls the cat's 'pinna reflex'. Repeated tactile stimulation to the cat's ear first causes it to be laid back. If stimulation persists, the ear is fluttered; thirdly the cat shakes its head and when all else fails to remove the irritation, it brings its hind leg up and scratches. Clearly there is more involved here than the recruitment of a few extra motor neurons. Mechanisms which control patterns of movement such as ear-fluttering and head-shaking must be recruited. Perhaps all these mechanisms are activated in some way by stimuli to the ear but their thresholds are different. That for laying back the ear will have the lowest threshold, with successively higher ones for the other three patterns. Workers studying complex behaviour frequently rank the patterns they observe on a similar intensity scale of increasing thresholds. A similar threshold idea has been used by Bastock and Manning (1955) to explain the fact that a male fruit fly switches from one courtship pattern to another when courting a female whose behaviour remains constant.

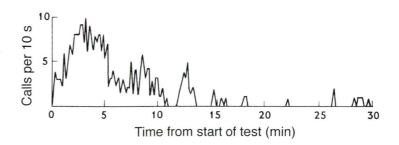

Figure 1.11 The 'warm-up' and subsequent 'fatigue' of alarm calling when a chaffinch is presented with a stuffed owl in its cage. The maximum rate of calling occurs after about 2.5 min. (From Hinde 1954.)

Inhibition

Inhibition operates at every level within the nervous system. As we have seen, nerve cells actively inhibit each others' transmission of information. Similarly, the prevention of one activity's occurrence while another is in progress constitutes inhibition at the behavioural level. In many ways, inhibition is just as important for the coordination of behaviour as excitation and to see why, we can again turn to Sherrington's work on reflexes.

Muscles are commonly arranged in antagonistic pairs, such that one flexes a portion of a limb and the other extends it. Clearly, it would be impossible to both extend and flex the same limb at the same time: Sherrington showed that excitation of one member of a muscle pair is accompanied by inhibition of its antagonist. Such inhibition is not absolute, and an inhibited muscle does not simply go limp. Once it is stretched by its antagonist then its own 'stretch reflex' (see p. 31 for a fuller description) will tend to make it contract. Although the antagonist muscle may override it, it will take up the slack, so to speak, in an active fashion. Much finer control of movement is possible if muscles can be made to work against one another in this way. Mutual inhibition allows them to take the lead in turn during limb movements, and to alternate flexion and extension of the limbs. Sherrington found that it is not only antagonists on the same limb which inhibit each other, but also muscles located on opposite limbs and which have antagonistic effects during locomotion. When the flexors of one limb are contracting, the flexors of the opposite limb are inhibited. Reciprocal inhibition of this type is one of the basic integrating mechanisms for walking and without it coordination of the different limbs would be impossible.

The role of inhibition in complex behaviour is superficially less obvious than that of excitation. We stimulate an animal and the conspicuous result is that it makes a response. But in so doing it has made a swift transition which requires the inhibition of its behaviour prior to the stimulus and other behaviour which it may be stimulated to perform at the same time. Sherrington saw reflexes as 'competing' for the final common pathway, i.e. the muscles whose action is common to several different reflexes. In an analogous way, we can see the different systems controlling patterns of complex behaviour like fighting, feeding and sleeping competing for the control of the animal's musculature. Such systems are obviously incompatible in the sense that only one behaviour can occur at a time. The 'integrative action' of Sherrington's title refers in part to the necessity for the nervous system to allocate priorities since there will often be conflicts. Which stimuli should an animal respond to; which can it safely ignore for the moment? Inhibition of action will be as crucial as excitation. For example, many animals require to take in food and water over the same time in order that digestion can proceed normally, but they do not try to do both at once! In a study on doves, McFarland & Lloyd (1973) gave *ad lib.* food and water to birds which had previously been deprived of both. The doves alternated their behaviour smoothly between bouts of feeding and drinking (see Fig. 1.12). Thus long before they had assuaged their hunger, they abruptly stopped feeding and switched to drinking, then after a short time they switched back again, continuing in this fashion until, some time later, they gradually ceased both activities. What is striking about their behaviour is that they did not 'dither' between feeding and drinking, spending only a few seconds on each. Rather, sustained bouts of each activity in turn inhibited the other. We do not know the exact mechanisms involved here but McFarland and Lloyd found that it was the time spent in feeding and drinking, not amount of food or water consumed, which seemed to determine the points at which switch-overs occurred. They called the process 'time-sharing' and it certainly suggests alternating inhibition of one activity by the other.

Sherrington found that when inhibition was removed from a reflex, it returned at a higher intensity than it had previously. Figure 1.13 shows this phenomenon, which Sherrington called 'reflex rebound', for the scratch reflex. We commonly observe that when a particular type of complex behaviour – for example, courtship – has not been elicited for some time, it has a

Figure 1.12 Cumulative number of food and water rewards obtained by a dove which was both hungry and thirsty. The dove could choose between pecking one key which yielded food or another which yielded water. Note how it starts by repeatedly pecking the food tray, then switches to the water key for several pecks before switching back for another bout of food pecking, and so on. (Modified from McFarland 1974.)

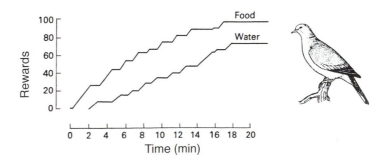

lowered threshold and is performed with high intensity when it is, at last, evoked. Vestergaard (1980) found that when laying domestic hens are kept on wire so that they have no substrate in which to dustbathe, they start dustbathing quickly and dustbathe in very much longer bouts when eventually given access to litter than do hens kept all the time on litter. It is possible that the system controlling dustbathing shows something akin to reflex rebound. Kennedy (1965) interpreted some aspects of the behaviour of aphids along these lines. The behaviour of winged aphids alternates between periods of flight and periods of settling and feeding on leaves. If an aphid settles on an 'unattractive' surface – an old leaf, for example – it does not stay long and soon takes off, but flies relatively weakly and soon settles again. Conversely, if it has settled on an attractive young shoot, it stays for a long period but, when it takes off, flies vigorously and for a long time.

By an elegant series of experiments, Kennedy was able to exclude any simple explanation for this relationship based on physical exhaustion during flight and recovery after resting and feeding on a young leaf. He suggested that there is mutual inhibition between the systems controlling flight behaviour and those controlling settling. As with reflexes, activation of the settling system may temporarily inhibit the expression of the flight system but, at the same time, gradually lower the threshold for flight. In Chapter 4, we shall discuss other evidence that the systems controlling complex behaviour do inhibit one another and the various other interactions that occur between different control systems.

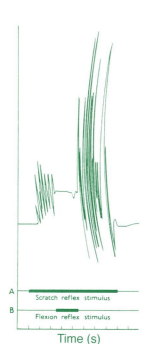

Figure 1.13 Inhibition of the scratch reflex by the flexion reflex. The stimulus denoted on line A evokes the scratch reflex, but this response is inhibited when the stimulus on line B evokes the flexion reflex. The moment B is removed the scratch reflex returns, and much more vigorously than before – an instance of 'reflex-rebound'. (Modified from Sherrington 1906.)

Feed-back control

Very commonly, reflex or complex behaviour consists of a steady output of some activity which has to be held at a given level. When we 'stand at ease', our body is evenly balanced over the pelvic girdle and easily corrects for any slight jostling we may receive. To do so, the muscles of the legs and back must be held at a constant level of tension and if shifted away from this level, they must correct to bring the body upright again. Analogously, animals under normal circumstances maintain a very constant body weight and they eat and drink sufficient for their needs at regular intervals. If a surplus is available they do not overeat. In times of scarcity, they spend a higher proportion of their time in searching for food and consume more when the chance arises to replace any deficit.

Both these examples show us behaviour acting as a **homeostatic** system ('homeostasis' means literally 'same state') and serving to preserve the *status quo*. In the first case, this was achieved through reflex systems controlling the leg and trunk muscles; in the second case, it was achieved by a series of more complex systems regulating the search for food, feeding and satiation. In both cases the operation requires that the end result (posture and balance whilst standing, state of nutrition) is monitored in some way. When it deviates from a set value a signal is sent to the control mechanisms to correct the imbalance and bring the end result back to the set value again. This idea is shown diagrammatically in Fig. 1.14 and its application to feeding and drinking behaviour is discussed in more detail in Chapter 4.

The homeostatic control of posture is understood quite thoroughly at a neurophysiological level. In some cases we know the paths of the neurons involved and can actually identify the structures which function as parts of the control system. One such is illustrated in Fig. 1.15, which represents a typical muscle on the limb of a mammal, such as would be involved in maintaining posture. Motor neurons which have their cell bodies in the ventral horn of the spinal cord run to the muscle; it is their activity which determines the tension developed by the muscle. In parallel with every skeletal muscle, and embed-

Figure 1.14 Diagrams of simple open- and closed-loop control systems. The output from the behavioural system is affected by disturbance factors (the crossed circle represents interaction.) In open-loop control (a) if the output is affected by disturbance no correction takes place. In the closed-loop arrangement (b) the results of the disturbance feed back to affect the input to the behavioural system. The dark segment of the crossed circle represents an interaction between the feedback mechanisms and the input, which tends to bring back the output to its original value. The feedback system produces its own output, which is proportional to the disturbance and changes the input to the behavioural system both to the right amount and in the right direction.

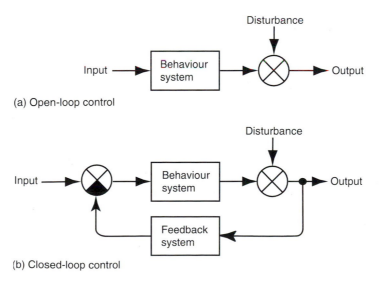

(a) Open-loop control

(b) Closed-loop control

ded within its fibres so that they contract and relax with it, are muscle spindles. These are specialized sense organs for recording the degree of tension in the muscle. Their sensory nerves run back to the spinal cord and, entering through the dorsal root, synapse with the motor neurons to the muscle. Thus a loop is closed which forms the basis of the stretch reflex, already mentioned.

When a muscle is stretched by the contraction of its antagonists, the muscle spindles are stretched also and their sensory fibres increase their rate of firing, stimulating the motor neurons so that the muscle contracts. It is easy to equate the units of the closed-loop control system in Fig. 1.14 with those of this real muscle mechanism. The output (state of tension in the muscle) is affected by a disturbance (being stretched by other muscles), and a feedback mechanism (muscle spindle) records the change and feeds back to change the input (motor nerve) and restore the original input. This is a simplified picture of the real situation which, in fact, includes other regulatory mechanisms allowing for very fine graded control over the muscle contractions involved both in the maintenance of posture and in movements, but it serves to illustrate the reality of feedback control at a reflex level.

Not all behaviour involves feedback control. When a movement must be made very rapidly, there is simply not time to modify the movement while it is in progress. The flip of the toad's tongue towards the cockroach is just such a case. The strike of the mantis is very similar. It moves towards a fly and orientates its body very slowly and precisely (operations which certainly involve

Figure 1.15 A simplified diagram of some of the neural pathways involved in the stretch reflex. Further explanation is given in the text.

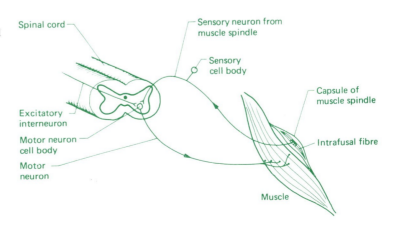

Spinal cord

Sensory neuron from muscle spindle

Sensory cell body

Capsule of muscle spindle

Excitatory interneuron

Motor neuron cell body

Intrafusal fibre

Motor neuron

Muscle

feedback control), but once aimed, the strike is an all-or-nothing movement. If the fly moves after the strike is initiated, this makes no difference to the movement and the mantis strikes in the wrong place. Such behaviour which occurs without feedback is said to be under 'open-loop' control (as opposed to the 'closed-loop' control of a homeostatic system). The difference between open and closed systems is shown in Fig. 1.14.

So, despite the many levels at which the behaviour of animals is studied, we can see that there are certain principles – excitation, inhibition, summation, facilitation and feedback control amongst others – that appear to be common to many different levels. Studying single neurons and studying the behaviour of whole animals may require very different techniques and, in many cases, very different concepts. Nevertheless, it is valuable to keep these common principles in mind. It is often possible to break down complex behaviour patterns into smaller units, some of which are immediately equatable with reflexes. However, we cannot always explain behavioural observations using reflex terminology, nor is there any point in trying to do so. There *are* differences in complexity and these often require different types of approach, as we have seen for the cockroach versus the sage grouse.

Diversity and unity in the study of behaviour

It will be clear by now that the study of animal behaviour presents us with a great diversity of subject matter, levels of analysis and questions to be answered. At one end, neuroethology merges into cellular physiology and biochemistry. At the other, the study of animal groups moves into evolutionary theory and ecology. Psychologists, ethologists, physiologists and behavioural ecologists all contribute to this diversity, so much so that it may be tempting to think that there is no sense in which there is a single 'study of animal behaviour' at all.

There are, however, two very positive reasons for regarding it as a single subject of rich diversity rather than as a collection of isolated disciplines. The first is that the same questions, particularly those of mechanism, survival value and ontogeny, can be asked at all levels. It is as important to ask 'why' (in the survival value sense) an animal has a nervous system with sensory, motor and interneurons connected in particular ways as it is to ask why male sage grouse display. It is as important to ask about causation (how does it work?)

when looking at the way in which a bird assesses the amount of food in its territory as it is when recording from a single nerve with a microelectrode. Studying adaptation without looking at mechanism tends to become sterile, since it is essential to understand the constraints imposed by what the animal's body can do before we can understand the evolutionary significance of what it actually does. Studying mechanism without asking about adaptation tends to give lists of facts with no context. Evolution through natural selection is the unifying theme of all biology and the study of animal behaviour, at all levels, is at its best when questions about mechanism and questions about evolutionary significance are asked in parallel.

The second reason for not compartmentalizing the subject and trying to see it as a whole is that the different levels of analysis, although distinct, are never totally separate. Many of the same concepts that have been used and tested at one level have been found to be useful at other levels. The behaviour of animals is, ultimately, due to the workings of nerves, muscles, hormones and sense organs. Equally, however well we understand the workings of those parts of the body at a physiological level, our understanding is incomplete unless we understand how they interact to produce complex behaviour. The most fruitful studies have been those in which the boundaries between the different levels have been breached and in which 'outside the skin' and 'inside the skin' have not been the demarcation lines of a particular investigation.

Some of the best examples of how our understanding has benefited from just such a multi-level approach have come from studies of behavioural development, where we can often begin to link up the building of a nervous system with the emergence of a behavioural repertoire. Thus although – quite arbitrarily – Tinbergen put development as his fourth question for ethologists, it is the one to which we choose to turn first.

2　The development of behaviour

One of the most remarkable features of living organisms – plant or animal – is the way in which a single-celled zygote, the fertilized egg, is transformed by cell division, cell differentiation and cell movement into the adult form, far more complex and often millions of times larger. Development (often called embryology when it describes the progress from egg to a young but free-living stage) remains one of the most challenging fields of research in all science.

The zygote contains all the information necessary to build a new organism, provided that it can develop in and interact with a suitable environment. When we study the development of behaviour, we must obviously concern ourselves with some aspects of embryology – for example the way in which the basic framework of the nervous system is laid down – but we need to go far beyond this. A young animal's behaviour may continue to develop long after it is independent and it is perfectly reasonable to argue that in some animals behavioural development continues throughout life. Thus learning (which we shall concentrate on in Chapter 5) might well be regarded as a form of development and young animals sometimes learn a great deal as they develop. However, here we shall concentrate on other types of behavioural change which often take place both rapidly and dramatically during the early part of life.

At the outset, we must recognize that young animals are not simply part-formed creatures, inadequate stages on the path to adulthood: they have at all times to be fully functional animals capable of behaving effectively in their own world. During their early stages of growth some animals may be protected inside an eggshell or a uterus, or watched over by attentive parents, but others are free-living and have to look after themselves entirely. Young animals may emerge as miniature adults, gradually growing in size, but their behavioural responses must change as well to keep pace. Young cuttlefish begin and remain as carnivores, but at first they can kill only tiny crustacea which are ignored as prey when the cuttlefish have grown. They move on to larger and larger food and the behaviour patterns employed for detecting and catching prey have to change accordingly with growth towards the adult size. The behavioural and morphological changes can be even more dramatic, for some young animals lead a life totally different from that of adults. Tadpoles swim and breathe like fishes and are herbivores before metamorphosing into land-living, carnivorous frogs or toads. Aquatic, filter-feeding rat-tailed maggots (so called because they breathe air through a long snorkel tube at their rear end) metamorphose into flower-feeding hoverflies (see Fig. 2.1). For these life histories, young and adult each require an almost totally separate repertoire of behaviour.

Such changes mean that development often has to generate patterns which operate for only part of an animal's life and then disappear. Provine (1976) has described the particular synchronized movements by which cockroaches break out of their individual eggshells and then the protective case which packages a batch of eggs together. These movements involve a series of reversed waves of contraction along the body from tail to head and they are seen only on this one occasion. They appear, precisely timed, at the close of early development in the egg and serve to launch the young cockroach nymph into its next stage of growth; this done, they are never elicited again. The feeding behaviour patterns of the aquatic larvae of hoverflies, just mentioned, must have a rather longer life than this, for they have to carry the larva through

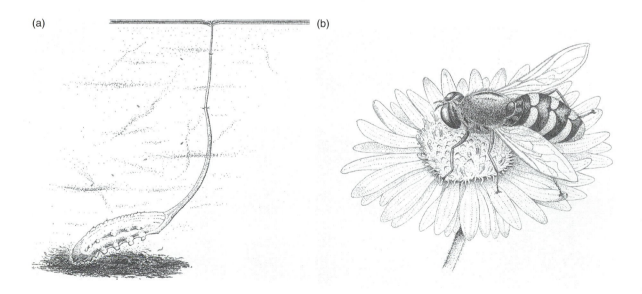

(a) (b)

Figure 2.1 (a) A 'rat-tailed maggot', an inhabitant of stagnant water where it browses on detritus or the bottom mud, breathing by means of an elongated 'snorkel' linked to the insect's tracheal system. After a pupal stage it metamorphoses into (b) a flower-feeding hoverfly, whose environment and behaviour are totally different. (Drawings by Nigel Mann.)

its whole growth period until it pupates, but then like the cockroach hatching movements, they disappear. Animals with complete metamorphosis have, in effect, to be capable of organizing two different lives, carrying two sets of genes which organize two developmental programmes, interacting with two different environments. Just sometimes the two stages of life can combine. Figure 2.2 shows a nice example of 'cooperation' between young and adults

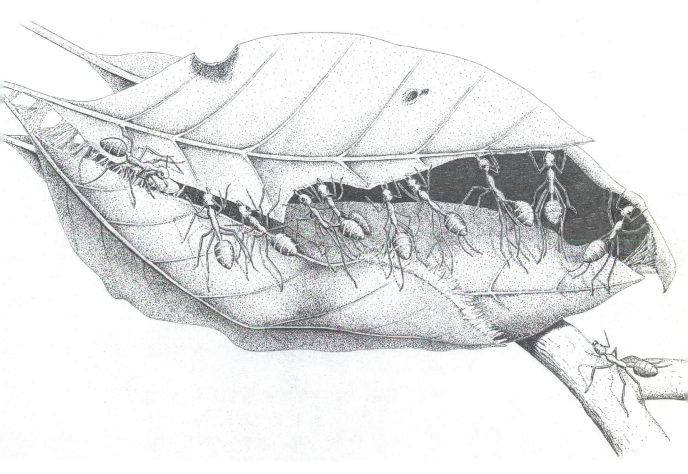

Figure 2.2 Collaboration between the generations! Weaver ants constructing their nest by bringing two leaves together and then binding the edges to join them firmly. Adult workers cling to the open edges with jaws and feet to pull them even closer together, whilst on the left another worker holds one of the colony's larvae and, by moving it from side to side like a living bobbin, induces it to spin silk which binds the leaf edges together.

within a colony of weaver ants. Using adult behaviour patterns, clinging on with jaws and clawed feet, worker ants bring together the two opposing edges of a leaf to form a sheltered nest space within it. They then seal the edges together by bringing up larvae and inducing them to spin silk across the gap – a pattern the larvae had to evolve to spin their cocoons prior to pupation. There are many other examples of the divergent lives of young and adult animals: Oppenheim (1981) provides an excellent review of this aspect of behavioural development.

We shall be describing in a later section (p. 62) some of the developmental processes in the nervous system which mediate the changeover from juvenile to adult behavioural systems. Here we may note that specialized infantile behaviour patterns do not always disappear but may return in a slightly different context. Baby meerkats (an African mongoose) go limp and behave passively when an adult seizes them by the scruff of the neck. This reflex facilitates their being moved without injury. Adult female meerkats similarly relax when seized in a neck bite by the male during copulation. The neural basis for the reflex remains beyond infancy and we have other examples which reveal that even if they are not used again, some juvenile reflexes remain. In a series of elegant experiments, Bekoff and Kauer (1984) showed that the unique movements made by the chick to break out of the eggshell (beating the beak against the shell and strong thrusting movements of the legs which serve to rotate the body) will be performed again by young chickens several weeks after hatching if they are gently pushed back into a huge artificial eggshell and then placed in the posture of the chick at hatching. Probably one factor which contributes to the retention of the neural circuitry underlying this pattern is that elements of it are involved in adult locomotor behaviour. We shall later discuss other examples of this kind of 'neural recycling' from insects (see p. 66).

The progression, sudden or gradual, which marks the transition from young to adult, is not always made along a single pathway. In this chapter we shall be considering a number of ways in which behavioural development can proceed and how far young animals can compensate for deviations from the normal path. However, few animals have one single end point, one fixed pattern of adult behaviour. Some animals become males, others females, and their behavioural repertoires may be very different. Female honey-bees develop into queens or workers, grasshoppers of certain species may stay solitary or become swarming locusts. In these cases we know that there are clear-cut genetic or environmental triggers which send subsequent development along one of two or more alternative paths. The presence or absence of a Y chromosome in the zygote of mammals determines the sexual pathway, although as we shall see, subsequent 'sign posts' along the path are not genetic (p. 68). The length of time that royal jelly is provided to the growing larval honey-bee determines whether it develops into a queen or a worker (p. 383). The desert locust exists in two forms, or 'phases' as they are termed; it is either a highly gregarious, black and orange locust or a solitary-living, pale green grasshopper. These two are so totally distinct in both behaviour and appearance that when first described they were put into different genera! The switch is brought about in part behaviourally, in part it can be through egg quality. Solitary nymphs change their colour and behaviour if they are crowded together, whilst the eggs of females which have been crowded, even if they started adult life solitarily, hatch into nymphs with gregarious tendencies (Roessingh *et al.* 1993; Islam *et al.* 1994; Fig. 2.3).

Not all triggers are so clear cut. Males of some vertebrates fight to defend territories to which females are attracted and then mate with them. Other males may not fight and hold a territory of their own, but rather hang around submissively in another male's territory as so-called 'satellite' males. Such behaviour in the breeding season is found in animals as diverse as sticklebacks (Fig. 2.4), ruffs and white rhinoceros! Although at first sight satellites may appear to be unsuccessful, close observations show that they certainly do manage to mate with females – often when a territory holder's attention is concentrated on fighting off other territorial males. In other words, becoming a satellite – at least for a time – may be an adaptive strategy for adult life.

With both males and females, we now know of many cases where newly independent young birds or mammals have two alternative reproductive strategies to follow: they can disperse, find mates and reproduce themselves, or they can stay behind with their parents and help the latter to rear another set of offspring. We shall discuss this kind of choice in later chapters on evolution and on social behaviour.

Identifying the events, genetic or environmental, which bias such development one way or the other, is now an active area of research. In addition to genetical and physiological considerations, ecological factors often turn out to be crucial for such life-history 'decisions'. Nor will they necessarily be permanent decisions, for further change and development may be possible as ecological conditions allow, or as animals get older and individual experience enables them to adapt.

Figure 2.3 The solitary (above) and gregarious phases of the grasshopper locust (*Schistocerca gregaria*). The former is cryptically green to brown in colour with prominent eye stripes, and has 5 or 6 larval stages. The gregarious phase is yellow and black with less conspicuous eye stripes, and always completes its development in 5 stages. There are many behavioural differences between the two phases. (Photographs by Steve Simpson.)

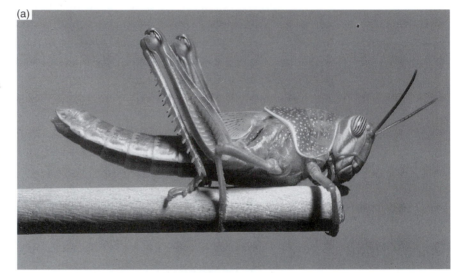

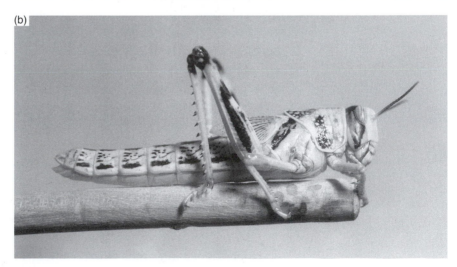

Instinct

Whatever our level of study, the **adaptiveness** of behaviour is one of the most dominant features that we observe. Whether we are looking at negative feedback control processes in the muscular control of a limb movement or the nest-building patterns of a bird, whether we are observing young animals or adults, the beautiful match between form and function is striking. Of course animals do make mistakes and may appear clumsy at times, particularly when they are put into unnatural situations, but for the most part their behaviour is beautifully matched to their way of life. They respond appropriately to the features of their world and thereby feed themselves, find shelter, mate and

Figure 2.4 Two alternative mating strategies for male ten-spined sticklebacks. The black male in full nuptial colours has built the nest and led the female, swollen with eggs, towards it. He 'shows' her the entrance, inducing her to enter and spawn after which he will pass through the nest and release sperm to fertilize the eggs. But below is another male; cryptically coloured, he appears very similar to a female. He has moved stealthily around the territories of nest-building males and approaches with the female. He may slip into the nest after her `and fertilize the eggs before the territory holder has a chance to do so. (From Morris 1952.) Morris recorded that this reproductive strategy, which he called 'pseudo-female,' often resulted when competition between males for territorial space was intense. It was not always successful but probably offers more subordinate males a chance to breed and, perhaps, also a chance to inherit a territory if the dominant male is killed.

produce offspring. How can behaviour acquire this near perfect match to an animal's mode of life?

This question has fascinated people for centuries, because we have always been observers of animals. Walker (1983) and Sparks (1982) discuss the history of our thinking about animal behaviour. The manner in which different ages and cultures have explained behavioural adaptativeness has depended to a large extent on how they viewed the relationship of human beings to animals and to the world around them. In other words it has been treated as a metaphysical question, not just as a biological one. At various times and in many different cultures, animals have been regarded as sentient beings just like ourselves and with similar goals and feelings, not unlike Brer Rabbit and Brer Fox in J.C. Harris's *Uncle Remus* tales! Such attitudes tend to stultify the objective study of animals for their own sake. We come closer to some semblance of biological reality with the highly influential views of the seventeenth century French philosopher and mathematician, René Descartes. He certainly recognized that animals had special characteristics of their own. As a devout Christian of his time he saw them as altogether below us in the scale of things, with human beings very much set apart from brute creation. Yet he had to recognize that the brutes did cope remarkably well with the problems of their lives. How could they do this, lacking souls or the power of reasoning which enabled humans to survive and prosper? Descartes supposed that animals were able to respond adaptively because they operated using **instincts** endowed by the Creator, which automatically provided them with the correct response, requiring neither experience nor thought. We deliberately choose the term 'automatically' because Descartes did indeed view animals as little more than automata, blindly driven along pre-set but adaptive paths.

The concept of instinct or instinctive behaviour (the terms 'innate behaviour' or 'inborn behaviour' are effectively synonymous) is still a familiar one. Instinct is often described as patterns of inherited, pre-set behavioural responses which develop along with the developing nervous system and can evolve gradually over the generations, just like morphology, to match an animal's behaviour to its environment. It might be defined in a negative kind of way, as that behaviour which does not require learning or practice, but which appears appropriately the first time it is needed. This definition immediately suggests its converse and the other familiar way in which behaviour can become matched to circumstances. Animals may not have any pre-set responsiveness, but be able to modify their behaviour in the light of their individual experience. They can learn how to behave appropriately and perhaps practice or even copy from others to produce the best response.

Expressed in these terms, we appear to have come up with a clear dichotomy, rather as Descartes intended: instinct (nature), in which adaptation occurs over generations by selection, or learning (nurture), where the same end result – responding appropriately – is engendered afresh within the lifetime of each individual. As we shall see, studies of the way in which behaviour develops do not support such a clear division, and suggest that the categories themselves are inadequate. Nevertheless, it is useful to look briefly at the nature of behavioural adaptiveness across the animal kingdom with its huge diversity of forms and lifestyles, for some pattern is discernible.

Instinct and learning in their biological setting

Pre-set behaviour which requires no learning or practice is obviously going to be advantageous for animals with short life spans and no parental care. Many insects lead almost totally solitary lives with no overlap between the generations and precious little social contact of any kind. Mason wasps (genus *Manobia*) construct a series of cells inside the hollow stems of plants. A female wasp emerges from her cell and has a brief moment of interaction with another wasp when she mates with a male. Thereafter she is totally on her own. She selects a hollow stem and builds a partition of mud mixed with her saliva at the inner end. She then lays an egg attached to the roof of the stem, close to this partition. Next she hunts for caterpillars which she lightly paralyses with her sting and, after repeated excursions, provisions the cell with 5–8 of them; these are the future food supply for the larva when it hatches. This done, she builds another partition sealing off the egg with its food supply and, laying a second egg beyond this, provisions a second cell, seals that off and so on. In this way 8–10 cells may be constructed in line along the cavity of the stem until the female, reaching the outside end, plugs this with mud. She then moves off to seek another stem and constructs more cells (Fig. 2.5).

The female wasp lives only a few weeks and carries out this elaborate series of behaviour patterns in total isolation. She could not possibly achieve this tight schedule if she had to acquire everything from scratch by trial and error and she has to rely on pre-set, unlearnt responses. This conclusion is given further emphasis by the fascinating observations made by Cooper (1957) on this particular wasp species. Examining a stem in which the larvae have pupated prior to their emergence as adults, reveals that all the pupae are orientated with their heads facing the open end. Making the correct choice of end is a matter of life or death, for although sometimes the emerging adults could turn round in the narrow stem, they do not do so but move on ahead breaking through the partitions. Normally the outermost pupae (although they derive from the later eggs of the series of cells) emerge first, leaving a clear passage, as it were, for their siblings from deeper in the stem. Adults emerging from artificially reversed pupae struggle on inwards through the deeper cells and accumulate at the blind end (Fig. 2.5).

How can a larva, about to pupate, make the correct decision? Cooper's experiments showed clearly that there was no possibility that they detected light, or used gravity or oxygen concentration as a cue. They rely on information left behind by their mother. As the female wasp retreats outwards building the partitions between the cells, the inner side of each is, of necessity, left as rough mud whilst the outer sides she smoothes into a concave form. Cooper experimented with artificial grass stems in which he used glass paper to make partitions, rough on one side and smooth on the other, and he could show unequivocally that it is just these characteristics of concavity and smoothness versus roughness which the larva uses: it pupates with its head towards the latter (Fig. 2.5). Information is thus passed from one generation of wasps to its offspring and it must be encoded genetically in a way that allows the larvae

Figure 2.5 (a) Cells of the mason wasp (*Manobia quadridens*) in a hollow stem. Note the form of the mud partitions, rough and convex on the 'outer' side, smooth and concave on the 'inner'. The larvae have all pupated with their heads towards the rough wall. (b) An artificial stem of hollowed out dowelling made by Cooper (1957). The partitions were reversed in orientation and, as a result, emerging adults have headed 'inwards'. They gather at the blind end where one wasp still remains in the pupal stage – also headed inwards.

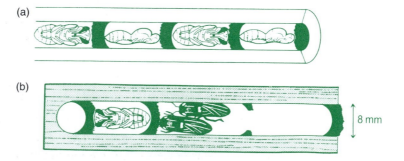

and the adult female to develop appropriate behaviour. Again we must recognize that neither the mother wasp's actions nor the response of the larvae can rely on experience.

For the most extreme of contrasts we may compare the mason wasp's life history with that of the African elephant (Moss 1992). The elephant's life span is comparable to our own and the key element of elephant society is the matriarchal group led by a mature female with her daughters and their offspring (Fig. 2.6). A baby is born into a group where every individual knows all the others from long experience in their company. It is nourished and closely protected for several years by its mother and other relatives. Slowly it acquires the adult repertoire of feeding behaviour, learning how to select food and how the group migrates around its home range to match the seasonal changes of vegetation and water supply. Females do not become sexually mature until about 20 and puberty is even later for males, who leave the group when mature to lead more solitary lives. The behaviour of individuals and of groups varies considerably according to their history. For instance, Ian and Oria Douglas-Hamilton (1975) record how one group of elephants at Addo in South Africa are abnormally nocturnal in their habits and, whilst fearful, are also unusually aggressive towards humans. This behaviour can be traced back to an attempt to annihilate the Addo population by shooting, in 1919! None of the elephants alive at that time can still be there, but their descendants have acquired and transmitted the behaviour which enabled a few to survive nearly 80 years ago.

The mason wasp which must rely on pre-set instinctive behaviour and the elephant which can learn at relative leisure and even transmit information from one generation to the next, represent two extremes on the behavioural scale. In fact, our descriptions are greatly over-simplified, for the wasp can and must learn many things during its brief life – the exact locality of each of its nest stems, for example, so that it can return to them after its hunting trips. The young elephant possesses some tendencies, such as those for feeding and reproduction, which are instinctive even though it may have to learn how to direct them. Certainly all animals whose complexity exceeds the annelid worm level show both types of behaviour and each has its own special advantages. One advantage of learning over instinct is its greater potential for changing behaviour to meet individual changing circumstances. Such a consideration is obviously more important to a long-lived animal than to an insect which lives only a few weeks. A further relevant factor may be body size, because highly developed learning ability requires a relatively large amount of brain tissue, insupportable in a very small animal. Usually body size and life span are positively correlated to some extent and most large animals live longer than small ones.

Figure 2.6 A 'family group' of African elephants. The old matriarch leads the group across dry grassland of the Amboseli reserve in southern Kenya. Behind her follow sisters and daughters with their offspring of all ages. Males will leave the family as they become sexually mature around 10–15 years of age, but they are unlikely to be able to mate with females until much later. They move in small bachelor groups, continuing to grow, and are unlikely to breed until they are 40 or 50 years old. (Photograph by Karen McComb.)

Apart from these physical constraints it is clear that natural selection can operate to match different degrees of learning ability to an animal's life history. The two most advanced orders of the insects, the Hymenoptera (ants, bees and wasps) and the Diptera (two-winged flies), are comparable in size and life span. The Hymenoptera, in addition to a rich instinctive behaviour repertoire, show an extraordinary facility for learning, albeit of a specialized type, and this plays an important role in their lives. During her brief three weeks of foraging life, a worker honey-bee will learn the precise location of her hive and the locations of the series of flower crops on which she feeds. She will move from one to another of these in the course of a day's foraging because she also learns accurately the time of day at which each is secreting the most nectar. Comparison with the Diptera yields a striking contrast. Certainly some flies *can* learn. Hoverflies learn something of the position of the flowers they visit, and houseflies tend to return to the same place to settle in a room. Using some ingenuity, it has proved possible to get the fruit fly (*Drosophila*) to learn some simple discriminations. But for the most part the dipteran memory is brief and their learning powers very limited. Unlike the Hymenoptera, they do not reproduce in fixed nests to which they must return regularly. Their short lives are largely governed, with complete success, by inherited responses to stimuli which signal the presence of food, shelter and a mate.

The characteristics of instinct and learning

We mentioned earlier that instinct and learning are not much more than shorthand labels at an early stage of trying to understand what actually happens during the development of behaviour. Superficial considerations may suggest that whereas learning results in flexible patterns of response, instinctive behaviour is characterized by rigid, stereotyped responses and patterns of movement. Yet common observation will tell us that learnt patterns can be just as stereotyped as instinctive ones. In his delightful book, *King Solomon's Ring*, Lorenz (1952) describes how water shrews learn the geography of their environment in amazing detail. If at one point on a trail they have to jump over a small log, this movement is learnt with such fixity that they continue to make the jump in precisely the same fashion long after the obstacle has been removed. All mason wasps build the partitions between cells in the same way and all kittiwake gulls show the same 'choking' display (Fig. 2.7) on their cliff nest sites. Neither of these patterns requires learning for their development. But all chimpanzees of one community in Tanzania show one specialized mutual grooming pattern (Fig. 2.8) which is not seen in other communities and is certainly copied afresh by each generation of young animals. Rowley and Chapman (1986) describe how occasionally natural cross-fostering occurs between two very distinct parrots, the galah (*Eolophus roseicapillus*) and Major Mitchell's cockatoo (*Cacatua leadbeteri*), when a female galah lays eggs in a cockatoo's nest hole (see Fig. 2.9). Galahs reared in such nests emerge strikingly abnormal in their behaviour. They retain some galah-like calls, but have also the cockatoo's calls; they have also aquired the cockatoo's manner of flight and choice of diet. So even such species-specific stereotyped behaviour patterns may develop through learning and practice in some cases.

Such an example illustrates some of the problems that arise if one tries to force behaviour into categories, such as instinct *or* learning. There is a natural enough temptation to do this because instinctive behaviour is such a constant feature of an animal species: unless their environment is grossly changed, they all tend to develop a repertoire of the same patterns of behaviour and perform them in the same way. The cross-fostered galahs mentioned above must make us more cautious, but we know parrots to be long-lived and very intelligent animals. They probably are rather exceptional and we know many cases in which behaviour patterns typical of feeding, courtship and aggression do develop in the absence of learning.

Nevertheless, it has become clear that the dichotomy suggested by Descartes is not by itself going to be sufficient to account for the behaviour of animals. If the question is posed, 'where does this or that behaviour come from'? or, 'how can it be so well adapted to circumstances?' it has to be accepted that simple, either/or, nature/nurture answers are inadequate, however attractive they may seem. Our path to understanding is by an examination of how genetic and environmental influences interact to shape the behaviour of animals as they move from egg to adult.

One way of attempting this, which used to be popular amongst ethologists – see, for example, Lorenz(1966) – is the 'isolation' or 'Kaspar Hauser experiment'. Kaspar Hauser was a youth who appeared on the streets of

Figure 2.7 The 'choking' display of the kittiwake. The bill is wide open revealing the brilliant reddish-orange gape and tongue. As the body inclines forward the bird makes a rapid series of jerky up-and-down movements of the head and neck. (After Tinbergen 1959.)

Nuremberg, Germany on 26 May 1828 with confused stories of his origin and upbringing. He was obviously mentally disturbed and displayed bizarre behaviour. He died in 1833 of wounds, almost certainly self-inflicted. Kaspar Hauser attracted the attention of various philosophers and occultists of the day and was sometimes regarded (for no fathomable reason) as an example of

Figure 2.8 Chimpanzees of the Kasoge community in western Tanzania raise arms and clasp hands for mutual grooming of each others underarms. (Photograph by W.C. McGrew and C. Tutin.)

Figure 2.9 A mixed brood of two species of Australian cockatoos at about 45 days from hatching and ready to fly. Three Major Mitchells are on the left with their foster sibling, a galah, the latter's mother having laid an egg in the Major Mitchell's nest. All are marked with conspicuous wing tags which allowed them to be followed in the wild. The galah NO spent his first 2 years living in a flock of Major Mitchells. (Photograph by Graeme Chapman; from Rowley & Chapman 1986.)

the totally untutored human being. He was claimed to be an isolate, revealing native human characteristics unchanged by experience. Consequently his name was sometimes used as a convenient label for experiments in which animals are reared in isolation.

Such a procedure attempts to keep animals solitarily, out of all contact with others from as early an age as possible. When mature, their responses to a variety of stimuli are tested and compared with those of animals reared normally. As we have seen, the life histories of many insects are effectively natural isolation experiments. Rather few other animals have been tested under really rigorous conditions, but some fish and birds have been shown to perform

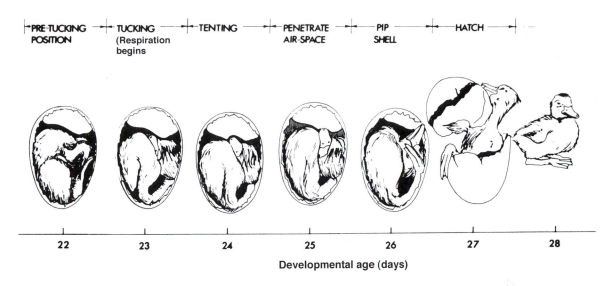

PRE-TUCKING POSITION	TUCKING (Respiration begins	TENTING	PENETRATE AIR-SPACE	PIP SHELL	HATCH

Developmental age (days)

22   23   24   25   26   27   28

Figure 2.10 The final stages leading up to hatching in ducklings. They hatch after 27 days of incubation but for several days before this the bill first presses up against the air space in the egg (tenting) and then penetrates it. The duckling begins breathing air and vocalizing 2 or 3 days before hatching. It is these calls which affect the development of its subsequent responses to the mother's calls. (Modified from Gottlieb 1976.)

various feeding, sexual and alarm patterns of behaviour quite normally following isolation. This are interesting data but, however easily isolation experiments enable us to eliminate the possibility that animals learn how to do something, they offer only a very restricted view of what happens during development. Isolation can tell us only what factors are *not* important for the development of behaviour; it tells us nothing of what *is* involved and some factors may not be at all obvious. We ought to begin our experiments by asking from what, exactly, are the young animals being isolated?

Gottlieb (1971, 1993) has carried out a series of investigations into the way in which young ducklings of various species come to respond correctly to the calls of their mother and how this choice is affected during their early life. One might regard the young bird, developing inside its eggshell, as another natural isolation experiment but this is too simple a conclusion. For example, young mallard hatched from an incubator respond more strongly to the call notes of mallard ducks than to other related calls, and do so the very first time that they hear them, a preference we would call instinctive. But experiments reveal that one factor necessary for this instinct to develop is an embryo duckling's ability to hear the calls – quite unlike those of the mother – which it and other ducklings make while still inside the eggshell (see Fig. 2.10). This 'instinctive preference', once established in the newly hatched ducklings, remains stable in isolated birds, but if they are reared socially as would be normal in nature, then they become more responsive to other calls that they hear and may abandon their early choice. Bolhuis (1995) provides a number of other examples showing that various experiences, as apparently unrelated to vision as stroking, or being put into a running wheel – both in total darkness – can affect the visual preferences of young birds. It causes them, a few hours later, to express a preference for a realistic model mother bird (again we would call this instinctive) over a simple, artificial visual stimulus such as a wooden cube. Such experiments illustrate that instinctive responses can have a fairly complex developmental history. The details of the history may vary even between quite close relatives (something we shall also find for bird songs later in this chapter). Gottlieb (1983) found that young wood duck (*Aix sponsa*), which also hatch with a preference for the characteristic descending pitch calls of the mother, subsequently *lose* this preference unless they continue to hear the calls of other ducklings around them; their own calls are not sufficient.

To fall back on simple concepts of learning or practice to account for such changes would require one to extend such terms beyond any level of usefulness. We must accept that behavioural development may involve processes and factors which are not easy to categorize.

Genetics and behaviour

Any account of behavioural development must include some consideration of genetic factors because genes consitute one source of information which is present from the very outset of life. Certainly genetics immediately springs to mind when thinking about instinctive behaviour and it has been common enough to refer to such patterns being 'inherited', most especially when discussing their evolution. We have every reason to conclude that behaviour has often evolved alongside morphology to achieve adaptiveness.

While there can be no doubt that genes are involved in the development of behaviour, it is not usually a straightforward matter to investigate how they act. We now know very precisely how genes code for the structure of the proteins produced by cells, but this is basically all that an individual gene can do. Other processes, whose functioning depends on the properties of other proteins, then operate to deliver the end product – a muscle, a bone or a kidney. It is easy to lose sight of gene action with such complex processes. The environment in which genes operate also affects these developmental processes. For instance, the cytoplasm of many eggs is variable in its nature across different regions and this affects how the embryo develops, which end becomes the head, which the tail and so on. How can we 'translate' these processes into the terms of behaviour and its development? It is made extremely difficult because the 'distance' between the genes and the end product is so large and the nature of this end product is so diverse. We cannot expect the development of, say, a courtship display, an alarm call and learning ability to have much in common. Even a highly complex organ like a kidney is obviously made up of proteins and their metabolic products and we can watch structures moving into place as the embryonic kidney differentiates. This is far from the case for even the simplest of behaviour patterns. The kidney is a constant – it is always there to be examined – but a behaviour pattern appears only under particular circumstances: indeed, it may never appear in the lifetime of a particular animal. We can only really speak of a potential to behave in this way or in that.

There is a further difficulty, because any form of genetic analysis depends on us having suitable genetically based variations for our material. Common sense suggests that genes must be involved when we observe instinctive behaviour emerging fully fledged, as it were, at the first performance in every individual of a population. But paradoxically such constancy of development offers no rigorous way of investigating if and how genes are acting.

We can take an analogy from morphology. It is very easy to work out how genes operate to produce the colour of the eyes in *Drosophila* or in human beings. Eye colour varies and the pattern of inheritance of the variants leaves no doubt that they are under simple genetic control. What eye colour you develop usually depends entirely on what genes you are carrying. This means that we can readily proceed to study how genes affect the synthesis of pigments, at what stage in the development of eye colour they act and so on. However, if we are interested not in the development of eye colour, but of eyes themselves we have no such opportunities. Apart from certain gross abnormalities, all *Drosophila* and all humans develop eyes. Genes *must* be involved at all stages in the complex growth and differentiation which lead to the formation of the eye cup and the retina, but unless we can isolate suitable mutations we can say virtually nothing about how they act. So it must remain with most of instinctive behaviour.

For the reasons outlined above, many would argue that it is not possible to speak of the genetic control of behaviour itself, but only of behaviour's structural substrate, the nervous system, which has the potential to organize the behaviour. This is fine, but of course we have have only limited ways of relating the structure of the nervous system to the behaviour it produces. All the detail of the instinctive patterns will be hidden in the minute detail of neural interconnections and their physiological functioning. (Glance back again to Fig. 1.6!)

Nevertheless we have the clearest possible evidence that genes do play a powerful role in the formation of just such interconnections. Doves and pigeons develop their species-specific cooing song patterns in the absence of any experience, even when deafened so that they cannot hear themselves. Interspecies hybridization is frequently possible and, although the hybrids themselves are sterile, the males display and coo. When they do so, the patterns of their calling are grossly distorted, full of cooing which is jumbled and fragmentary (Lade and Thorpe 1964). The genes the hybrids inherit are mixed up and the songs with them.

Other interspecific hybrids reveal a more subtle type of gene-based disturbance. Dilger (1962) worked with small parrots of the genus *Agapornis*, of which there are a number of African and Madagascan species which readily breed in captivity. They all have a unique method for gathering nesting material, using their sharp bill to perforate large leaves parallel to an edge and then tearing off a strip of leaf which they carry back to their nest hole as building material. At one point in this sequence of behaviour patterns there is a divergence between species. Fischer's lovebird (*A. personata*), carries leaf strips back to its nest singly, held in its bill; the peach-faced lovebird (*A. roseicollis*), interposes another stage. As each leaf strip is torn free, the bird elevates its rump feathers, turns round and tucks the strip amongst them, then lowering them so that the strip is held fast. The bird repeats this part of the sequence a number of times, finally flying back to the nest with several strips in place (Fig. 2.11).

Dilger hybridized these two species; as with the doves the hybrids are sterile, but they perform all the usual nesting behaviour. In aviaries the parrots readily used newspaper instead of leaves as nesting material. The hybrids all showed normal piercing and strip-tearing behaviour, and we must assume that the genes received from each parent species were, at the very least, sufficiently compatible to allow normal development of the neural mechanisms behind this pattern. The sequence continued and it appears that peach-faced behavioural control now took over because a hybrid would always raise its rump feathers and, turning round, tuck in the strip. So far, so good, but at this point things went wrong. Instead of following through in the manner of its peach-faced parent, the hybrid did not open its bill at the end of the tucking behaviour. It hung onto the strip, more like the Fischer's parent pattern, and so, as its head returned to the front position, the strip was drawn back out of the feathers and the parrot was left standing with a strip still held firmly in its bill! It could then have reverted to Fischer's-style single strip carrying, but never did so. Instead it opened its bill, dropping the strip and then returned to the paper again and began another strip-tearing sequence, which ended in the same abortive fashion. Dilger found that it took *months* of repetition before some sequences began to be successful – always when the hybrids managed to inhibit dropping the strip at the end, and flew back to the nest. However, they always went through the tucking pattern first. We know parrots are intelligent birds which can rapidly learn fairly complex tasks, but it was 2 years before they were successful in nearly all sequences and even then their heads would turn round slightly before they flew off with their strip – revealing a last persistent trace of the tendency to tuck.

Figure 2.11 A peach-faced lovebird tearing strips of paper as nest material and then tucking a strip into the feathers of its back to carry them to the nest. (From Dilger 1962.)

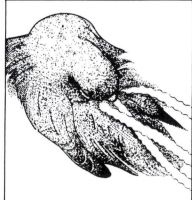

One could scarcely ask for a more vivid demonstration of the force of genetic instructions operating on the development of particular neural mechanisms and thus persistently biasing an animal to perform a particular behaviour pattern. The huge numbers of neurons and the complexities of advanced vertebrate brains make it impossible to get much closer to gene action, but given simpler material, it is sometimes possible to identify how genes operate on their neural substrate. Even if its development still eludes us, we may be able to understand how genes may regulate the way different behaviour patterns are brought into action at the appropriate time. The big marine mollusc *Aplysia*, the sea hare, is hermaphrodite and individuals usually mate when they encounter one another. Following mating, *Aplysia* switches its behaviour from feeding to egg-laying, which involves a series of specialized stereotyped patterns not seen at any other time. Using the techniques of molecular genetics, Scheller and Axel (1984) have been able to identify a gene and its product which, acting on specific nerve cells, causes them in turn to release 'egg-laying hormone'. This acts at several sites, neural and muscular, in *Aplysia*'s body and activates the neural circuitry controlling the egg-laying movements.

Thus in one sense, we can understand the genetic control of *Aplysia*'s behaviour in great detail but, as mentioned above, to understand the genetic control for building the neural circuitry during development requires much more evidence. We can get some few steps along this complex path by examining the effects of gene mutations which alter the way the nervous system is built; there are now plentiful examples. For obvious reasons some of the best known ones come from *Drosophila* and mice. *Bar* reduces the number of facets in the eye of *Drosophila* to a narrow strip; *no-bridge* causes a gap in the interconnections of the central complex, a major structure in *Drosophila*'s brain; *staggerer* mice have gross abnormalities to the Purkinje cells of their cerebellum, whilst those carrying the mutant *waltzer* have defective inner ear structure.

In all these cases, and many others like them, we can observe changes to behaviour which correspond to the neural defects. Not suprisingly, they are rather gross changes for the mutants do act rather like hammer blows to a complex and sensitive piece of machinery. Nevertheless we can gain useful information about how genes may operate to underpin the development of behaviour's neural basis. It may be obvious enough that *Bar* flies have defective visual responses and *reeler* mice – the name describes the behavioural effect – lack good balance, but in both cases we find that it is not just one target organ which is affected. *Bar* flies lacking eye facets also lack large areas of neural tissue in their optic lobes, the centres controlling visual information. The formation of eye facets triggers a range of developmental processes downstream, as it were, leading to normal development of the neurons which form the optic lobes. *Reeler* mice exemplify a different kind of developmental sequence. They seem to have the potential to form a normal cerebral cortex, but the cells that should migrate out to form connections fail to do so and remain piled up beneath the cortical layer (see Barinaga 1996).

Drosophila carrying the *no-bridge* mutant show rather specific changes to behaviour. They can fly normally and see normally but, tested in a laboratory set-up, they cannot follow a moving target in flight in a normal way. Indeed, a number of other mutants affecting the central body of the brain (which is the target of *no-bridge*) show related problems, which suggests that this body is one controlling centre for the orientation and control of locomotion (Ferrus & Canal 1994).

Here, then, we are just beginning to build up a picture of how the normal alleles of mutants we identify act so as to control processes which lead to the formation of the brain. With one very special subject, the tiny (2 mm) nematode *Caenorhabditis elegans*, it is possible to go much further thanks to remarkable developmental studies by Brenner and his group (see Brenner 1974). Nematodes have a particularly deterministic type of development such that their eventual cellular structure is very fixed. *C. elegans* adults are made up of exactly 959 cells and the nervous system takes up 302 of these! Mutants are known which affect many stages in the construction of this neural network and usually corresponding changes to behaviour can be identified, mostly concerning how the worm moves and its responses to chemicals in the environment (see Fig. 2.12). Partridge (1983) provides a clear summary of the behavioural aspects of this work.

Caenorhabditis has a very simple repertoire and our problem in linking genes and the development of the nervous system with behaviour is far greater in complex animals. We may be able to identify neural mutants with 'sledgehammer' effects but we have little or no idea of their implications for the elaborate patterns of behaviour studied by ethologists and psychologists.

Figure 2.12 (a) A culture of the tiny (2 mm) nematode, *Caenorhabditis elegans*. (b) Tracks left by *C. elegans* on an agar plate. The normally elegant sinusoidal patterns (hence its specific name) may be distorted in mutants which affect the nervous system and hence aids their detection. (c) Tracks of one mutant, the dominant Rol-6, which actually affects the structure of the cuticle rather than the nervous system. It affects the nematode's behaviour in a conspicuous way. (Photographs by Mark Blaxter.)

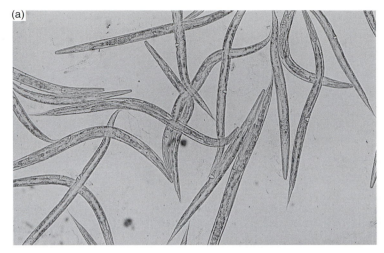

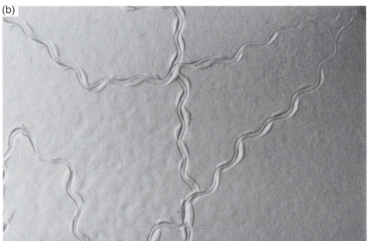

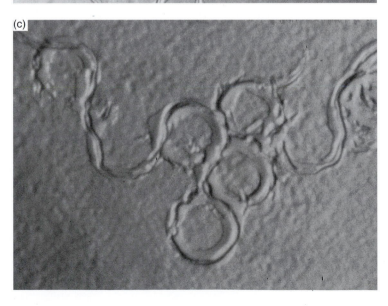

Perhaps our best hope for progress along these lines lies in the remarkable technical advances of modern molecular biology. It is now almost routine to isolate genes and generate multiple copies of them to discover what they produce in the cell. It is sometimes possible to identify patterns of gene production in embryos and hence to deduce where and at what stage particular genes are acting. It might become possible to locate gene action within the developing nervous system, but the fine scale required is very demanding.

In fact, a step towards discovering where in the central nervous system genes must act to affect behaviour has been achieved with *Drosophila* using more simple genetical techniques. Benzer and his colleagues (Hotta & Benzer 1976) have made use of some extraordinary genetic manipulations that are possible in this insect to produce 'mosaic' flies – flies that are genetically different in different cells of their bodies. The most interesting mosaics are *gynandromorphs,* individuals in which some cells are male, some female. About one third of *Drosophila*'s genetic material is carried on the X (sex) chromosome, of which males carry one copy and females two. If, during cell division in a female embryo, an X chromosome is lost from one of the two daughter cells, this cell and all its descendants from then on will carry only a single X and hence be male. Loss of an X is quite common in certain genetic stocks and can occur at any point in development. Thus the proportion of male tissue can vary from 50% (the X is lost at the first cell division of the zygote), to a minute patch of male cells only (see Fig. 2.13). So, if there are

Figure 2.13 Diagrammatic representations of mosaic or gynandromorph fruit flies. Female cuticle is represented dark, male cuticle light. The varying proportions result from the different stages of development at which an X chromosome is lost from one daughter cell following cell division in a female embryo. This cell and its descendants will become male. The top left hand example is exactly divided into male and female halves: the X chromosome must have been lost from one daughter cell at the very first division of the fertilized egg.

genes on the X which affect male sexual behaviour and these mutate, and, as is usual, the mutants are recessive, it may nevertheless be possible with gynandromorphs to identify which cells in the fly's nervous system must be male if it is to show masculine behaviour.

Male fruit flies vibrate their wings as they face females during courtship. They produce two types of vibrating 'song', so-called pulse song and sine song. Only if the mesothoracic ganglion has genetically male cells carrying the right genes will the insect produce pulse song. Though all its other central nervous system be male, if just these neurons are female in genotype, no pulse song is produced. This shows that the presence of genes *in particular cells* can determine how the fly behaves. Such a remarkable result is a consequence of the way insects are organized: each cell determines its own sex genetically. We could not have such results in vertebrates where sexual behaviour develops as a result of hormones secreted by the gonads which affect all the cells of the body.

It will be argued, and with justification, that this kind of information tells us little about the way genes act in the actual processes of development itself, although once again the new molecular techniques may help eventually. It will not be easy and for the moment we have only glimpses of how genes operate. One such glimpse comes from the work of Gwadz (1970) on wild populations of mosquitoes. The females of different populations of these mosquitoes become sexually receptive at different times after emergence. The females of one strain, called GP because it was originally collected at Gunpowder Falls, Maryland, are receptive to insemination quite early – a mean of 38 hours after emergence. Females of another strain, called TEX because it was originally collected at Austin, Texas, take much longer to accept males – with a mean of 120 hours (Fig. 2.14). Hybrids between the two strains (either GP/TEX or TEX/GP) have an intermediate time to insemination, with a mean of about 54 hours, but with a slope more similar to the GP strain. When the F1 hybrids were backcrossed to the GP and TEX parental strains, the results were compatible with the idea that early receptivity was due to a single, autosomal, semi-dominant gene. Gwadz also showed that late-emerging females of the TEX strain could be made receptive much earlier by application of the juvenile hormone, normally produced by one of the insects' endocrine organs, the corpora allata. Given early hormone artificially, they behaved like GP females. Thus the genetically controlled difference between the two strains may well be be due to changed timing of hormone production and this results in changes to the rate at which female receptive behaviour develops.

So here is an example of rather simple genetic control over a rather simple type of behavioural development. Such clear-cut results are not likely to be the rule. The development of most behavioural traits will be far more complex and the processes involved are almost certain to be affected by many genes. Furthermore, they will interact in ways that are not always straightforward. This means that the variation between individuals in the way their behaviour develops is not discontinuous, as with the female mosquitoes and hence amenable to simple genetic analysis, but continuous. For instance, mice cannot be classified as 'aggressive' or 'not-aggressive'; rather, they show widely varying degrees of aggressiveness. Continuous variation of this type requires the analytical methods of quantitative genetics and even then they can give us only a rather general overview of how behavioural potential is organized genetically.

The best source book for a wide range of behaviour genetics is still Fuller and Thompson (1978). Keverne (1994) reviews the modern molecular

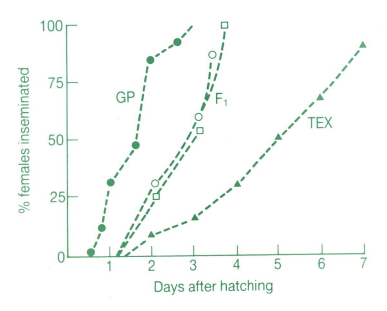

Figure 2.14 Onset of receptivity to insemination for two parental (GP and TEX) strains of mosquito, and the F$_1$ hybrids. (Open circles show GP/TEX hybrids and squares show the reverse TEX/GP hybrids.)

genetics in relation to behaviour whilst Hay (1985) provides an attractive introduction to the field.

For the moment we have to accept that there is very incomplete information on how genes control the development of the capacity to perform behaviour patterns. However, once the capacity is established we have better material and a better understanding of how genetic changes are involved in its modification in the course of evolution (see Chapter 6, p. 325). Meanwhile, our discussion of the processes of behavioural development can continue along other lines.

Development and changes to the nervous system

External events as well as genetic factors affect the way the nervous system develops. Gottlieb's work, described above, showed clearly that the auditory system of mallard ducklings requires experience – perhaps one might call it 'priming' – with certain types of sound if it is to develop normally and maintain its special responsiveness to the mother's calling. Specific external factors affect this behaviour through some effect on the developing nervous system. Obviously the development of behaviour must often run parallel to the complex process of building a nervous system. As with the study of gene mutations affecting the nervous system, sometimes we can deduce the changes underlying behavioural development only from observing changes in the physiological responses of neurons. Sometimes we can observe relevant structural changes to the nervous system directly.

To take a neurophysiological example first, it has been possible to study very closely the changes underlying the acquisition of visual reponsiveness in some mammals. It turns out that the process shares some features analogous to the development of auditory responses in Gottlieb's ducklings. Thus the eyes of many mammals develop all their basic structure in total darkness, but do not develop their normal functioning unless 'primed' with certain types of visual experience. Blakemore and Cooper (1970) reared kittens in complete darkness except for brief periods in a very visual environment (Fig. 2.15a). From 2 weeks of age when their eyes opened they spent an average of 5 hours per day looking at a pattern of either horizontal or vertical stripes. When first brought out into the normal world at 5 months the kittens were, understandably enough, disorientated. However, they quite quickly managed to cope but their vision remained abnormal. This was revealed dramatically when a 'horizontal' and a 'vertical' kitten were put together. If the experimenters presented them with a rod held vertically and shaken the 'vertical' kitten moved forward to play with it, but the 'horizontal' kitten ignored it. When the rod was turned to the horizontal, the 'horizontal' kitten immediately took interest and approached whilst the 'vertical' kitten no longer responded – it behaved as if the rod had suddenly disappeared.

Subsequently, Blakemore and Cooper used neurophysiological techniques to look at the functioning of neurons in the kitten's visual cortex. (This is the part of the brain which receives, in coded form, the visual information from the cat's retina.) Each of these neurons has a 'preferred orientation', which means that they fire when lines or edges at that orientation are presented to the eye. In a normal cat the orientations are distributed all around the clock, but vertically or horizontally reared kittens showed a very strong bias towards the corresponding orientation (Fig. 2.15b). Thus

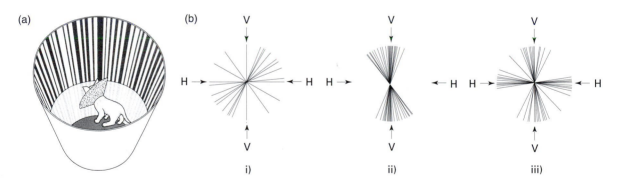

Figure 2.15 (a) Apparatus used by Blakemore and Cooper to restrict the visual environment of kittens. The kittens wore a black collar so that they couldn't see their bodies and stood on a glass plate inside a high cylinder whose walls carried vertical or horizontal stripes of varying thickness.

(b) Diagrams – so-called polar graphs – of the preferred orientation of visual cortical neurons of kittens. Each line represents one neuron and they are arranged according to their preferred orientation around the marked vertical and horizontal axes. (i) Neurons from a kitten reared in total darkness; (ii) from a kitten after 33 h exposure to vertical stripes in the apparatus illustrated in (a); (iii) from a kitten after 50 h exposure to both horizontal and vertical stripes, but no other visual experience. (Modified from Blakemore & Cooper 1970.)

horizontal kittens had no units responding to vertical or to lines or edges within 20° either side of vertical, they literally cannot see a rod when it is held vertically.

Such experiments reveal clearly how the early environment, sometimes before hatching or birth, sometimes later, affects the physiological functioning of the nervous system and hence shapes the development of fully functioning behaviour. At the level of morphology too we can regularly observe that behavioural development takes place in parallel with normal growth processes both of the nervous system and the rest of the body. Thus the emergence of sexual behaviour in vertebrates depends on the growth of the gonads which begin to secrete hormones. Very often the nervous and muscular systems of young animals have to go through further growth or differentiation before they reach adulthood and there are attendant behavioural changes.

For example, young birds can often be seen making vigorous flapping movements with their wings while still in the nest and it is commonly supposed that they are practising flying. Similarly, human parents often support young babies on their legs and encourage them to 'practise' walking. In fact there is no evidence that the early development of bird flight or human walking are affected in any way by such activities. Over a century ago Spalding (1873) showed that young swallows, reared in cages so small that they could not stretch their wings, flew just as well when released as normally reared birds. By the age of 18 months, when the majority of children are walking, their skills are very uniform whether they walked first at 9 months or at 15 months. In both cases it is the maturation of the central nervous system and its coordination with muscular development which count. Practice, of course, eventually adds all the finer points of skill – young fledglings are clumsy fliers – but the basic pattern emerges without it.

Because the developing embryos of amphibia and birds are relatively accessible it has been possible to investigate more closely the growth of their nervous systems and behavioural changes. As development proceeds, the increasing complexity of an embryo's structure is paralleled by an increasing repertoire of behaviour, both spontaneous and in response to external stimuli. Oppenheim (1974) describes work on chicks showing how movements appear quite suddenly just at the stage when the requisite connections between growing nerve fibres are made. Sensory fibres grow in to the spinal cord from the muscle spindles and then can provide information on the state of contraction of a muscle (Chapter 1, p. 32). These fibres must form their central connections before that muscle can take part in coordinated movements with others.

The gradual nature of development in crickets enabled Bentley and Hoy (1970) to study one example of neural growth in great detail. Crickets go through a series of 9–11 larval or nymphal stages (instars) which become increasingly similar to the adult with each successive moult. The final moult involves the biggest changes, with the full development of wings and adult genitalia. Since they have no wings, cricket nymphs do not fly, but some elements of the flight pattern can be detected as early as the 7th instar. Nymphs at this stage will adopt the typical flying posture (see Fig. 2.16) when suspended in a wind tunnel. All the muscles necessary for flight are present from an early stage, albeit reduced in size. Bentley and Hoy inserted fine wire electrodes into some of these muscles of the thorax and recorded action potentials from them. Since the relationship between input from motor neurons and the output from the muscles is relatively simple in insects, the muscle activity gives a very accurate picture of the nerve activity responsible for it. Bentley and Hoy could detect signs of the rhythmic nerve impulses characteristic of flight from 7th instar nymphs, but it was incomplete and not sustained. It became more complete with each successive instar; new elements

Figure 2.16 A 7th instar nymph of the field cricket (*Teleogryllus commodus*), adopts the flying position when suspended in a current of air, although it has no wings. The antennae point ahead, the fore- and mid-legs are drawn close to the body and the hind legs are held straight back parallel to the body. All the features are identical to the position of the adult in suspended flight. (Drawn from photographs in Bentley & Hoy 1970.)

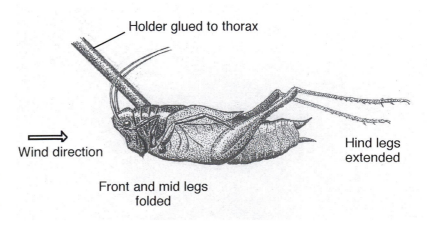

Holder glued to thorax

Wind direction

Front and mid legs folded

Hind legs extended

could be identified as becoming active, presumably as functioning synaptic contacts were established. Although timing varied between individuals, the units always developed in the same sequence until, by the last instar before the final moult, the entire pattern was complete. At this stage, then, the maturation of the nymph's nervous system is finished and it awaits only the development of wings. Substantially the same story is true for the cricket's song, another activity which requires the wings of an adult (they are rubbed together rapidly to produce sound) for proper expression. For both flight and singing Bentley and Hoy found that they could not elicit neural activity from the thoracic centres of nymphs unless they made lesions in the brain. This presumably removed inhibition, which normally prevents the nymphal muscles from being stimulated into useless activity until the rest of the body's development has caught up.

The crickets and their relatives show gradual development towards the adult form which provides a gradually growing morphological framework in which to build up the neural basis of adult behaviour. Insects with complete metamorphosis have to reorganize their nervous systems quite radically during the pupal stage. In experiments similar to those on crickets, Kammer and Rheuben (1976) could detect patterns of activity in the developing muscles of the pupae of some large moths which indicated that the neural basis for both flight and warming up (shivering) movements have developed before half-way through the 21 days of the pupal stage. Elsner (1981) reviews these and other studies in more detail.

Studies on the changes to the nervous system of insects during larval growth and metamorphosis have revealed some remarkable parsimony in their development (Levine 1986). Insects sometimes modify and re-shape the same neurons for use in both larval and adult forms. Levine and Truman (1985) have shown that in the hawk moth *Manduca sexta,* all the motor neurons supplying the abdominal muscles are used again. Different as the two body forms are, the abdominal movements of the caterpillar and those of the adult moth may not be totally dissimilar. Some of the motor neurons can stay innervating the same muscles, but others lose their specifically caterpillar connections at pupation and grow new processes to innervate newly formed adult muscles (Fig. 2.17). Most larval sensory neurons supplying sensory hairs die away at pupation but one group come to innervate specialized trigger hairs on the abdomen of the pupa. They trigger a reflex, unique to the pupa, which causes it to flex suddenly in a movement which defends it against predators.

And so behaviour patterns mature, flourish and disappear, correlated with growth, modification and death of the underlying neurons. It all looks very much like the running of a pre-set programme of development, but this is a description, not an explanation. Close examination of neuro-embryology can help us to understand what 'pre-set' implies and this certainly exemplifies how neurobiology is an aid to studying behavioural development itself.

Again, insects are excellent material for studying the way in which a nervous system is built up during development. For example, in grass-hoppers the axons of sensory neurons from the developing legs grow out from cells in the sensory hairs and make connections with the central nervous system. Bentley and Keshishian (1982) have shown that the first axons to grow this way – the pioneers, as they call them – use special cells

Figure 2.17 How three motor neurons of the moth *Manduca* are 'remodelled' during metamorphosis. Individual cells can be identified regularly in all moths and injected with a cobalt dye, which reveals the rich arborization of the dendrites. The top row shows the relatively simple dendritic pattern of each cell in the larva – cell body and axon to the larval muscle are clear. After metamorphosis into the adult each cell has a greatly enlarged dendritic field, which has grown into areas not penetrated in the larval state and the axons have, in fact, lost their target larval muscles and now innervate new muscles in the adult moth. (Redrawn from Levine 1986.)

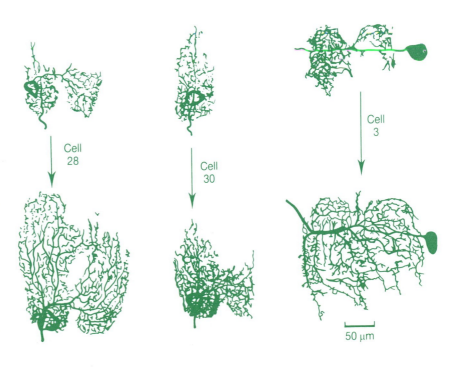

Cell 28

Cell 30

Cell 3

50 μm

67

situated at intervals along the route as 'guide posts'. As the limb grows and lengthens, the guide posts are stretched apart and later axons do not use them but grow along the line already established by the pioneers. Thus, although the final result looks uniform – sensory axons running smoothly into the central nerve cord – the actual developmental histories of those neurons have not been uniform. There is also considerable flexibility in the way in which detailed synaptic connections are formed as sense organs develop and their sensory axons reach their central destinations. Which cells they connect with and how they do it depends upon what other cells are in the vicinity and what sort of inputs are arriving from the sense organs. If inputs are abnormal or the cells with which they would normally have synapsed are missing, they may be able to compensate, at least to some extent. There must be a genetically based programme for development but it is much more likely to lay down ground rules rather than a specific point-to-point blueprint. Murphey (1985, 1986), Stent (1980; 1981), Keshishian & Chiba (1993) and Katz & Shatz (1996) provide interesting discussions of such evidence which is certainly relevant to the development of behaviour itself and its genetic basis. We shall meet similar concepts of flexibility, predispositions and sensitive periods when we come to discuss the development of bird song (p. 103), where apparently similar behavioural end points may have been attained by markedly different paths.

Hormones and early development

It is appropriate to discuss here some of the fascinating discoveries about early sexual development, because they involve clear developmental changes to the nervous system with major behavioural consequences. Sexual development, as we mentioned earlier, is a clear case of alternative pathways, along one or other of which an animal's development, both morphological and behavioural, is switched by an early 'triggering event'. Because being male or female is such a pervasive difference in ourselves and the animals we are most familiar with, it is important to keep some broader biological perspective and remember that a good number of invertebrates are hermaphrodites and have only a single developmental pathway to becoming both male and female. Even more strange in some ways, a number of fish species retain the capacity to switch paths and behavioural factors are involved in this change (see Francis

1992). For example, in the coral reef fish *Anthias* all individuals develop into females, but some transform later into males. They normally live in mixed groups and if a male dies, one of the females changes sex. This change is very rapid: it takes only a few days for the previously female fish to develop male features and begin making sperm in its gonads rather than eggs. But even earlier, within hours in some cases, its behaviour begins to change and it is treated like a male by other females in the group (Shapiro 1979; Fig. 2.18). As if to demonstrate that all variations are possible in nature, the anemonefish – also a coral reef species with a remarkable symbiotic association with large sea anemones – does the opposite! It normally begins life as a male, mating with a larger, more dominant female. If she dies, or is removed, the male changes into a female and another, hitherto non-breeding male takes up the fertile male role (Godwin 1994).

For most vertebrates the developmental switch to one sex or the other is permanent. In some reptiles it appears to be controlled by the temperature at which the eggs develop. In Mississippi alligators, Bull (1980) finds that shaded nests, where temperatures are slightly lower and incubation slightly longer, have a high proportion of males whilst nests in the open sunlight of the river-side produce mostly females. Birds and mammals do not rely on such environmental triggers to achieve a balanced sex ratio. With them the switch is a genetic one, depending on the segregation of one pair of sex-determining chromosomes in the gametes. (In mammals we now know that it is one gene carried on the non-paired part of the Y chromosome which induces the embryo gonad to become a testis; Goodfellow & Lovell-Badge 1993).

The fish which can change sex demonstrate that they inherit the potential for both types of sexual behaviour. The same must be true for mammals and birds, where the two sexes share almost all their genes. The 'problem' for natural selection has been how to make them develop differently. In morphology we can see that sometimes it has not been worth differentiating between the sexes. Male mammals develop functionless mammary glands and nipples; some male marsupials develop the bones which support the female's pouch. For those aspects of morphology and behaviour where sexual differentiation, is vital, the switch is a hormonal one. Once the gonadal sex – ovary or testis – has been determined, so is the pattern of hormone secretion. Birke (1989) and Short & Balaban (1994) review the complex story of sexual differentiation

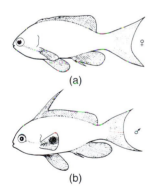

Figure 2.18 (a) Female and (b) normal male of the coral reef fish *Anthias sqamipinnis*. The males are brightly coloured and have a conspicuous dark spot on the pectoral fin. The females are a uniform orange–gold colour; they begin to show colour changes towards the male type within 3–6 days of the removal of the male from a group. (Drawn from photographs in Shapiro 1979.)

which has many interesting variants across the mammalian groups. Here we must summarize the main events as they affect behaviour.

In mammals, the differentiation of the genitalia and of those neural mechanisms responsible for initiating sexual behaviour is not determined at fertilization, but much later in development. Both males and females inherit a brain which can mediate both masculine and feminine behaviour; which pattern becomes dominant depends on whether the brain receives a pulse of hormone. The embryonic or newborn testis, unlike the ovary, has a brief period of activity and its hormone is picked up by particular regions of the hypothalamus. This is a small region of the brain much involved in regulatory activities of all kinds, and which controls the functioning of the pituitary gland that lies attached to and immediately below it. (The anatomy of these structures crucial for the control of sexual behaviour is described in more detail in Chapter 4, p. 214).

If the developing hypothalamus picks up hormone during a certain brief period, it is switched along masculine lines; if not, it becomes feminine. If females are given tiny quantities of hormone at the right stage, their behaviour is masculinized. It is interesting to note that either the male or female sex hormones, testosterone and oestrogen respectively, will serve to masculinize male or female embryos – they are both steroid hormones and closely related in chemical terms. Denying male embryos their pulse of hormone at the normal time or removing the testes before they secrete, or injecting a chemical which inhibits the action of testosterone, results in de-masculinization. We may note that this is a clear case of a critical period in development (see p. 97), for hormone treatment earlier or later has no effect. The die is cast and, for example, a male de-masculinized by early castration will not have his masculinity restored by large doses of testosterone given later in life. When adult, masculinized females show changed sexual behaviour, sometimes dramatically so. Thus in some rodents such as rats and mice, masculinized females will mount and thrust on receptive females and eventually go on to perform all the behaviour patterns which are shown by males when they ejaculate; they also show masculine type aggressive behaviour, (Manning & McGill 1974). In addition, such females are 'de-feminized' in that their vagina fails to open, they lack oestrus cycles and are consequently sterile. Female primates can readily be masculinized, and as with the rodents, it is not only specifically sexual behaviour which is altered. They are more aggressive and as infants show more of the 'rough-and-tumble' patterns of play which are characteristic of males. However, unlike rats and mice, such females show little sign of de-feminization for they remain fertile and can reproduce. Actually it is rather difficult to generalize about the effects of early hormone treatment on behaviour, because the extent and nature of the critical period varies between species. In sheep, for example, it extends for some weeks during mid-gestation and there are different critical periods for masculinization of genital development and for sexual behaviour itself. By closely time-limiting a female foetus's exposure to hormones, it is possible to produce sheep whose apparent genital sex and their sexual behaviour do not match up. In primates it is also the middle period of gestation which is critical but in rats and mice it is very late in development and in fact can be extended until a day or two after birth.

Careful observations have shown that the masculinization of females is not simply an artificially induced condition in animal behaviour laboratories. Female mice produce 8–10 pups after a 3-week pregnancy in which the foetuses have been developing in a row along each horn of the uterus (Fig. 2.19). Each has its own placenta but it appears that some mingling of hormones must occur between adjacent foetuses. Vom Saal and Bronson (1980) found

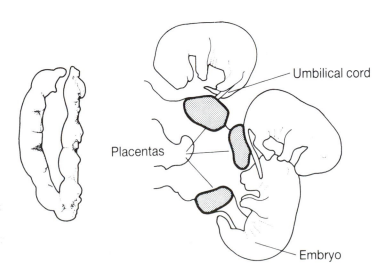

Figure 2.19 Left: the two uterine horns of a female mouse about two-thirds of the way through pregnancy. The rows of developing foetuses can be clearly made out, strung out like peas in a pod. Right: mouse foetuses dissected out to show each with a separate, disc-shaped placenta attached to the wall of the uterus. No matter how separate they appear, some diffusion of testosterone must occur between adjacent foetuses (see the text for further explanation.)

Umbilical cord

Placentas

Embryo

that, at birth, male foetuses had three times the level of circulating testosterone in their blood as did females – this is presumably a result of the pulse of secretion by the testis which we know occurs around birth. Having delivered mouse pups at full term by Caesarian section, they noted the position each female had occupied in the uterus. They compared females which had developed between two male foetuses (2M females) with those who had had two female neighbours (0M females) and found a number of significant differences. Most interestingly, they found significantly more testosterone in the blood of 2M females than their 0M sisters and they could correlate this effect – presumably due to leakage between adjacent blood systems – with a number of morphological and behavioural differences which extended into adult life. 2M had slightly masculinized genitalia, had longer and less regular oestrus cycles, showed more aggression towards other females and in defence of their young, and were less attractive to males than 0M in a choice situation. This does not necessarily mean thay are at a disadvantage. Vom Saal and Bronson suggest that under crowded conditions, the increased aggression shown by 2M females may improve their survival and that of their young. It is probably a case of swings and roundabouts in fluctuating environmental conditions. Such naturally arising variation within a litter may in fact be advantageous for the parents. Consequently, natural selection may not act so as to insulate female foetuses better from their brothers' hormone.

And so it appears that some of the normal variation we expect to find between females may have its origin in the chances of foetal life. For centuries it has been known that if a cow gives birth to twin calves of different sexes, the female calf very frequently has malformed gonads and uterus and hence is sterile. This so-called 'freemartin' is an androgenized female resulting from her sharing some blood circulation between her placenta and that of her brother in their mother's uterus. In human beings, too, there are clear effects when female embryos are exposed to male hormones during pregnancy. These are not detectable when dizygotic twins are of different sexes but happen when a single female foetus has a rare genetic abnormality which leads to a condition known as Congenital Adrenal Hyperplasia (CAH). The adrenal glands of the foetus develop abnormally and release a masculinizing hormone into the bloodstream. When CAH girls are born their genitalia are often somewhat masculinized but this can be corrected with surgery and they can grow up to be fully functioning females in every sense (see Fig. 2.20). However, it is fascinating that the effects of masculinization can be detected in their behaviour as children. Ehrhardt and Baker (1974) have shown that CAH girls are more 'tomboyish'. For example, they indulge in more rough play, prefer boys to girls as playmates and tend to reject dolls as playthings. Now, we know for certain that such preferences can be greatly influenced by parents and the particular culture in which children grow up (see Money & Ehrhardt 1973). However, it seems very probable that the CAH girls were indeed treated as girls by their parents and that the changes to their behaviour were spontaneous.

Ehrhardt and Baker are cautious in their interpretation of how early hormones may act. It might bias infants towards specifically male patterns of behaviour; but in humans these would always be interpreted by a developing child in ways which reflect the 'male role' as presented by its culture. Of course, it may not be a specific effect at all. Tiefer (1978) has suggested that early androgen may do no more than increase metabolic rate and hence energy levels. An especially active girl infant will be most likely to end up playing with other active infants and they will be boys; hence the other male-orientated patterns follow. We may note that although the behaviour of CAH boys appears qualitatively unchanged, they do have raised energy expenditure.

The early pulse of hormone is picked up by the hypothalamic region of the brain. Here the pre-optic nucleus (so called because it is a discrete cluster of neurons situated just anterior to the point where the optic nerves enter the brain) is a particular centre of hormone concentration. The pre-optic is far larger in male rats than in females and this difference reflects the hormone received in early life. Oestrogen or testosterone added to cultures of brain cells from the hypothalamus cause a great surge of growth but they have no such effect on cells from other regions of the brain. Presumably the differentiation of the pre-optic nucleus is one of the maturational changes leading to the organization of male or female sexual behaviour. Certainly, this region also concentrates hormone during periods of sexual activity when the animal is mature. Recent studies have revealed that the brains of males and females, of mammals and other vertebrates, develop slightly differently in a number of ways and it has sometimes been possible to offer functional explanations

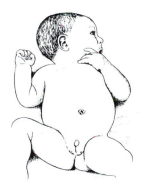

Figure 2.20 Masculinized appearance of the external genitalia of a baby girl whose adrenals have been hyperactive during foetal life, the CAH syndrome. Sometimes surgical correction is advised but such girls can grow up as fully functioning females. Similar masculinization can result from treatment of a pregnant mother with progesterones. These used to be given when miscarriages were threatened but can act as the steroid pulse which sets sexual development along more masculine lines. (From Keverne 1985.)

relating this to the behaviour of the two sexes (see Breedlove 1992; Jacobs 1996).

We tend to think of such developmental effects as part of the growth process which ends with the attainment of maturity. With a few exceptions, such as the extensive powers of regeneration in the urodele amphibia, it has been generally accepted that the nervous system of vertebrates shows no further growth or cell division in adults. However, recent work with canaries reveals seasonal changes in those parts of the brain responsible for the control of singing. Figure 2.21 is a simplified diagram representing a longitudinal section of a song bird brain showing the main centres controlling singing. On the motor output side there are two higher nuclei, HVc (which might well be assumed to stand for 'Higher Vocal centre', but actually relates to its neuro-anatomical name, hyperstriatum ventrale, pars caudalis!) and RA (*R*obust nucleus of the *A*rchistriatum) feeding down to the nucleus of the XIIth cranial nerve which controls the bird's voice box, the syrinx. Nottebohm (1989) and

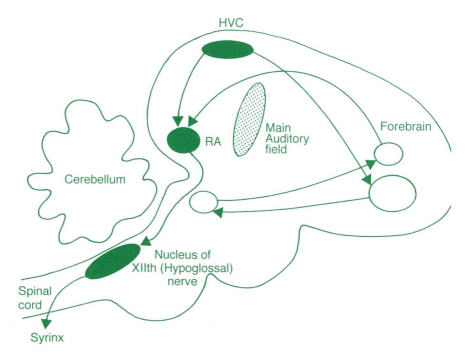

Figure 2.21 A simplified diagram of the main neural centres concerned with the control of song in a bird such as the canary. The brain is depicted in longitudinal section, anterior to the right. The dark-shaded nuclei are those, referred to in the text, controlling the descending motor pathway for song production. The unshaded nuclei, especially those of the forebrain, are known to be involved with song learning. Main nerve tracts linking the nuclei are indicated by arrows. (Modified from Catchpole & Slater 1995.)

74

his group (see also Bottjer & Arnold 1984; Balthazart & Ball 1995) had shown some time ago that HVc and RA were much larger in male canaries than in females. Further, they found that the size of these centres fluctuated markedly through the year. At the end of each breeding season they diminished greatly in size, only to enlarge again early in spring, as the canaries began to take up singing again. These cyclical changes are under the control of testosterone and by injecting a radio-labelled form of the amino-acid thymidine, which is taken up only by newly forming cells, Goldman and Nottebohm (1983) produced convincing evidence that the recrudescence of HVc was due to its incorporating newly formed nerve cells. This conclusion has since been amply confirmed – Catchpole and Slater (1995, chapter 2) give a full account of recent work – which proves that, at least in some advanced vertebrates, neural death can be followed by neurogenesis under hormonal control.

These changes are associated naturally with the canary's annual cycle of singing and the learning of new song types each breeding season (see below, p. 103). Such neural changes are quite distinct from, for example, the differentiation of the pre-optic nucleus in rodents, for not only are they reversible and continue into adult life, but they can be shown by female canaries as well as males. Testosterone injections lead to the growth of the appropriate brain regions of adult females and they begin to sing. (Canaries may be rather exceptional in this, for in some other birds adult females cannot be induced to sing in this way.) These remarkable results illustrate the value of a comparative approach which makes use of the diverse specializations that can be found if we look beyond the conventional set of laboratory animals. Song birds are proving to be really useful for linking given volumes or numbers of neurons with 'amounts' of information (i.e. how many song variants are learnt) being stored and hence give a very direct approach to memory processes (see Chapter 5). The work with canaries is also forcing a new look at mammalian brain growth and is a good example of the combination of behavioural, anatomical and physiological studies – real neuroethology.

Early experience and the diversity of parental behaviour

So far we have been concentrating on behavioural development which proceeds along guidelines parallel to and influenced by the normal growth processes of the young animal. In the course of such maturation the young animal will be interacting with its environment both animate and inanimate, but not all individuals will undergo identical experiences. We must now turn to examine some results of such early experience.

A great deal of work on a variety of animals, especially birds and mammals, has shown how sensitive they are to events when young. We have already discussed Gottlieb's work on the effects produced by the experience of sounds young ducklings hear while still in the egg. Auditory responsiveness in the egg may have surprising results. Vince (1969) showed that embryo quails respond to the clicking sounds made by the other members of the clutch. Remarkably, this communication between embryos serves to synchronize their hatching. Slow developers are accelerated by hearing the calls characteristic of advanced embryos and, to a lesser extent, advanced embryos are slowed down by the calls from less advanced neighbours. In nature this synchronization may be important, because it will reduce the dangerous period when there are both chicks and unhatched eggs in the nest before the mother quail can lead the whole brood away.

In mammals, the embryo is more insulated from the external world but, of course, more directly dependent on its mother's physiological state. If female mammals are stressed during pregnancy this may have significant effects on the development and behaviour of their offspring. They may become more fearful in their approach to novelty and their sexual development may be affected (see Sachser & Kaiser 1996, for an example from rodents and Clarke *et al.* 1994, for one from primates). Similar effects can be produced by a variety of treatments to young animals after birth. Mild electric shock, cooling, or even the simple act of lifting them from the nest and then replacing them can affect their subsequent behaviour as adults. Rats which have been handled during infancy mature more quickly – their eyes open sooner and they are heavier at a given age than unhandled controls. In addition, they appear less 'emotional' in frightening situations. By this we mean, for example, that when placed in a large, brightly lit arena they move around and explore more than unhandled rats, which spend more time crouching and cling to the edges of the arena if they do move.

Such effects and the way in which early experience produces them are not fully understood, nor is it a very easy subject to follow because, as with many complex questions, different studies do not always yield consistent results. The nature of emotionality and its relationship to fearfulness and other stress-related responses is certainly not straightforward. Archer (1973, 1979) provides critical reviews of the various measures of emotionality used with rats and mice. Probably the reason we can detect any effect at all from such apparently trivial manipulations is that we are measuring them against the background of the laboratory-reared animal which is clearly deprived of much stimulation. Rather than emphasizing how early handling 'adds' something we ought, perhaps, to recognize that laboratory rearing takes something away. In the wild it is common for a mother rodent to abandon her nest and shift the litter to a new site, sometimes more than once. Several mothers may share a nest at times and the environment around the nest, when the young begin to emerge at about 12–15 days of age, is usually far more complex than a laboratory cage. A number of studies have found that rats and mice which grow up in 'enriched' environments show improved learning in a variety of situations. For example, Joseph (1979) reared rats under two conditions, one group were kept in standard laboratory cages, the other group were given a variety of 'toys' – cups, ladders, corks, bottles, etc. – and these were changed regularly. Also their cages were moved around within the animal house room every other day, whereas the control group's cages remained in one position as is normal practice. The enriched rats proved to be markedly superior in maze learning and they also explored new environments more actively. There are a number of such studies and we know something of the neural changes which underlie such effects on behaviour. It has been possible to show that the brains of enriched animals are slightly heavier than those of unstimulated controls, they have a richer pattern of interconnections and they contain higher concentrations of key enzymes concerned with neural functioning (Rosenzweig 1984).

The 'extra' stimulation of young animals can certainly act on them directly so as to change their subsequent development but, where very young animals are concerned, there will also be effects through their mother. Such 'enrichment' experiments break into the normal intense and almost continuous communication system that operates between a mother rat or mouse and her growing litter. The pups provide a variety of stimuli and in particular have a range of calls, the most important of which are ultrasonic and far above the limits of our hearing. Disturbing the pups in any way – as when they are handled – leads to an increased rate of calling and the mother responds (see Cohen-Salmon *et al.* 1985, for an introduction to such work). A litter of pups may get scattered and if, for example, some towards the outside of the nest get chilled, their outburst of calling will accelerate the mother's retrieval of them back to the nest cup. As the pups get older their normal rate of calling decreases, they move around voluntarily and the responses of the mother are less intense. The behaviour of both mother and developing young are reasonably balanced so as to promote successful weaning. Elwood (1983) and Fleming & Blass (1994) have reviewed a range of work on parent–offspring interactions in rodents.

Whatever the influence of parents on young, we must remember how much variation there is in the amount of contact between them, even within a single group like the mammals. The details of the developmental pathways towards behavioural independence must be equally diverse. Marsupial infants are born after a very brief gestation period and they are nothing more than externalized foetuses at first, fused to the mother's nipples. Their sensory capacities are so restricted at this stage that mother–infant interactions can only be through the milk supply, but even when they become more independent they are constantly with the mother, carried on her body and often in a pouch (Tyndale-Biscoe 1973; Fig. 2.22a, b). This represents a far higher degree of contact than the part-time attention received by rats and mice. At the other end of the scale are rabbits, where Zarrow *et al.* (1965) showed that the mother visits her litter only once per day and then for only a few minutes. Milk is pumped into the babies, they are briefly groomed and then the nest is covered and left; the mother herself has another nest chamber elsewhere. Presumably this behaviour helps to minimize the chances that predators will find the nest by picking up the mother's scent. A similar explanation is suggested for the yet more extreme maternal behaviour of tree shrews – small insectivore-like creatures whose relationship to other mammals is uncertain. Martin (1968a) (see also D'Souza & Martin 1974) observed pairs breeding in captivity and found that they build two separate nests, in one of which the adults sleep while the female produces her litter, usually of two, in the other. The young are visited briefly on every second day; they suckle rapidly and become distended with milk (Fig. 2.22c). The mother then leaves them with the very minimum of grooming and typically does not return for 48 hours.

* As mentioned, the taxonomic position of the tree shrews has been the subject of debate for many years. Although they were often associated with the Primates, Martin (1968a, 1990) has discussed their many distinctive features. He further speculated that, however different they may appear, the striking similarities in maternal behaviour between rabbits and their relatives (Order Lagomorpha) and tree shrews might suggest a relationship (Martin 1968b). It is thus very intriguing to find that recent molecular evidence (Graur et al. 1996) has shown affinities between these two groups and places them rather separately from the other mammals. It could be that both inherited their pattern of maternal care from a common ancestor and that, in this case, behaviour has been more stable than morphology over evolutionary time.

Martin discovered that the young are capable of grooming and licking themselves from the day of birth. They stay alone in the nest until they emerge at 33 days of age.*

For a further contrast we may compare the tree shrews, whose maternal attention, although brief and intermittent, extends over 4 or 5 weeks, with the extraordinary specializations of some seals. These large mammals come ashore only to breed and for some arctic species their time on land, or more often on floating pack ice, is very hazardous. Pack ice is always liable to break up and move off in sea currents and even if it remains stable, mother and pup are threatened by foraging polar bears. Perry and Stenson (1992; see also Riedman 1990) describe how hooded seals (*Cystophora cristata*) have adapted to this situation. The female hauls out onto an ice floe and almost immediately gives birth to a single pup, which is born with a layer of blubber beneath the skin. She remains alongside it continuously with frequent and intense bouts of suckling (Fig. 2.23) and the pup grows at a phenomenal rate: the seal's milk is some 60% fat. The mother then abruptly abandons the pup to return to the sea and mate; this whole process from birth to desertion all takes place within *four days*!

This is a very remarkable type of development for a mammal and it forces us to recognize that natural selection can produce a range of specializations in the pattern of development which are matched to other aspects of a species' life history. The influence of the mother on the earliest stages of behavioural development must be very different in the marsupial, rat, rabbit and tree shrew. Even close relatives can vary enormously in their stage of development and independence at birth: we can contrast the helpless blind naked young of rats and rabbits with the active young, fully furred and with eyes open, born to guinea pigs and hares. As we have seen, some seals are suckled for a few days only but Californian sea-lions stay close to their slow-growing offspring for 6–9 months. There are no predators menacing them on the rocky islets where they breed, and mothers can go off to feed between bouts of suckling.

Not surprisingly, it is with long-lived, highly social animals that parental behaviour is most richly developed. Our closest relatives, the primates, have some of the longest and closest associations between parents and offspring, lasting for 6 or 7 years and beyond in the great apes, as in ourselves. The rela-

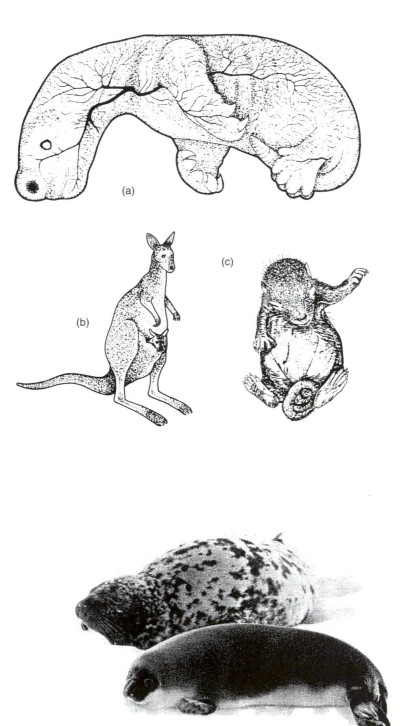

Figure 2.22 Contrasting mammalian young and contrasted maternal styles. (a) A newly-born kangaroo, minute in relation to the size of its mother, is effectively still an embryo. The forelimbs haul it into its mother's pouch where it attaches to a nipple. The prominent blood vessels run just beneath a moist skin and at this stage the young kangaroo may well breathe as much through its skin as through its embryonic lungs. (b) Many months later, as a large subadult it still retreats into the pouch when alarmed. (c) A newly born tree shrew, its abdomen distended with milk pumped in immediately after birth by its mother who will now leave it alone for 48 h with its other litter mate. It has only thin fur and its eyes are yet to open but it has well-developed limbs with fully clawed digits: young tree shrews groom one another from an early age. Compare with Fig. 2.23. (a and b from Tyndale-Biscoe 1973; c from a drawing by R.D. Martin.)

Figure 2.23 A hooded seal mother with her pup resting on Arctic pack ice where she gave birth. The pup is born with a layer of blubber and, after 4 days of intense lactation, the mother will leave it to fend for itself. (From a photograph by Kit Kovacs in Riedman 1990.)

tionship between a primate mother and her offspring is absolutely crucial for almost every aspect of behavioural development. Separation, particularly in the early months, can have very severe effects on the infants. In the experiments of Harlow and his group (Harlow & Harlow 1965) young rhesus monkeys were isolated at birth and reared artificially. Although they grew well enough they were behaviourally crippled and, when subsequently put with other monkeys, showed almost none of the normal social responses. In particular they showed totally inadequate sexual and parental behaviour. Male isolates did not respond normally to receptive females; if female isolates became pregnant and gave birth, they totally ignored their offspring.

This is another illustration of the subtlety of behavioural development. If we watch a young monkey with its mother, it is not easy to identify in its behaviour those processes which are contributing crucially to the infant's social development, yet they must be there. The work of Hinde's group, well described in his book (1974) followed in great detail the normal development of rhesus monkeys reared by their mothers living in small groups. They could trace the gradual growth of independence due in part to the infant, in part to the mother. At first the infant is scarcely ever out of contact. The mother rarely allows her infant to move beyond arm's reach and even when it leaves its mother to explore it returns frequently, using her as a secure base from which to investigate new objects and new situations. Gradually, in parallel with the infant's increasing independence, the mother becomes less solicitous and even begins to reject some of the infant's approaches (Fig. 2.24).

Knowing the normal course of development, Hinde and his co-workers then went on to study how it was disturbed by forced separation of mother and infant. The levels of deprivation they studied are far less drastic than those imposed by Harlow. In a typical set of observations a baby rhesus and its mother, living in a group, were watched regularly for some weeks. Then when it was 6 months old – well able to feed itself – its mother was removed from the group for a few days. Far from being isolated, the baby was usually 'adopted' by other females of the group and given a great deal of attention. Nevertheless its behaviour showed a marked change, distress calls increased, it moved around less and spent long periods in a very characteristic hunched posture. When the mother was returned, there was usually an instant reunion and the infant spent much more time clinging to her than just before separation. The pattern of their relationship was different from that normal with a 6–month infant and it took several weeks to recover.

Several fascinating conclusions, with obvious human parallels, have emerged from systematic studies of this type. For instance, the infants worst affected by the separation were those whose relationship with their mother before separation appeared to be less good. From our point of view, in this discussion of early experience and development, we may note how profoundly the normal smooth adjustment of mother–infant relations was affected by even a brief interruption. The effects on the young monkey were both general and persistent, for Hinde could detect even years later that monkeys which had been separated were more fearful in strange environments.

Just as within other mammal groups, so observations on primates in the wild have revealed striking differences between species in what we would call child-rearing practice. Thus mother langur monkeys are quite permissive in that they allow other females to take their babies from time to time. Rhesus monkeys are far more restrictive, scarcely allowing their babies to leave their side for several weeks and controlling outside contact for much longer.

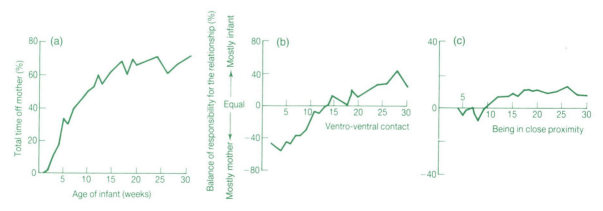

Figure 2.24 Three facets of the changing mother–infant relationship in rhesus monkeys taken from the work of Hinde's group. (a) The increasing proportion of time infants spend off the mother (i.e. not in contact with her body, although it is often very close) as they grow. (b) and (c) are more complex measures. (b) The ratio of responsibility for ventro-ventral contact, i.e. whether mother or infant initiates or terminates bouts of clinging; (c) is a similar measure for close proximity between them – being within arm's length effectively. Further details are given in Hinde 1974.

However, a close relative of the rhesus, the Barbary macaque of North Africa, behaves quite differently. Mothers allow other individuals, males as well as females, to hold their babies even within a few days of birth. Deag and Crook (1971) described how young babies can be seen moving about in the group not obviously under the care of any particular adult. It is a common observation for all primates that newborns and young generally are objects of intense interest to other members of a group, particularly young females. 'Aunting' behaviour has frequently been reported from primates and is also an important part of parental behaviour within the matriarchal groups of African elephants. Lee (1987) has shown that young elephants do best if they grow up in a group where they have not just their mother's attention, but that of other females and juveniles as well.

Even within a single species mothering styles can vary substantially, depending on circumstances. Altmann's (1980) long-term study of baboons describes the behaviour of young adults, and suggests how differences in the social, exploratory and feeding behaviour of young animals relate to the type of mothering they have received. Are there then 'good' and 'bad' mothers and ought not natural selection to eliminate the latter? Altmann's conclusion is that in a varying and uncertain environment, restrictive or permissive mothers may sometimes be 'good', sometimes not. Perhaps natural variations in the pathways of behavioural development are not, over a lifetime of reproduction, disadvantageous but quite the opposite, a point we have already made (p. 72) when considering variations in the development of females within a mouse litter.

We need to recognize that the interests of parents and young do not always coincide. Trivers (1974), in an influential review, referred to **parent/offspring conflict**. The latter wish to get a lot of help from their parents before they are forced to face the world alone. The parents, at least of those species which reproduce more than once in their lives (e.g. most birds and mammals) have to balance their investment in the current set of young with how such investment will leave them placed to invest in the next clutch or litter. They may well need to husband their resources and, as we have seen, some of their behaviour may be directed towards accelerating the independence of their offspring. We return to this issue in Chapter 6, for it has important implications for the evolution of behaviour.

Play

For anyone who has enjoyed watching kittens or puppies grow up, the most conspicuous aspect of their behaviour is play. They spend long periods playing with each other and with objects but, although adults may continue to play intermittently, it does tend to fade out with age. It is commonly suggested that play has a role in the development of adult behaviour. Indeed, there is an enormous literature on the importance of play in infancy and childhood for the normal development of human beings (e.g. Bruner *et al.* 1976).

In animals we can try to argue from analogy, but with care. If we observe that any animal spends significant periods of time engaged in a certain activity, we would, as a first assumption, conclude that it is important for survival. Play is a very energy-consuming activity in ourselves but there is some debate as to how far this is true of other species. Estimates are that between 5 and 20% of the 'surplus' energy available, i.e. that not required for metabolism and for growth, is expended in play. Even at the lower figure, this is not a trivial amount and play clearly imposes a cost. The cost is sometimes a physical one; young desert bighorn sheep (*Ovis canadensis*) cavorting in play through their harsh environment are often pierced by cactus thorns; young gelada baboons (*Theropithecus gelada*) playing on the steep cliffs of Ethiopian gorges have been seen to fall 5 or 10 metres and limp painfully away! Another significant cost arises because vigilance is bound to be reduced when young animals are playing and this will make it potentially dangerous. For example, Harcourt (1991) finds that young fur seals spend about 6% of their time in play swimming with others. They are preyed upon by sea-lions and over 86% of such casualties take place when they are playing in this way. Caro (1995) noted that their play certainly made cheetah cubs more conspicuous to him as a human observer and presumably also to lions and spotted hyaenas which are known to kill them.

We may be convinced of the importance of play for behavioural development, but a comparative survey across the groups reveals that, apart from a few fragmented anecdotes, there is no evidence of play outside the mammals and a few bird species. Even here it is not universal: for example, young rats show conspicuous chasing and wrestling behaviour, reasonably described as play, but for mice there are fewer records. It is among the ungulates, carnivores and primates that play becomes so conspicuous. Each group has its own typical repertoire; ungulates butt each other, chases and mock fights are common among primates, whilst carnivores show elaborate stalking, leaping and all aspects of prey-catching behaviour. The behaviour is often repeated again and again and has an exaggerated form, perhaps easier to recognize than to describe. Adults may join in and there are sometimes special postures which indicate that an individual wants to engage in play (see p. 159). All this may give the impression that play is clearly distinguishable from 'real' behaviour, but this is not always true. It can be quite difficult to identify play with certainty, particularly as animals get older, and we have all observed that animals themselves may have the same problem, for mock fights may change rapidly into real fighting as one partner really gets hurt.

When discussing parental care we had to accept that, although there could be little doubt that the interchange between mother and offspring contributed to the healthy development of the latter's behaviour, it was difficult to identify exactly how it operated. The same problem is met even more forcibly with play. As we watch mock fights among kittens (Fig. 2.25) or the amazing high-speed chases and wrestling bouts of young monkeys, it is difficult not to believe that such behaviour plays a part in the development of skills which are going to become vital for survival in adult life. Apart from the obvious physical skills, numerous other functions have been suggested for

Figure 2.25 Two kittens face and threaten one another in play. (Drawing by Melissa Bateson.)

play – gaining knowledge about potential prey species, gaining knowledge of the social group, exploration of the environment and so on. Yet whatever common sense may seem to tell us, it is certainly not easy to show convincingly any effects of play itself on subsequent adult behaviour.

Although we may not understand how parental care operates we can at least demonstrate the effects of its deprivation, as in some of the work just described. However, it is very difficult to deprive a young animal of the opportunity to play without depriving it of many other things, such as social contact, which we already know have drastic results anyway. One natural experiment comes from Lee's (1983) work on vervet monkeys in East Africa. Her observations extended over several seasons and included periods of severe drought. She noted that the play which was such a conspicuous feature of the infants and juveniles during normal seasons, and which occupied a great deal of their time, virtually disappeared during drought. All animals, including the young, managed to survive only by constant searching for food. When Lee compared young adults which had grown up during the dry period – and hence had been deprived of play opportunities – with the rest, she could detect no differences in their behaviour. Baldwin and Baldwin (1973) found a rather similar situation in some troops of squirrel monkeys where there was great variation in the amount of play, and in some troops almost no play was ever seen. Nevertheless it was hard to identify corresponding differences in their social behaviour as adults.

The ease of keeping domestic cats and the very conspicuous play behaviour of kittens makes them better experimental subjects. Additionally, they develop rapidly and their period of most intense play is over by about 3 months. Bateson and his collaborators devised a simple type of enclosure (Fig. 2.26) which enabled them to observe the behaviour of a mother cat with her litter and to quantify developments in play as the kittens grew up. They could distinguish two fairly distinct phases of play, the first from about 4 to 7 weeks was largely 'cat-orientated' play with each other and with their mother. From 8 weeks onwards more attention was directed towards objects in their environment, playing with ping-pong balls, a small stuffed dog etc. It was noticeable that in single-sex litters male kittens showed more object play than females, but this difference tended to disappear when brothers and sisters grew up together.

Mothers often take part in play sessions and their influence can be detected indirectly as well. If early weaning is forced upon a mother by administering a drug which suppresses milk production or simply by keeping her on short rations, the kittens respond by eating more solid food. This is an obvious result, but the signalling of early weaning by reduced milk supply also leads to the kittens accelerating the onset and increasing the amount of their play (Martin & Bateson 1985; Bateson *et al.* 1990). This certainly suggests that play is not just a casual activity for the kittens, but something they need to do in order to function well later as independent animals.

Perhaps the kittens' object play is related to special feline skills. With some members of the cat family there is one very conspicuous type of play which does seem particularly directed towards the development of prey-

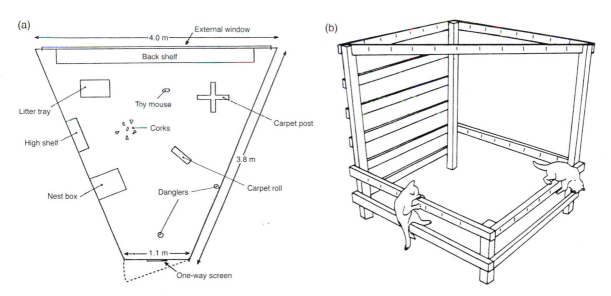

Figure 2.26 (a) Observation room used by Bateson's group for the study of a mother cat rearing kittens and the development of their play. Several of these rooms are fitted together around a central observation space fitted with one-way screens. Various objects to stimulate play are provided during testing; a carpet post for scratching and a high shelf onto which the mother could retreat out of the kittens' reach, at least for the first few weeks. (b) Wooden 'climbing frame' which was placed in the room for a half-hour session each day to test the development of the kittens' locomotor play and climbing skills. (From Martin & Bateson 1985.)

catching. We have all noticed that domestic cats bring home live prey which they proceed to 'play' with, letting them loose only to pounce on them again, often repeatedly and to our distaste. In the natural context, such behaviour falls into place, for mother cats loose live prey close to their kittens and may thus familiarize them with suitable prey types. It also gives them repeated opportunities to practise capturing a relatively crippled mouse, for instance, before they have to tackle a fully functioning one for themselves. Caro (1980) has been able to demonstrate directly that such behaviour does provide some pay-off. Under well-controlled conditions he found that experience of particular types of prey – birds, mice or fish – at 12 weeks of age does improve the predatory skills of cats tested when 6 months old. However, the effects were not dramatic and unexposed animals probably soon catch up. In the wild, Caro (1995) has made extensive observations of the play of young cheetahs which occupied about 3–4% of their time, but again it was not possible to identify any unequivocal results as they grew up. We may note that sometimes the most important predatory skill a mother can impart to her offspring is restraint. The play of young cheetahs has been observed to render useless the careful preparatory stalking of gazelles by their mother (Caro 1987).

More comparative evidence will help us to obtain a reasonable assessment of the role of play and help us to set it into the context of behavioural development. It unlikely ever to have only one function in any animal, and the balance of different functions is bound to vary between species. Fagen (1981) and Martin & Caro (1985) provide good reviews of this perplexing but fascinating field of research.

Imprinting

The effects of early experience may be fairly general in their nature, affecting a wide range of behaviour. Some of the examples we have been discussing might come into that category. However, there are circumstances under which particular early experiences have rather sharply defined results. Of course, whether we call the results of an experience 'general' or 'specific' depends to a great extent on how closely we look for effects. It is very easy to label an effect as specific if only a few aspects of behaviour are studied. Nevertheless some examples are quite striking: playing a tape recording of song to a young bird may affect the song it sings when it becomes sexually mature months later

(p. 107), but as far as we know nothing else in its behaviour is changed by this experience. Some of the results of 'imprinting' appear to be equally specific.

Imprinting cannot be precisely defined. As we shall use the term here it refers to various behavioural changes whereby a young animal becomes attached to a 'mother figure' and/or a future mating partner. It was Konrad Lorenz who introduced the topic to most behaviour workers, by experiments with geese in which he got broods of goslings to follow him and treat him as their mother figure (Fig. 2.27). A good deal of the work on imprinting has been carried out using birds like geese, ducks, pheasants and chickens, which have precocial young (i.e. those that can see, walk and feed themselves from hatching and do not stay in the nest). Imprinting may be measured by the amount of attention paid to the mother, time spent close, latency to approach, time spent following if she moves, and so on. This type of response to the mother figure is usually called 'filial imprinting' to contrast it with 'sexual imprinting', not measurable at the time but only later in life when we observe how its early experience affects an animal's choice of sexual partners at maturity.

Imprinting usually takes place soon after hatching or birth and often

Figure 2.27 Konrad Lorenz, one of the founding fathers of modern ethology, with a brood of imprinted goslings. (Photograph © Erbengemeinschaft Lorenz.)

results in a very fixed attachment, difficult to change. Lorenz described it as a unique form of learning which, unlike other forms, was irreversible and restricted to a brief 'critical period' just after hatching. It clearly does involve a learnt association between a particular stimulus and a response, such as we shall be considering in more detail in Chapter 5 on learning. The context in which we observe it, so early in life, does make the results of imprinting very distinct, but now few would wish to consider the actual process of imprinting as different from other types of associative learning. It seems more useful to concentrate on the role which it plays in development. Because imprinting occurs before anything else has been acquired by learning, it is often a very clear and identifiable event and this has made it extremely useful for studies of the neural basis of learning and memory (Chapter 5, p. 313).

As Lorenz suggested, the range of objects which can elicit approach and attachment in young birds is very large. His original observations were most dramatic because he got broods of greylag goslings to imprint on himself. There seems to be no limit to the range of visual stimuli for imprinting. Birds have been imprinted upon large canvas 'hides' inside which a person can move, down through cardboard cubes and toy balloons to matchboxes. Colour and shape seem to be equally immaterial (see Figs. 2.27, 2.28 and 2.29); auditory stimuli are also attractive to many young birds and will often enhance attachment to objects close to their source. Sound is very important to induce following in mallard ducklings, for example. Wood-ducks nest in holes in trees, and the mother calls to her young from the water outside the nest hole. They are induced to approach before they have ever seen her properly.

With visual stimuli, movement helps to catch attention but it is not essential: a stationary object will attract young birds provided it contrasts with its background, so will flashing lights. Bateson and Reese (1969) have shown that within a few minutes, day-old chicks and ducklings will learn to stand on a pedal in order to switch on a flashing light, which they then approach. The rapidity with which they acquire this response suggests that the light has reinforcing properties even before the young birds are imprinted to it. They actively seek such stimulation and such attraction forms the basis for the imprinting attachment. The young animal comes prepared or biased to respond to conspicuous objects, to approach them, to learn their characteristics – visual, auditory or olfactory – and to form an attachment.

For all the wide range of potentially attractive stimuli, we know that some young animals do not come to the imprinting situation with no biases at all. As mentioned above (p. 52), given certain, rather non-specific early experience, chicks will rapidly develop a preference, made at first sight, for a realistic mother hen over other imprinting objects. Hampton *et al.* (1995) show that

Figure 2.28 (a) Apparatus used to study the following responses of young birds. Different models can be attached to the long arm, which rotates slowly, moving them around the circular runway. (After Hinde 1974.) (b) The end-point of such a procedure: a young duckling firmly imprinted on a wooden cube. It stands close to it at all times; if the cube moves, it follows.

(a)

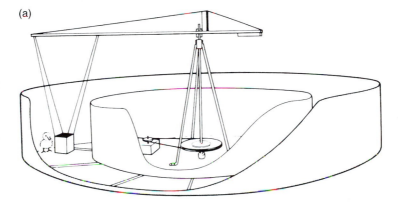

(b)

this preference develops over a few hours in dark-reared chicks which are handled or exposed to the calls of a mother hen; non-exposed chicks do not develop it. In nature, one supposes, some of the relevant experience will almost always be there to bias the chicks towards approaching and attaching to their mother, rather than other conspicuous objects around them. Similar biases, however aquired, may be involved in sexual imprinting situations, as we shall discuss later.

Imprinting-like phenomena are also clearly involved in the social development of mammals. It is common knowledge that the younger we take and rear a litter of wild mammals, the easier it is to tame them. Orphan lambs reared by humans follow them about and often show little attraction towards other sheep ('Mary had a little lamb . . .!'). This may not be just a filial attachment; a form of sexual imprinting sometimes occurs and zoo staff know to their cost that hand-reared animals are often useless for breeding when they are mature. Whilst visual and auditory stimuli are dominant in birds, the behaviour of mammals is very much dominated by their sense of smell and it is not surprising to find that their early olfactory experience often affects their choice of a mate. In guinea-pigs, mature animals are more attracted by others whose scent matches that which was present in the nest during the time when they were being reared (Carter & Marr 1970). With mice, the situation is more complex and greatly affected by the genotype of the animals, but experience of scent is always a key factor on which they base their choice (D'Udine & Alleva 1983; Lenington & Egid 1989).

Both of the criteria which Lorenz claimed made imprinting a unique form of learning – irreversibility and restriction to a brief critical period – have had to be modified in the light of later research. Certainly we observe that fidelity to the parent figure is the rule in nature, but we must remember that the original stimulus for attachment is constantly or at least repeatedly present and there are few other distractions. Irreversibility, were it to be established, would be almost unique, certainly for learning in vertebrates. It would mean that 'primacy' – the object learnt first, would take precedence over 'recency' – that learnt most recently, and with conventional learning the reverse is usually the case. A number of studies have yielded inconsistent results on this point and Bolhuis & Bateson (1990) have been able to unravel some of the factors which might help to explain why.

They kept chicks individually in small cages with brightly coloured objects of two contrasted types. One was a blue sphere with white squares on it, the other a red cylinder with white stripes (Fig. 2.29a). After 3 days' exposure the chicks were tested for filial attachment and all, as expected, showed a powerful preference for the object they had been reared with. Then they were exposed to the other, non-preferred model, sometimes for the same period, sometimes for a shorter time, it actually made rather little difference. At a second test after such exposure, all the chicks showed reduced preference for their first object and a marked shift towards the second, more recently

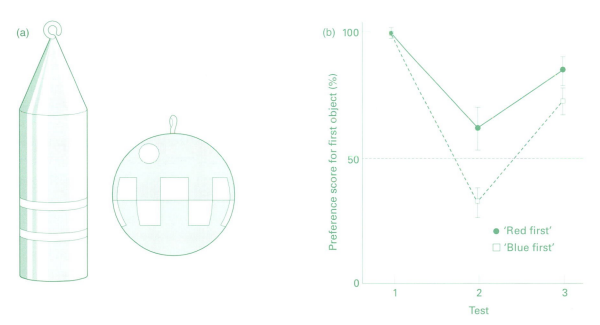

Figure 2.29 (a) The two objects, used in Bolhuis and Bateson's experiment described in the text. The ball was painted blue with white squares, the cylinder was red with white stripes.

(b) The preference scores for Bolhuis and Bateson's chicks. Test 1 followed 3 days of exposure either to the blue ball or to the red cylinder. All showed complete preference for their familiar object. Each group was then kept with the *other* object for 3 further days up to test 2. They showed a marked change in degree of preference, larger for the 'blue first' group than 'red first'. Lastly, kept with *both* objects prior to test 3, both groups shifted their preference back towards their first object. Further explanation in text. (a and b from Bolhuis & Bateson 1990.)

observed stimulus object. They had certainly not shown irreversibility in any complete sense. However, in the final phase of the experiment Bolhuis and Bateson either kept the chicks isolated, with no object present, or both objects were put into their cage. At a third preference test which followed, all the chicks reverted back to a clear preference for their *first* object, although not up to the original level (Fig. 2.29b). So primacy did have some power over recency, but not an absolute one.

Lorenz's second distinguishing characteristic for imprinting concerned its restriction to a brief 'critical period', or perhaps less exactingly, 'sensitive period', following birth or hatching. Certainly everyone would agree that imprinting only manifests itself in really young animals, but the reasons for this are not entirely simple. Our perception of the sensitive period depends to a considerable extent on how exacting are the criteria we use. Taking filial imprinting in birds as an example, if the criterion 'follows a moving object at first exposure' is used, then many precocial birds will satisfy it for 10 days or more after hatching. However, if one takes as a criterion 'the formation of a lasting attachment following a single exposure to an object', then the sensitive period may appear much more limited. Figure 2.30 illustrates some results using both types of criteria. Those from Ramsay and Hess (1954) measured how well mallard ducklings discriminate and stay close to a model 5 to 70 hours after a single 30-minute exposure to it. Boyd and Fabricius (1965) simply scored birds which followed a model on their initial exposure.

Guiton (1959) proved that the length of the sensitive period was not independent of external factors because, whilst young chicks kept in groups ceased to follow moving objects 3 days after hatching, chicks reared in isolation remained responsive for much longer. He could also show that the socially reared chicks did not remain inert and ignore their surroundings until exposed to a moving object; they became imprinted upon one another. This phenomenon may be important in the natural situation. Boyd and Fabricius (1965) point out that mallard ducklings do not normally leave the nest until the second day after hatching, well past the peak of the experimentally demonstrated sensitive period. By this time some of them may be imprinted upon each other. Even if only a few of the brood actively approach and follow the mother bird as she leads them to water, the brood will act as a group and stay together because the rest will follow the maternally imprinted ducklings.

Reviewing the evidence suggests that the process of imprinting involves, first, that the young bird seeks and approaches a conspicuous object in its immediate environment. It rapidly attaches itself to the object, learning its characteristics more fully and this knowledge serves as a familiar reference point from which it can then discriminate against the unfamiliar things in the environment. Now it often reacts fearfully towards the unfamiliar and it is the ability to make the discrimination which effectively brings the sensitive period to an end.

With young mammals also we have to try to distinguish between a change in responsiveness that occurs as a result of previous experience and a change which occurs anyway by some form of growth or maturation process. We can certainly point to clear sensitive periods at particular turning points in their behaviour. Scott and Fuller (1965) summarized their extensive work on dogs (see also Scott 1962). They found that there is a period from about 3 to

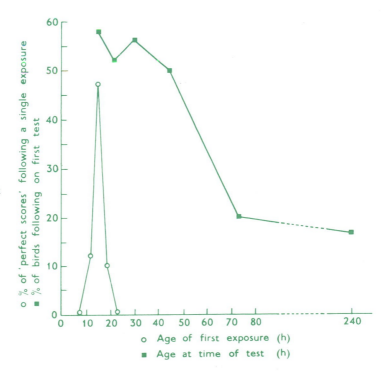

Figure 2.30 Two ways of expressing the sensitive period for filial imprinting. The line joining the open circles shows the very sharp peak obtained when mallard duckings were tested for following and discrimination of a moving object after a single exposure to it at different ages. The line joining the black squares shows the much broader peak obtained by scoring the percentage of duckings that follow a moving object in their first exposure to it. (Data from Ramsay & Hess 1954; Boyd & Fabricius 1965.)

Figure 2.31 Twin orphan lambs of the Soay breed fostered on to a Blackface ewe by tying half of her dead lamb's skin over each of the orphans. (Photograph by M.L. Ryder.)

10 weeks of age during which a puppy is forming normal social contacts. If isolated beyond the age of 14 weeks the young dogs no longer respond and their behaviour is very abnormal. A very short exposure at the height of this sensitive period is sufficient for them to form a normal relationship with human beings. Dogs, like some sexually imprinted birds, seem perfectly capable of accepting both humans and their own species as social partners.

Restricted periods during which an experience must occur if it is to be effective are also observed with adult mammals. The time around birth requires a crucial switch in the responses of the mother, who must accept her newborn and start lactation. We know with sheep and goats that the mother must be able to smell and lick her newborn within an hour or two of birth if she is to accept them and lactate. Beyond this period, all young are rejected. Farmers have long understood that if they are quick, an orphan lamb can sometimes be cross-fostered to a ewe whose lamb has died. They present the orphan covered by the skin of the dead lamb, which the ewe had 'accepted' during her sensitive period (see Klopfer & Gamble 1966; Vince 1993; Fig. 2.31). Discussing the analogous stage in rodents, Rosenblatt and Siegel (1983) describe the behavioural and physiological events surrounding parturition. They point out that the high arousal of the female facilitates the development of interactions between her and the newborn. A brief contact at this stage has been found to change the behaviour of females for weeks subsequently, even though there is no further opportunity to interact.

The argument about sensitive periods is not merely academic, for it relates to contrasting views of behavioural development. Is it a totally continuous process with interactions between the growing animal and its environment possible at every stage? Or is it more like embryonic development, where we know that certain events must take place within a critical period if they are to take place at all? Once the critical period is past, embryonic cells may no longer be competent to respond. The development of behaviour probably exhibits both sets of characteristics and much will depend on a particular animal's life history. With short-lived animals which must grow rapidly and are sexually mature at a few months of age, behavioural development is highly compressed and events like imprinting come to take on an all-or-nothing, irreversible appearance. (When we come to discuss the development of bird song on p. 103 we shall certainly find evidence of sensitive periods beyond which experience has no effect.) But the infancy of some mammals is very prolonged and development is far more a continuous process – we have only to think of the work with rhesus monkeys which was discussed earlier on p. 81. The behaviour of the young monkey gradually develops over months or even years. Experience has effects at all stages, some transient, some permanent, but rarely appearing as irreversible since there is usually time for later experiences to change the direction of development.

Imprinting has always attracted the attention of psychiatrists because there is no doubt that human infants are extremely sensitive to early experience of many kinds. In particular, Bowlby (1969, 1973) developed a theory of the attachment of a baby to its parent which drew extensively from ethological work. He suggested that the period from 18 months to 3 years was especially important and that separation from, or lack of an adequate parent figure at this time led to a greatly increased risk of psychological disturbance in adolescence and later life. The idea of a restricted sensitive period in human development, with its implicit suggestion of irreversibility, has been widely criticized (e.g. Clarke & Clarke 1976; Rutter 1979). Nobody would deny that maternal separation in childhood may have long-term effects on human children but inevitably the situation is not as clear cut as with simpler animals. Separating a human baby from its mother will involve a whole series of changes to its life, almost always deleterious, and it is difficult to link up cause and effect very closely.

Sexual imprinting

Lorenz found that the early experience of his young geese and ducks affected their choice of sexual partner when they were mature. Subsequent work with a wide variety of birds and mammals has confirmed this (Bateson 1976). As with filial imprinting, there seems to be virtually no limit to the range of objects which can provoke attachment. Male and female turkeys have been sexually imprinted on human beings, cockerels imprinted onto cardboard boxes which they courted and attempted to mount, whilst we have already referred to the sexual imprinting of many species of mammal kept in zoos and artificially reared by human foster-parents.

Some of the most complete experimental work has used Estrildine finches, in particular the zebra finch and the Bengalese finch. They are particularly useful because they breed readily in small cages and have a very rapid life cycle: they fledge and become independent at about 5 weeks from hatching and will breed themselves soon after. Immelmann (1972) carried out a range of experiments to investigate the effects of cross-fostering between these two species. He placed a single zebra finch egg in a clutch of the Bengalese finch and allowed the Bengalese parents to rear the whole brood. Subsequently, cross-fostered zebra finch males were isolated until they were sexually mature. Immelmann then gave them a choice between a zebra finch female and a Bengalese finch female. A cage was divided into three parts by two transparent partitions with a continuous perch running through them. A male was placed in the central part with a female zebra finch on one side and a

Bengalese female on the other. The results were unequivocal: males directed their courtship towards the Bengalese finch female. This preference was all the more striking because when a zebra finch male was put in with the two females, the zebra finch female usually responded at once with all the usual conspecific greeting calls and perched as close to the male as the partition allowed. The Bengalese finch female was, at best, neutral and usually showed avoidance as he approached her (see Fig. 2.32).

Such results are very dramatic and they parallel some of Lorenz's original observations in which, for example, hapless cockerels attempt in vain to attract female mallards which they avidly court from the bank whilst the ducks are swimming some way off shore! From a developmental point of view, one wants to ask questions similar to those raised for filial imprinting. Is sexual imprinting as irreversible as it may appear from the above observations? Does it too have a sensitive period and if so, is this coincident with that for filial responses and brought about in a similar way?

Once again, studies of zebra finches have helped to provide answers. As we have seen, sexually imprinted male zebra finches initially prefer to court

Figure 2.32 A male zebra finch (on the left), reared by Bengalese finch foster-parents, shows full courtship song and posture, with raised feathers, to a Bengalese finch female. (Photograph by C. Ten Cate.)

females of their Bengalese foster-parent species. In some of Immelmann's original experiments such imprinted males were then forced to pair with zebra finch females. Uninterested at the outset, in the absence of any choice the males eventually gave in and bred with their conspecific partners. They were allowed to raise one or two broods of young together. Then the males were once again tested for their preference in the three-part cages. In almost all cases the lure of the foster-parent Bengalese females remained as strong as ever – another dramatic result which suggests how resistant sexual imprinting is to change.

However, all is not as simple as it seemed from these early experiments. Later work (Immelmann *et al.* 1991; Kruijt & Meeuwissen 1991) has shown that the crucial factor in determining such apparent irreversibility is the first preference test, when the young, but fully adult males were put with both species of female to see which they preferred to court. If cross-fostered males were force-paired with conspecific females *without* first being allowed to court foster-species Bengalese females (which we know they certainly would have done), then the adult experience of pairing with zebra finch females does greatly reduce, or even eliminate, the preference for those of the the foster-parent species.

So it would appear that the experience of taking part in courtship with foster-parent females serves somehow to 'stamp in' or consolidate the effects of the sexual imprinting which is initiated during the period of dependence as a nestling. We must recognize, then, that even adult birds are still susceptible to experience which affects their sexual preferences and there is not the rapid closure of a sensitive period which we see for filial imprinting.

We have too little comparative evidence to be able to judge how typical the zebra finch developmental story is. However, with birds that develop more slowly, such as domestic chickens, there is also evidence that any sensitive period for sexual imprinting is certainly later and probably more extended than that for following a parent figure. Vidal (1980) exposed male chicks to a moving model for a period of 15 days at one of three stages of the young bird's early life: 0–15 days, 16–30 days or 31–45 days. They were then kept either with a hen or isolated until full sexual maturity at 150 days when they were given a series of choice tests. The strongest sexual preference for the object over a normal hen was shown by those birds which had been isolated after exposure at 31–45 days, the group which showed the least following or other signs of filial imprinting.

Few other types of bird or mammal have been tested properly, but sexual imprinting of a similar type certainly occurs in pigeons, geese and ducks (as Lorenz showed), in gulls and probably in parrots. Note that it *cannot* occur in cuckoos and cowbirds, whose young are always reared by foster-species. Such nest parasites must develop their sexual preferences in other ways, presumably by some inherent recognition of 'own species' characteristics.

We have already described (p. 91) the work with domestic chicks which suggested that here, too, there are signs of a preference for a mother hen figure which was revealed when they saw her for the first time (Hampton *et al.* 1995). If such preferences are there even in non-parasitic species, they could be the basis for successful mate-recognition as parasitism began to evolve. Some early experiments with zebra finches (Immelmann, 1972) suggested that males imprinted more rapidly and more firmly on their own species than on foreign ones although, as we have seen, the effects of foster-parenting are fairly dramatic. However easily overridden by experience, could there be some degree of inborn recognition at work? Again, careful developmental studies can reveal some environmental factors which can affect the responsiveness of nestlings.

Ten Cate (1982) made a close study of the parental behaviour of both zebra finches and their Bengalese finch foster-parents. He found that zebra finches interact more intensively with their offspring than do Bengalese and suggests that this increased attention may account for more rapid imprinting to zebra finch parents. Further, the young themselves are affected by the other nestlings they are reared with. Thus a single zebra finch foster-chick in a Bengalese brood is more likely to develop the very strongest preference for a Bengalese sexual partner. One reared with both zebra and Bengalese finch siblings will show a less marked preference (Ten Cate *et al.* 1984). Experiments of this type help to unravel the interactions between inherited tendencies and social experience which operate during development, other examples of which we shall discuss further in the section on bird song.

We should not be suprised that sexual and filial imprinting have very different time scales and characteristics because functionally they have such different jobs to do. Attaching to a parent figure rapidly and avoiding other figures once this has been accomplished is a key to survival in many young animals. But this phase of simple attachment for protection soon passes. Acquiring the preferences and predispositions it will need for a successful adult life can take place later and, at least in long-lived animals, can be rather more leisurely.

Bateson (1979) has argued that it makes sense to delay the onset of sexual imprinting until young birds are beginning to develop their adult plumage, because siblings will then become better models. In chickens, Vidal's result fits in with this idea and the timing of the sensitive period in quail and in mallard also agrees. Bateson (1980) has good evidence that a rather precise and intense form of learning is taking place which may require a longer period to become established than does filial imprinting. He suggests that during this period the young bird is learning the detailed characteristics of its mother and siblings as a basis for its subsequent choice of sexual partner. The bird chooses some individual which is neither too similar (and therefore likely to be a close relative) nor too dissimilar. He calls this a choice for 'optimal discrepancy' and he has tested the hypothesis with Japanese quail. This species shows great variation in the details of their head-markings – variation which has been shown to be genetically based and therefore a good indicator of relationship. Birds are reared in groups and then given a choice between three birds of the opposite sex: (i) a close relative with a head pattern very similar to the test subject, (ii) a bird with different markings but of the same basic colour, and (iii) a bird with white plumage (Fig. 2.33). The choice of which bird to approach was consistently for the second type. Choice for optimal discrepancy could lead to a reasonable balance between avoiding inbreeding, but also avoiding outcrossing too distantly with the risk that a genetic constitution which has been selected to match local conditions is broken up. We have concentrated on the effects of sexual imprinting for males, but Bateson has found similar effects for female quail. In the zebra finch, too, Sonnemann and Sjolander (1977) found that although the results are not so striking as with males, females definitely are affected by cross-fostering and choose to perch close to Bengalese males rather than males of their own species in three-part cage tests.

We have already noted that brood parasites like cuckoos cannot develop sexual preferences through imprinting on parents or siblings. Natural selection can operate to shape the pathways of development but Bateson's findings with quail do suggest possible reasons why sexual imprinting, with its gradual learning of species and individual characteristics, has evolved in animals whose life history allows. Apart from providing a mechanism for optimal outbreeding, imprinting also removes any constraints on rapid changes of appearance, size, shape, colour or markings if these become advantageous for any reason. The learning system can immediately follow the changes, something which would be very difficult for a system which involved inherited responsiveness to one type.

Bird song development

Imprinting has just provided us with an example of how an inborn tendency to approach conspicuous objects is linked with learning of their characteristics. We have also mentioned previously how a motor pattern, such as bird flight, may mature to a basic level in the absence of practice, but that all the finer skills of flying are added later with practice.

Some of the most beautiful examples of the interweaving of inherited

Figure 2.33 Apparatus used for testing courtship preferences of adult Japanese quail. Stimulus birds of various types are put singly into the inner compartments, which are brightly lit from above. Their outward-facing windows are fitted with one-way screens, so that although the stimulus birds cannot see into the unlit outer part of the apparatus, the bird making its choice can see through into each compartment. A sensitive platform in front of each window records automatically when a bird is standing on it. Those familiar with that city will understand why Bateson called this 'The Amsterdam Apparatus'. (After Bateson 1983.)

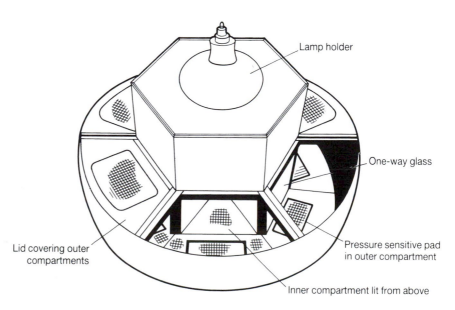

Lamp holder

One-way glass

Pressure sensitive pad in outer compartment

Inner compartment lit from above

Lid covering outer compartments

and learnt components during development come from studies of bird song. Song can be regarded just like any other behaviour pattern – it is a controlled sequence of muscular activity which in this case we perceive as sound. Detailed analysis of song is in fact easier than for most motor patterns, as it can be recorded and then played into a sound spectrograph to yield a chart on which marks of different densities show how much energy was emitted at the various sound frequencies over time. Some examples of sound spectrographs of bird song are shown in Fig. 2.34. With practice they can be interpreted rather like a musical score and they show up variations in sound pattern which are hard to detect by ear alone, and impossible to describe in words.

Bird vocalizations exhibit an astonishing variety, from the briefest of

Figure 2.34 Sound spectrographs of three closely related species of European warbler. The three are very similar in plumage and share the same general habitat, hence the songs probably serve, in part, as a mechanism of species isolation (see p. 363). These warblers were first clearly distinguished as separate species by the eighteenth century English naturalist Gilbert White, who published *The Natural History of Selborne* in 1789. He separated them by their songs.

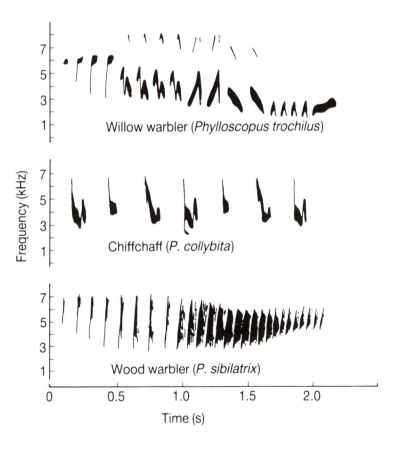

Willow warbler (*Phylloscopus trochilus*)

Chiffchaff (*P. collybita*)

Wood warbler (*P. sibilatrix*)

Frequency (kHz)

Time (s)

call notes through to elaborate songs lasting several minutes. Song, as we popularly understand it, is characteristic of one of the most abundant and advanced groups of birds, the Oscine families of the Order Passeriformes or perching birds (see Campbell & Lack 1985, for a concise account of this group). This order includes some of our most familiar birds, the finches, warblers (both Old World and New World) and thrushes. The ways in which they acquire their songs have proved to be a rich source of material for investigation. In their diversity they provide us with opportunities to study some important functional and evolutionary aspects of behavioural development. Catchpole & Slater's (1995) excellent book provides the best introduction to what they refer to in their subtitle as 'biological themes and variations'.

The songs which have proved most useful in developmental studies are those of intermediate complexity, such as those of the finches. The European chaffinch produces a 2–2.5-second burst of sound consisting of several sequences of repeated notes of different types. As we listen to chaffinches singing we notice immediately that individuals sing several variants of the song – all clearly recognizable as belonging to that species but differing in detail. Neighbouring males often have several songs in common but each bird will usually have its own individual mix of song types. Figure 2.35 illustrates some of these variants from a study by Slater & Ince (1979).

Such variability raises intriguing developmental questions; what are its origins? It has been known for some time that chickens and pigeons produce all their characteristic calls when isolated and even if deafened. Could the song of chaffinches be similarly resistant to environmental influences; could the variants even be genetically based?

To answer these questions Thorpe and his collaborators (Thorpe 1961) took young birds from the nest immediately after hatching and reared them alone in sound-proof chambers. Next spring the males began singing in isolation; their song was recognizably chaffinch but was much simpler in form and showed little variation either within or between individuals, (Fig. 2.35, bottom). Following our discussion of the validity of isolation experiments on p. 000 we may note that here such an experiment gives a very clear answer to a first stage question. The different song variants are not genetic but due to some influence from outside. It was an obvious guess that this external source was the song of adult birds which the young bird heard both in the nest and early in its first spring, and this proved to be the case.

Work by Marler and Tamura (1964) with an American finch, the white-crowned sparrow, allows us to go into more detail. This bird has a wide range on the Pacific coast and, like the chaffinch, its songs vary. But they vary in a more systematic way, in that birds from a given geographical area tend to sing similar song variants so that there are clear local 'dialects' (Fig. 2.36). If young males are taken immediately after hatching and isolated, as were the chaffinches, then no matter which region they come from, they all eventually sing very similar and simplified versions of the normal song. Obviously they must pick up the local dialect by listening to adult birds and modifying their own simple song pattern accordingly. Marler and Tamura found that this learning process usually takes place during the first 3 months of life and thus

Figure 2.35 At the top, three examples of chaffinch song. They share the same basic pattern but are very different in detail although they are all umistakably chaffinch to the human – and certainly to the chaffinch – ear. One male will commonly have several such variants in his repertoire and share some of them with birds in neighbouring territories. The lowermost song is that sung by a chaffinch brought up from the chick stage in complete auditory isolation from others. It has chaffinch timbre and rhythm but is obviously far simpler in form and, in particular, lacks the 'terminal flourish' so typical of the normal songs. (From Ince *et al.* 1980; Thorpe 1961.)

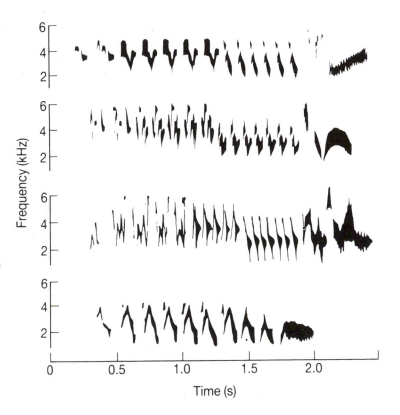

before the bird has ever sung itself. Males captured in their first autumn and reared alone begin to sing for the first time the following spring, and produce a recognizable version of the local dialect. There are clear indications of a sensitive period for learning dialect characteristics. Up to 3 months of age, isolated males can be 'trained' to sing their own or other dialects by playing tape-recorded songs to them, though the results of such experience do not show until the birds begin to sing themselves some months later. Beyond 4 months of age the birds are unreceptive to any further training by tapes and their songs, when they begin singing, are not affected. Here, then, we appear to have a simple, inherited song pattern which is sensitive to modification by learning but only during early life. The young birds 'carry' the memory of the songs they hear and reproduce them when they first sing.

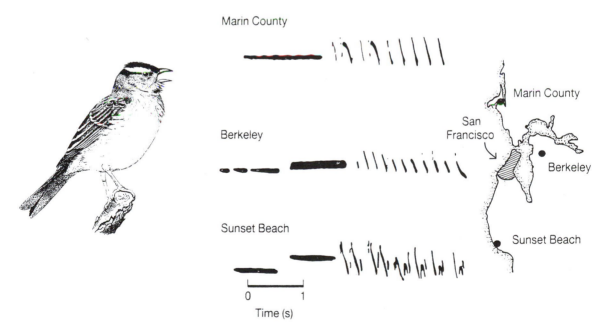

Figure 2.36 The song 'dialects' of the white-crowned sparrow around San Francisco Bay, USA. The examples given are typical of their area and, unlike the chaffinches of Fig. 2.35, there is much less song variation within geographical areas than between them. The use of the term 'dialect', as in human geographically based dialects, seems justified. (After Marler & Tamura 1964.)

Experiments by Konishi (1965) took the analysis still further and have enabled us to qualify the conclusion that simple song is inherited. Remembering our earlier reflections on what isolation experiments can tell us about development, we must note that there remains one song which the isolated sparrow can still hear – its own. If a young fledgling is deafened just as it leaves the nest it will subsequently sing, but it produces only a series of disconnected notes. These are quite unlike the song of isolated birds which, though simple in form, would still be recognizable as 'white-crowned sparrow' to an ornithologist. The bird has to be able to hear itself in order to produce this simple song pattern. Consequently, it is more accurate to say that it is not the capacity to produce the simple song which is inherited, but rather some kind of neural 'template' representing this song, against which the bird matches the notes which it produces and adjusts them to fit. It requires auditory feedback if it is to realize this inherited potential.

In many song birds, the way they first begin to sing in spring perhaps reveals this feedback control in actual operation. For example, chaffinches begin the breeding season singing a 'subsong', a soft rambling pattern of notes varying in pitch and length (Fig. 2.37). As this is repeated the notes become louder and less variable, always approaching closer to the final pattern which presumably corresponds to that held in the template.

With the white-crowned sparrow, Konishi found that if he deafened young birds after they had been 'trained' during their sensitive period with normal song but before they had themselves sung, their subsequent song resembled that of birds deafened as fledglings. They need to hear themselves in order to match up the song they produce with that which they have stored in their memories. Presumably the songs they have heard as fledglings modify their inherited template so that now it conforms to the more complex characteristics of normal adult song. Once the birds have matched their own output with this and sung the adult song, they can go on singing normally even when deafened. At this stage song development comes to an end in the white-crowned sparrow; after its first spring the bird is no longer susceptible to further experience and keeps much the same song pattern for the rest of its life.

The results of these experiments are summarized in Fig. 2.38. One final experiment illustrated there must now be mentioned. Playing tape recordings of other bird species to the young white-crowns during their sensitive period has no effect and the birds' subsequent songs sound like those of isolated males. It appears that the young birds are highly selective in what they will learn. Young chaffinches react similarly and Thorpe (1961) pointed out that this can scarcely be a result of limitations in its sound-producing organ or syrinx. The related bullfinch and greenfinch have syrinxes of almost identical structure. These species have only a poorly developed natural song but are good mimics and will learn to reproduce the songs of many other birds. Some clue to the chaffinch's limitations is provided by the song of the one species it will learn to imitate reasonably well from tape recordings. This is the tree-pipit, whose song to the human ear also has a chaffinch-like tone, though it is very different in pattern. The chaffinch may have an inherited tendency to

Figure 2.37 The development of song within a young male chaffinch in its first breeding season. Subsong is quiet and rambling, often quite lengthy. As the bird continues, some signs of phrasing are heard and louder and more structured songs begin – so-called plastic song. It justifies this name because it precedes a final version, which is completely structured with fixed length and pattern. After this stage, no further change takes place and the song is said to be 'crystallized'. Each male usually develops two or three different crystallized songs (Fig. 2.35) and, after its first season, does not change this repertoire, although early in spring it will run rapidly through these preliminary stages before crystallizing again into the familiar forms. Note how subsong is completely different from the song of isolated birds (Fig. 2.35, bottom). This latter is preceded by subsong during development and is, itself, a crystallized song. (From Catchpole 1980.)

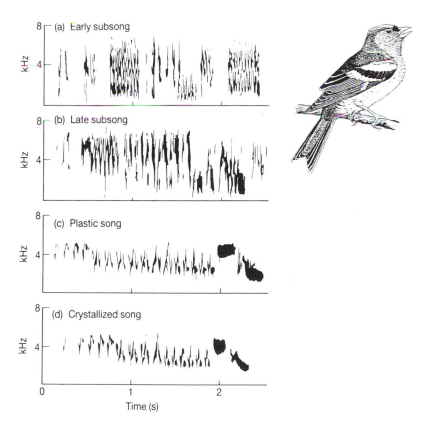

single out this tone from all others and to reproduce the pattern of notes. Marler and Peters (1977) have analysed in detail how in two closely related American species, the swamp sparrow and the song sparrow, young birds hearing tapes for the first time will unerringly pick out the rhythms of their own species' song to copy.

Certainly, as Kroodsma (1982) points out, such selectivity must be important in the song development of many species. A young male will have to concentrate on the song of its own species, often selecting it out from a variety of alternatives which can be heard in the nesting territory during its

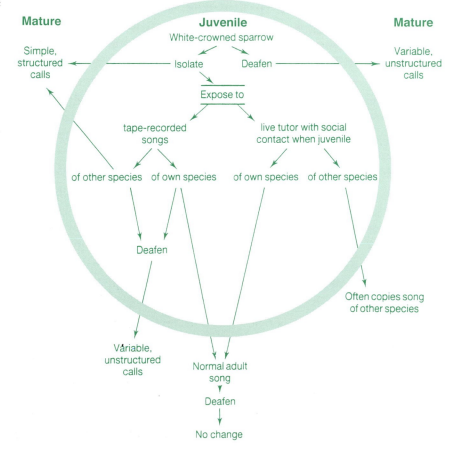

Figure 2.38 A diagrammatic summary of some of the results obtained by Marler, Tamura, Konishi, Baptista, Petrinovitch and others working on song development in white-crowned sparrows. Events inside the circle represent the juvenile phase before the bird has begun to sing itself. The results when it is mature and is singing are shown outside the circle.

sensitive period. Until recently this ability was commonly assumed to be another example of an inherited predisposition to recognize and learn particular things, in this case the characteristics of one's own species' song dialects. Recent research suggests that rather than thinking solely in terms of inherited responsiveness, we should also examine the nature of the learning situation.

Tape recordings offer only a sound stimulus and, as we have just described, they are remarkably effective within certain limits. But provide a young bird with a live song-tutor in the same or an adjacent cage and these limits are easily extended. For instance, isolated white-crowned sparrows have been shown to be capable of learning from live tutors well beyond the 50–55 day sensitive period for tape-recorded tutoring and, more strikingly, they will acquire the songs of other species of caged birds close to them and in sight of them, e.g. Lincoln's sparrow (*Melospiza lincolnii*) and the strawberry finch (*Amandava amandava*). They will certainly not learn these songs from tapes. Nor are such results confined to aviaries, for wild white-crowns have been recorded singing Lincoln's sparrow song in areas in California where the two species co-exist (Baptista & Morton 1988; Baptista and Petrinovitch 1984, 1986). Baptista and his colleagues suggest that it is the social interactions between tutor and pupil which stimulate learning and that aggressive, stressful interactions, which are often seen in aviaries and in the wild between adults and young birds of the same species, may be the most effective. Young males are often driven out of their father's territory once they are mature and this may serve to concentrate their attention on his song types in particular.

Social interaction as a means of focusing attention and thereby affecting the course of behavioural development, is likely to be an important general concept. We have already mentioned Ten Cate *et al.*'s (1984) work on imprinting between zebra and Bengalese finches, where the amount of attention given to nestlings affected their choice of sexual partner. High arousal, directing the attention of young animals, will accelerate learning in all kinds of social situations.

Catchpole and Slater (1995) describe the processes of bird song development as a theme and variations. We can examine only a small fraction of them. It is important to note that the songs of some species are far more elaborate and less proscribed than those of the finches we have been discussing, either in content or in time for learning and modification. In particular the thrushes often have songs of great beauty to us, which consist of a large number of elements, strung together in variable order to make up a song bout which may last for several minutes. These elements are renewed with each successive breeding season, some being retained, others jettisoned, whilst new elements are added. Many elements are copied from other species and even from non-avian sounds in the birds' environment. It is well known that some birds are superb mimics – starlings, mockingbirds, mynahs and lyre birds are famous for incorporating other birds' calls into their own songs. The developmental processes involved here must be significantly different from those we have been describing above, although the canary – a finch – also develops fresh song elements each year and the remarkable neural changes underlying this process (see p. 74) are probably shared by the thrushes and other elaborate songsters. This developmental variability probably reflects a varying balance of the functions which song serves in different species; carving out a territory, repelling rivals, attracting a mate, and so on.

One final discovery about bird song, with important implications for all behavioural development, is the fact that some birds begin singing for the first time using a much wider repertoire of elements than they will ever use again once their song has crystallized. Human infants when they begin babbling are said to incorporate all the utterances used in every human language across the earth, but they persist only with those that are employed by the language(s) used by the adults around them; the other sounds disappear. Marler and Pickert (1984), studying the songs of swamp sparrows across wide areas of the United States, found that the crystallized dialect songs from different areas are acquired by a rather analogous process. Young birds begin even their first sub-songs with a whole range of the species' repertoire. Taking their cue from the songs of local adults, they then gradually eliminate the elements which are not involved in the dialect of their area.

Presumably, for all the variations on the theme of development, the process of matching your own song to those of adults in the surrounding area is in some way reinforcing or rewarding to a young bird. We have rather left to one side the question of why so many passerine birds develop their songs through this often elaborate process of learning. Some birds have completely unlearnt songs – pigeons and doves, for instance – why not all? There are a number of possibilities, the balance of which will vary with each species' life history. One is that through such learning, with its inevitable imperfections, a group of adjacent territory holders will come to acquire a range of song variants, some of which will overlap, others being personal and distinctive. This gives the opportunity to distinguish individuals by their song repertoire. We can observe that the songs of neighbours with whom boundary disputes have been resolved tend to be ignored ('It's only old so-and-so'), but a new song variant attracts a rapid response ('I must repel this interloper'). Catchpole and Slater (1995) discuss the evidence for this, and other possible functions.

Whatever the the function, natural selection has to secure the development of the song within the individuals. Copying, as we mention above, must be rewarding. In one unusual example we have some direct evidence of the nature of such a reward, and it is the young male's potential mate who provides it. Cowbirds are brood parasites and therefore young males do not hear the songs of adults of their species during early life. West and King (1988) studied the origins of two of the geographical dialects in the United States, one eastern, the other southern. They discovered that, counter to all ordinary expectations, the songs of males reared in isolation are more complex than those of normally reared birds. Beginning with elements of both dialects, males crystallize down not to the song pattern of other males that they are exposed to, but to match the dialect of the area of origin of the *females* with which they are housed! The explanation for this remarkable result comes from close observation of pairs when the young male begins singing to a female. When the male sings particular elements from his repertoire, she responds with a brief 'wing-stroking' display – a rapid shuffling of the wings, similar to the copulation invitation display of a number of passerine birds. The male responds in turn by increasing the frequency with which he produces these rewarded elements and eliminating those others with no effect on the female. In such a fashion, the female's preference for her local dialect comes to be matched by the males of the locality. Song development in cowbirds is thus a natural form of the 'operant conditioning' much studied by experimental psychologists and which we shall discuss further in Chapter 5.

Cultural transmission as a form of behavioural development

For the most part living organisms are linked to succeeding generations by information passed on through the nucleus and cytoplasm of the germ cells.

Of course, animals and plants modify their physical surroundings throughout their lives and some of this may remain to affect following generations. Soil builds up in woodlands and alters the environment for seed germination, whilst animals as diverse as ants and rhinoceroses make trails which persist for years and are used by successive generations. Nevertheless the evolution of morphological or physiological traits normally requires inherited genetic variations to arise by mutation or recombination upon which natural selection can act.

Behavioural evolution has an extra and unique dimension because an animal may learn from its parents or others in its group. Later it may in turn act as a model from which its own offspring will modify their behaviour. Suppose one animal learns a completely new behaviour pattern which proves more successful for some purpose than the hitherto typical behaviour. This pattern may be copied and thence passed on to succeeding generations and gradually come to replace the old pattern, without any genetic changes being involved: a new 'culture' has evolved.

We are familiar with such cultural evolution in ourselves and many would argue that almost everything of importance in human behaviour is transmitted in this way from one generation of a society to the next. The various human languages provide an obvious example of a continuous cultural tradition which maintains different types of behaviour in different populations. Reviews of similar phenomena in non-human animals are given by Bonner (1980), Galef (1976) and Whiten & Ham (1992).

Clearly cultural evolution is possible only among animals with the ability to modify their behaviour by copying and practice. There are no systematic studies of such capacities, but it would be fair to assume that they are much more widespread than once thought. We shall discuss examples from birds below and it is also clear that rodents such as rats and mice can show observational learning. Rats have been shown to acquire the tendency to dig for hidden food from watching others doing it. Having learnt this, they can themselves become models for others to learn from, thus initiating the chain of cultural transmission (see Laland & Plotkin 1990; Laland *et al.* 1993). We are certainly not surprised to find other examples from primates, our closest relatives. Intensive, long-term observations on the Japanese macaques of Koshima Island have shown that some of the differences between the behaviour of different monkey troops are indeed of cultural origin. The monkeys are attracted to feed on maize, sweet potatoes and other foods artificially put out for them. The potatoes are often caked with earth which the monkeys used to rub off with their hands. One day observers saw a young female dip her potato into a stream and wash it. She persisted in this habit, which was copied, first by one of her infants and subsequently by nearly all the younger members of her troop. The behaviour spread more widely when young males, who never stay to breed in their native group, transferred out to join neighbouring troops. There is now a clearly distinguishable 'washing sub-culture' within the Japanese macaque population.

We have already seen (Fig. 2.8) that a distinctive and stereotyped pattern of grooming in chimpanzees can be passed on by copying. They also show similar group-dependent feeding habits. In some areas chimpanzees 'fish' for termites by poking a twig into the termite mound and letting the insects crawl up the stick when they can be eaten. Other populations use leaf stalks or grass stems for the same purpose (McGrew 1992; Fig. 2.39). The behaviour of the chimpanzees of the Tai Forest in the Ivory Coast differs strikingly from those at Gombe in Tanzania. In particular, the former show far more extensive hunting of monkeys for food and they also use a wider range of tools in foraging (see Fig. 2.40). It seems certain that almost all these differences are cultural in origin (Boesch 1994).

The manner in which the song dialects of birds such as white-crowned sparrows are passed between generations is a clear example of cultural transmission. As described on p. 106, the young male models the finer details of his song on that of his father and neighbouring males singing close by; he would pick up a different dialect and sing quite differently if he grew up in another area. Ficken and Popp (1995) gathered impressive evidence of the fidelity with which vocalizations can be copied and re-copied. They recorded the quite complex 'gargle' call of black-capped chickadees at the same site over a period of 19 years and found that certain variants had remained remarkably stable. Chaffinch songs persist less well, although variants can be traced over several generations and, to be fair, their song is more complex and usually lasts for something over 2 seconds compared with the half-second of the chickadee's call.

One of the best known examples of cultural transmission in birds is the

Figure 2.39 A chimpanzee trims a twig which it has selected into a suitable form for termite fishing. (Photograph by W. McGrew.)

habit of opening milk bottle tops by blue tits (Hinde & Fisher 1952). Starting in the 1920s, it was noticed that milk bottles delivered to doorsteps were being plundered by birds which then fed on the cream (Fig. 2.41). By plotting the spread of this habit across Britain, it seemed that it was 'invented' in three separate areas of London and then spread outwards apparently by imitation. However, Sherry and Galef (1984) subsequently showed that such cultural

Figure 2.40 A female chimpanzee in the Taï Forest Reserve, Ivory Coast, cracks open a nut of the tree *Coula edulis* – a much-favoured food. She uses a heavy piece of wood whilst the nut rests on a stone which thus serves as an anvil. Her one-year-old infant watches. (Photograph by Ch. Boesch.)

transmission could occur without imitation – that is, without a bird actually seeing another bird open a bottle. They showed that chickadees (which are closely related to tits) learnt to open containers with lids even if all they saw was just the opened container. Naïve individuals could learn what to do simply by seeing the results of behaviour, not the behaviour itself.

Another important area of cultural transmission is the recognition of danger. Young animals often learn what is dangerous by the responses of adults, particularly their alarm calls, in the presence of predators. Curio *et al.* (1978) showed that young European blackbirds gave alarm calls to objects such as a stuffed owl or even a plastic bucket if they had heard their parents' alarm calls while looking at them. Curio devised an ingenious arrangement whereby he could present adults and young with different objects at the same time, with neither being able to see what the other saw. So, while the adult would be giving an alarm call to a stuffed owl, the young bird who heard the call would be faced with a plastic bucket and thereafter associated buckets with danger! The example of the Addo elephants, given earlier in this chapter, shows that culturally transmitted fear can persist indefinitely.

By its behaviour an animal can modify its habitat or vote with its feet and

Figure 2.41 A blue tit opening a milk bottle to get at the cream. (Drawing by Nigel Mann.)

choose another habitat which suits it better. Thus it may cease to be a passive agent upon which the environment 'acts' and may actively change the nature of the selective forces which impinge on it. Cultural transmission can be a potent factor for change to behaviour, one which is both rapid and extensive. Cultural evolution is certainly not to be regarded as something quite separate from genetically based evolution via natural selection, for both processes will interact. The Japanese macaques, which began washing their food in streams, subsequently moved down to the edge of the sea to do so. Once there, some began swimming and started to eat seaweed – a whole new range of behavioural possibilities opens up. One can easily imagine how, given time, this might lead to various morphological and physiological adaptations to match. Monkeys could evolve insulating blubber and webbed feet! Cultural changes to behaviour might thus lead a species along completely new evolutionary pathways. These important issues are further discussed in Bonner's (1980) book and by Wilson (1985).

So, at the conclusion of this chapter on behavioural development, we link ideas on development with those on evolution, reasonably enough as we reflect on the question we began with: how does behaviour become so well adapted to requirements? Throughout we have had to emphasize the interactions between information provided in the genes and that acquired by diverse pathways from the environment, both physical and social. Animals never come to novel situations with a blank mind. The title of a stimulating review by Gould and Marler (1987) on this topic makes this point very neatly; they call it *Learning by instinct*. In many ways it would be logical to move on next to consider learning in its own right – it is, after all, a form of behavioural development. However, it is impossible to discuss learning without some detailed reference to the external and internal factors that determine how animals learn to modify their responses to novel situations which they may encounter. Accordingly, we shall return to the topic of learning after chapters on stimuli and motivation, the external and internal components of Tinbergen's causation question.

3　Stimuli and communication

Behaviour can be so well adapted to an animal's way of life that the animal seems to 'know' what to respond to, what to do and when to do it. Thus partridges and small rodents 'know' that the most effective defence against a hawk flying overhead is to flatten themselves on the ground and remain motionless since the hawk, which is very sensitive to movement, is least likely to see them. Most of us have experience of wasps that know about sweet drinks or mosquitoes that know where to find exposed human flesh.

Using the word 'know' in this context does not necessarily imply that animals are consciously thinking about what they are doing (although they may be, as we discuss in Chapter 5). Animals may 'know' things about their environments in the same way that a heat-seeking missile knows how to find its target or a computer knows how many mistakes you have made in the course of a game. In other words, quite simple unconscious mechanisms are capable of giving rise to what we might refer to as knowledge. When we ask causal questions about behaviour, we are asking what this knowledge is and how the animal uses it. For example, if we see a small bird or mammal taking evasive action in the presence of a hawk, we could ask, 'How does the animal discriminate between a dangerous hawk and a harmless gull flying overhead?' or, 'When it has detected a hawk, how does the animal decide whether to race for cover or to remain motionless where it is?'

Such questions about immediate causation are obviously closely linked to the ones about developmental history we discussed in the last chapter, but they are also usefully separated from them because they require rather different kinds of information to answer them. For example, two different species may both recognize a hawk by its distinctive shape, but one may have an innate (unlearned) ability to recognize it while the other may have learnt by watching the responses of its parents to different birds flying overhead. The causal mechanisms might be similar but the developmental routes quite different. Conversely, two species may each have an innate capacity for recognizing hawks but use quite different cues to do so.

In this chapter, we begin our study of how behaviour is caused and controlled by looking at the way animals initially detect what is going on in their environments. Do they see, hear, smell as we do? Can they touch and taste? Do they have sixth or even seventh senses that we know nothing about?

But we also have to go further and ask how animals then make sense of all the information they are receiving. Is the patch of red light that has just been detected a piece of food, a mate, a predator, or just the setting sun? Such objects may have some similarities, but the appropriate behavioural response for an animal to make will be quite different depending on which one it is, so the animal must be able to distinguish between them. We might think that the best way to do this would be to have more and better information and we do indeed find that animal sense organs are often superbly good at picking up fine details in the world. But this in turn leads to yet another problem. Paradoxically, sense organs are often too good in the sense that they provide so much information that the brain may not be able to process it all and an animal may have to ignore or throw away some of the sensory information provided by its sense organs. For example, sense organs can potentially provide information about many objects or events in the environment, such as grass waving in the wind. If these carry no significance for survival, the animal will take no notice at all while continuing to respond strongly to other types of event, such as an approaching predator.

The choice of what to respond to is not, however, always as clear cut as this. The animal may be receiving many different sorts of information at the same time, most of which may have some significance for its survival and reproduction. It may be receiving information about food and water as well as about its predator, and the fly that has landed on its back, and so on. It then has to choose between the different sorts of information its sense organs are providing it with and selectively ignore the rest, sometimes changing which information it makes use of at any one time. When there are no predators around, information about food may be responded to very strongly, but once the predator appears, sensory information about food now has to be ignored and the animal must start responding instead to the more immediate threat to its life. Decision-making of this sort is an inevitable outcome of receiving a large amount of sensory input, so that this present Chapter 3 on the sensory worlds of animals will lead naturally into Chapter 4 on motivation and decision-making. These two chapters together show how animals perceive, evaluate and then respond to their environments – the key parts of what constitutes an explanation of the causation of behaviour. We should not underestimate the complexity of the mechanisms involved. Animal bodies are far more complex than any man-made machine yet made and it is worth noting that scientists who design robots are increasingly turning to animal behaviour for ideas about how to make their machines perform tasks such as finding their way around their environment, tasks that animals successfully perform all the time.

We thus begin our study of how animal behaviour 'works' by looking at the way in which animal sense organs detect the host of smells, buzzes, flashes and other changes taking place in the animals' physical environment. Then we will look at how their sensory nervous systems and brains process this unpromising raw material into objects or events that the animals can recognize and respond to. In doing this, we shall see that many of the most important objects in an animal's world often turn out to be other animals and many of the most signficant stimuli to which they respond are in fact signals from other organisms. Understanding how animals detect and respond to stimuli will therefore, by the end of this chapter, have taken us into one of the most fascinating of all areas of animal behaviour: the study of animal communication.

What stimuli are and how they act

The term 'stimulus' means literally 'little goad' from the Latin verb *stimulare* – to goad, incite or arouse – and this describes very well what the term currently means when applied to examples of animal behaviour. An alarm call by one member of a flock of birds incites all the rest to take off. In this case the effect is immediate, the goad of the alarm call prodding the other animals into action. But at other times the action can be delayed and the effects of a stimulus can seem to accumulate gradually, changing the responsiveness of the animal over a period of time, so that the stimulus does not so much 'prod' as 'arouse'. The courtship of male doves leading to induction of hormone secretion in the female, which we discuss in more detail in Chapter 4, is a good example of this. The sight of a male dove courting leads to hormonal changes in the female which make her more ready to take part in nest-building. The stimulus of the male's courtship is only effective, however, if it is repeated many times. One bout of courtship has little effect. It requires several hours of courtship behaviour by the male, usually spread over a number of days, to bring the female into reproductive condition.

Often, however, it is impossible to make a sharp distinction between the immediate (prodding) effects of a stimulus and its longer term or arousing effects, because these overlap. The stimulus of sucking on the nipples by young rats, for example, triggers the final let-down of milk from the female, but the prior growth of the mammary glands has been accelerated by the cumulative arousing action of the pregnant female grooming and licking her own nipples.

Yet another effect that stimuli have on animals – which again differs from a straightforward prod – is to orientate their behaviour in particular directions. Animals respond almost all the time to the basic physical qualities of their environment – light, gravity, air, water currents and so on – and position themselves to be in correct relationship to them. Blowfly maggots crawl directly away from light when they move out of their food source to pupate. Fish rest with their heads facing into a water current. From an early age young rats, even in total darkness, show a 'righting response' which keeps them upright with respect to gravity. In each of these cases, the stimuli act continuously over a long period of time to orientate the response in a particular direction.

Orientation can also refer to the way in which stimuli control responses to more transitory features of the environment. Drakes direct their courtship displays according to the position of the ducks on the water. Some displays are given when lateral to the female, others when the male is directly in front of her. The experiment shown in Fig. 3.1 shows how a digger wasp, *Philanthus*, uses landmarks such as pine cones to orientate and find its burrow. The female wasp digs the burrow and provisions it with insects for her larva. She makes several trips to do this and as she leaves the nest on a foraging trip, she makes a brief circling flight to orientate herself and learn some of the landmarks surrounding it. If these are moved while she is away, she still uses them to orientate, or rather mis-orientate, her response on the return trip, since she will go to where the cones have been moved rather than to the real tunnel entrance.

At times, this orientation role for stimuli becomes even more dramatic, particularly when animals use them for long-distance migrations. Indigo buntings (*Passerina cyanea*) are small birds that breed throughout the eastern United States and migrate up to 2000 miles to winter in the Bahamas and Central America. They travel at night and orientate themselves by the ring of stars surrounding the pole star to tell them where north is. The pole star and the stars nearest to it appear to move much less than the other stars and so provide a relatively constant beacon to steer by, as humans have also discovered. Emlen (1975) showed how important these stars were to the orientation of the indigo buntings by putting them in a planetarium where he could control the pattern of stars they saw (Fig. 3.2). By selectively eliminating different parts of the night sky in the planetarium, he was able to show that as

Figure 3.1 The female digger wasp (*Philanthus triangulum*) builds a nest burrow in sand. While she was in the burrow, Tinbergen placed a circle of pine cones around the entrance. When she emerged, the wasp reacted to new situation by a wavering orientation flight (a) before flying off. Returning with prey (b), she orientated to the circle, although it had been moved during her absence. (From Tinbergen 1951.)

(a)

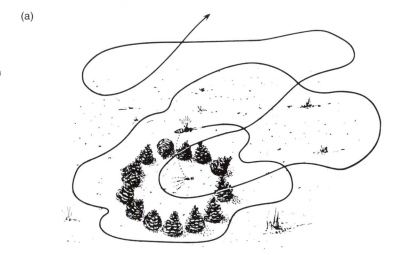

(b)

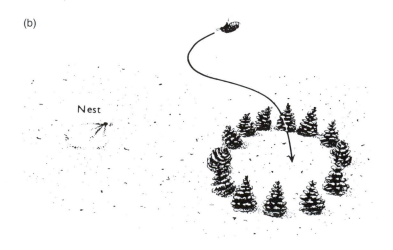

Figure 3.2 (a) The indigo bunting's summer and winter ranges. (b) The system used to test the buntings' responses to a planetarium sky. The birds stood on an ink pad and left a record of their activity on the blotting paper funnel. (From Emlen 1975.)

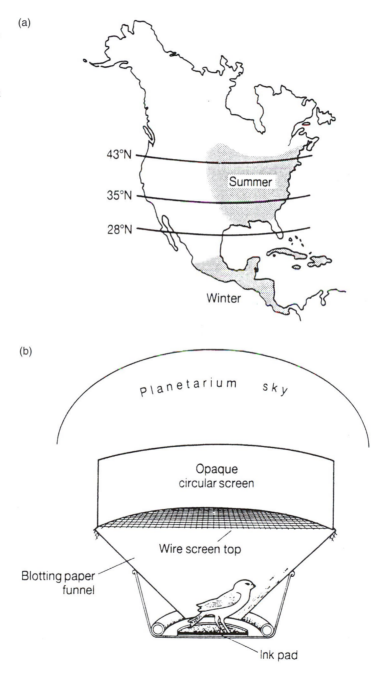

long as the birds could see the pole star and some of the stars within 35° of it, they orientated as they would under a real sky, i.e. northwards or towards it in spring and southwards or away from it in autumn. They used the same star patterns, but took up different orientations at different times of the year.

The sun is used for orientation by large numbers of animals including insects, fish and birds but, of course, using the sun to steer by is much more

Figure 3.3 Diagram showing the effect of 'clock shift' on the direction taken by birds released at noon due east of home. In (a) the birds steer by keeping the sun 90° to the left. In (b) their internal clocks have been shifted 6 h backwards so that when they are released at 1200 h they behave as though it is 6 h earlier and orientate directly away from the sun, as if it were 0600 h.

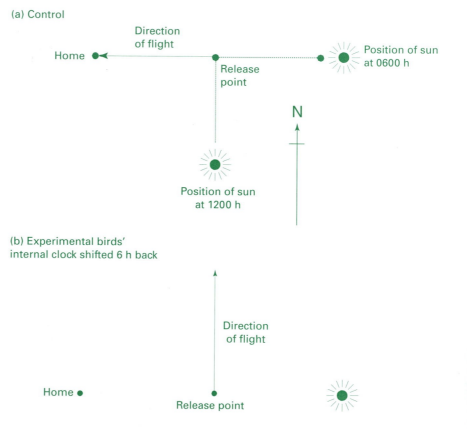

(a) Control

Direction of flight

Home

Release point

Position of sun at 0600 h

N

Position of sun at 1200 h

(b) Experimental birds' internal clock shifted 6 h back

Direction of flight

Home

Release point

complex than simply using it as a fixed guidance point as can be done with the pole star, because the sun appears to move across the sky during the course of a day. To travel in a consistently westerly direction using the sun, for example, would mean moving away from the sun in the early morning, keeping it on the left at noon and moving directly towards it as the sun set in the west (Fig. 3.3a). This means that the animal has to change its behaviour towards the same stimulus (the sun) depending on what time of day it is and this is only possible if it has some sort of internal clock.

Some dramatic experiments have been done to show that many animals, including honey-bees and birds, do have such a clock and can be fooled into setting off in quite the wrong direction if their clocks are reset (Hoffman 1958; Papi 1992). For example, the internal clock of birds can be reset by keeping them in a light-tight room in which artificial daylight is turned on and off at times different from real sunrise and sunset in the world outside. The birds can thus be exposed to the same total amount of daylight as they would outside but experience it either earlier or later in a 24-hour period. Such animals are said to be 'clock-shifted'. If an experienced homing pigeon is clock-shifted 6 hours backward – that is, it is kept in a loft in which the lights come on and go off 6 hours after sunrise and sunset in the world outside – its behaviour is predictably disturbed if it is then released outside. Imagine that the pigeon has been trained to fly westwards from a release site towards its loft and that real dawn and real sunset occur at 0600 and 1800 h, respectively. If the clock-shifted pigeon is released at noon the sun will be due south. In the normal way, it would fly 90° to the left of the sun at this time in order to get home. However, as it has been clock-shifted, its internal clock is telling it that it is only 0600 h and that therefore, to fly west, it should fly directly away from the sun. The clock-shifted pigeon is therefore expected to head off north, i.e. 90° in error of where its home loft really is. This is more or less what the pigeons are observed to do (Fig. 3.3b; Papi 1992), the errors made by the birds being roughly 15° for every hour of clock shift, which corresponds to the amount of horizontal movement the sun makes across the sky. Fortunately, the birds eventually realize their mistake, change course and find their way back to their loft, apparently using local landmarks when they get close to the loft.

This ability to correct for initial mistakes is, of course, very interesting in itself as it suggests that orientation in pigeons is not a straightforward response to one stimulus or even one stimulus combined with an internal clock, but a complex interplay between sun, internal clock and geographical landmarks. Braithwaite and Guilford (1991) showed that pigeons flew home faster if they had been kept in clear-sided boxes and allowed to view the surrounding landscape for 5 minutes before being released, than if they had been kept in opaque-sided boxes which prevented them from seeing where they were. Visual landmarks thus seem to be important to them, but as smells, sound and even magnetism are too, it is clear that there are many elements to the pigeons' ability to orientate (Papi 1992; Able 1993; Wallraff 1993).

On a larger time scale, Emlen showed that whether the indigo buntings used star patterns to orientate north or whether they used the same patterns to orientate south depended on their own hormonal state, which was in turn dependent on changing day length. This meant that the buntings responded differently to the same set of stars at different times of the year and could use them to guide both their spring migration south and their autumn migration north, depending on hormonal state. In this sense, some stimuli come from inside the animal itself.

As we saw in the last chapter, hormones have their effects because they act as stimuli on certain target organs. An example of this is the release of milk by the female rat in response to sucking by her young that we have already mentioned. The mechanical stimulation by the young as they suck induces the hypothalamus of the brain to release a hormone called oxytocin from the posterior pituitary gland. The oxytocin is carried in the blood and acts as a stimulus to the mammary glands to let down milk. Many other stimuli can also be said to be internal. There are receptors for detecting changes in blood pressure, the tension in muscles, the amount of salt in the blood, and so on. As we saw in Chapter 1, effective movement itself depends on information from within the body about the relative position of different parts of the body. There are receptors in joints, muscles and tendons, and in fact coordinated behaviour would be impossible if animals could not detect the relative position of their limbs or what was going on in different parts of their bodies. So the term stimulus, far from being just a simple 'trigger' that always leads to the same simple response, covers a wide range of different ways in which factors both inside and outside the animal's body are detected and evaluated over many different time scales.

Diverse sensory capacities

Every animal inhabits a world of its own whose character is inevitably shaped by the information it receives from its sense organs. As observers and interpreters of animal behaviour, we would be greatly handicapped if we had to rely solely on the evidence of our own senses. Fortunately, as we will see, modern technology makes it possible for us to overcome some of our own limitations and begin to enter the sensory world of other species.

For example, some animals are able to see ultraviolet light, i.e. light of much shorter wavelengths than our eyes are sensitive to. Bees foraging on flowers respond to colours and patterns that are invisible to us (and vice versa since we have a greater sensitivity to red or long wavelength light). Bees fly to the flowers of white bryony (*Bryonia dioica*) because the petals reflect large amounts of ultraviolet, although they appear to us as pale green, and provide little contrast with the leaves. Interestingly, not all the flowers that appear unpatterned to us do so to the bees. Many flowers have 'nectar guides' – radiating lines or blotches of contrasting colour on the petals that help the insect find the entrance to the flower where the nectar is. Some flowers appear, to us, to lack nectar guides altogether, but photographing them through an ultraviolet system (Fig. 3.4) shows that they do have patterns, but adapted to the eyes of bees, not humans.

Fish and birds, too, are also sensitive to ultraviolet light (Bennett & Cuthill 1994; Tovée 1995). The plumage of birds such as gulls and snowy owls, which we think of as white or dull grey, reflects ultraviolet light, possibly making them very conspicuous to other birds (Burkhardt 1989). Ultraviolet colouration may turn out to be an important component of bird communication, giving them a 'private channel' to each other from which mammals, including ourselves, are excluded.

Some snakes can 'see' infrared radiation, that is, the heat given off by their warm-blooded prey (Newman & Hartline 1982). Pit vipers (Crotalinae) and the distantly related pythons (Boidae) have pit organs or infrared 'eyes', small cavities on their heads (Fig. 3.5), which contain thousands of heat-sensitive nerve endings. When the warmth from the body of a mouse reaches

(a)

(b)

Figure 3.4 Flowers seen in two different lights. Photograph (a) records what the human eye can see: the flower appears to have no nectar guide patterns. (b) The view that approximates to what a bee sees, photographed through an ultraviolet system: a striking colour pattern is revealed. (Photograph by Thomas Eisner.)

the nerve endings, they excite the nerve fibres attached to them and the snake is alerted to the presence of a potential prey. Rattlesnakes are so good at detecting prey with their infrared eyes that they can accurately strike and kill even in complete darkness.

Another way in which our vision differs from that of other species is in our lack of sensitivity to the plane of polarization of light. Light reaching us from areas of blue sky is vibrating predominantly in one plane, known as the e-vector. The angle of this plane changes in a regular fashion with respect to the sun as it moves across the sky, so that animals that use the sun for navigation can potentially know where the sun is even if all they can see is a patch of blue sky (Brines & Gould 1982). Von Frisch (1967) showed that bees can indeed use the pattern of polarization of light to locate the sun's position even when it is obscured by clouds or a screen. Bees have a small dorsal region of the retina which has a row of analysers, each of which is maximally sensitive to a different e-vector direction. When a bee looks at a patch of polarized sky-light, the perceived direction of the e-vector is determined by which analysers in the array produce the largest output (Rossel & Wehner 1986). Birds and fish are also sensitive to the plane of polarization of light, as are squid, which

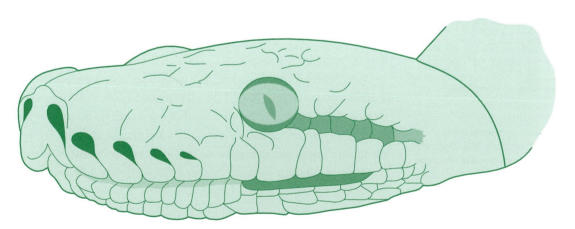

Figure 3.5 The head of the reticulated python (*Python reticulatus*), showing the heat-sensitive pits on the side of the snout behind the nostrils. A further series of pits are found along the lower lip, below and behind the eye. Each pit has a slightly different field of 'view' and its lining is highly sensitive to heat radiation.

appear to detect fish by responding to the pattern of polarized light that is reflected off their shiny bodies (Land 1981).

In the realm of hearing, too, we find some animals possessed of faculties far beyond our own and detectable only with sensitive instruments. Bats and the moths they feed on engage in a deadly battle in the night, using sounds far above our hearing range. About 70% of bat species, such as the horseshoe bats (Fig. 3.6) feed on insects and most of them use a form of sonar that enables them to detect flying insects even in complete darkness Bat sonar, like that of dolphins which also build 'sound pictures' of their world, works very like human sonar in that the animals emit pulses of sound and then listen to the echoes of those sounds after they have bounced off objects in the environment. Unlike the relatively slow sonar pulses of a submarine, however, bats can emit up to 100 pulses a second and then, remarkably, listen to the echoes coming back between each pulse. Different objects in a bat's environment distort the sound in different ways so that by comparing the original sound it made with the distorted echoes that come back to it, the bat gains information about the size, shape, speed of movement and surface properties of what is around it (Sales & Pye 1974; Simmons & Stein 1980).

However, because bats hunt insects that are in general small in size, they have to use very high frequency (i.e. short wavelength) sound because sound waves only produce echoes from objects when the wavelength of the sound is approximately equal to the diameter of the object in question. (We expect to hear an echo if we shout at a cliff face, for example, but not if we shout at a needle.) To obtain a reasonable echo from an object the size of a moth, bats have to use sound of very short wavelength, which means frequencies of 50–80 kHz. Because of the higher speed of sound in water, dolphins have to use even higher frequencies of sound – up to 300 kHz. We call such sounds 'ultrasonic' because the extreme upper limit of human hearing is about 20 kHz and we simply cannot hear bat echolocation calls at all.

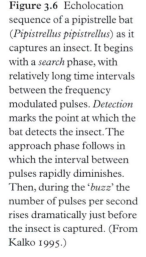

Figure 3.6 Echolocation sequence of a pipistrelle bat (*Pipistrellus pipistrellus*) as it captures an insect. It begins with a *search* phase, with relatively long time intervals between the frequency modulated pulses. *Detection* marks the point at which the bat detects the insect. The approach phase follows in which the interval between pulses rapidly diminishes. Then, during the '*buzz*' the number of pulses per second rises dramatically just before the insect is captured. (From Kalko 1995.)

But some moths and lacewings can. Noctuid or owlet moths (Noctuidae), for example, have ears on the thorax, consisting of tiny thin tympanic membranes connected to sensory cells. These simple ears can pick up the echolocating calls of hunting bats and enable a moth to take evasive action. If the sounds of the bat are very faint, indicating that the bat is still some distance away, the moth simply flies off in the opposite direction. But at close range, the moth goes into a series of erratic manoeuvres – flying in spirals or dropping suddenly to the ground – which makes it more difficult for the bat to catch it (Roeder 1967).

Some bats and owls also hunt in complete darkness not by emitting sounds of their own but by being extremely sensitive to sound made by their prey. The barn owl's hearing is so good, for example, that it can locate a mouse in complete darkness simply by homing in on the sound of the mouse rustling through leaves or even just chewing (Payne 1971). On hearing the sound, the owl turns its dish-like face (Fig. 3.7) towards the source, until it is pointing directly towards it, and then launches itself towards the prey. The owl uses the intensity differences between its two ears to home in on the sound, since when the two ears are being equally stimulated, the sound must be coming from a point directly between them. This gives it information about where the mouse is in a horizontal (left–right) plane. In addition, barn owls have asymmetrical positioning of the ear openings, with that of the left ear being above the midpoint of the eye and that of the right one below it. This gives them the ability to pin-point the mouse's position in the vertical plane as well (Knudsen & Konishi 1979).

Figure 3.7 The 'face mask' of the barn owl (*Tyto alba*) edged with short, still features which help to deflect sound into the ear openings. Not visible here, these are asymmetrically placed at the rim of the mask, the right opening slightly below the level of the eyes, the left one slightly above. Like the snake's pit organs, the owl's two ears have different fields of 'view'. (Drawing by Nigel Mann.)

Some animals can also pick up patterns of disturbance caused by objects moving near them without what we would conventionally call ears. Fish possess a 'lateral line' organ – effectively a row of little pressure sensors down each side of the body – that allows them to pick up mechanical vibrations in the water. Lateral line organs are most developed in fish that live in dark caves or the deep ocean where there is no light (Dijkgraaf 1962), but even fish with normal eyesight can, in complete darkness, collect detailed information about the presence of other fish in the school, predators and so on. Blind fish can keep up with the rest of the school but fish without a functioning lateral line system tend to crash into each other (Partridge & Pitcher 1980). Some caterpillars have hairs that can detect a flying wasp as much as half a metre away (Markl 1977) and, as we saw in Chapter 1, cockroaches use the tiny air movements made by the movement of a toad's tongue as it prepares to strike to enable them to escape.

It has been known for a long time that spiders will attack a tuning fork touching their web. This is because the vibrations of the tuning fork resemble the vibrations produced by prey caught in the web and struggling to get free. Spiders' webs are very good transmitters of information about both the nature and the location of a mechanical disturbance and a spider can use the vibrations to home in on its prey. Movements of as little as 1 nm (10^{-9} m) may be picked up by the spider (Masters & Markl 1981; Masters & Moffat 1986).

Smell is a faculty that is much more highly developed in some animals than it is in humans. Female silk moths (*Bombyx*) emit a scent that is extremely attractive to any male within a wide radius. The males locate the females entirely by their odour and are able to pick up the scent of a female when it is diluted to only 100 molecules of sex odour substance per cubic centimetre of air. Dogs can smell a variety of substances at concentrations one thousand to ten thousand times lower than humans can (Marshall & Moulton 1981). Male garter snakes can follow chemical trails left by females and can even determine the direction in which the female was going. As the female moves along she pushes against certain objects, thus leaving more scent on the side in her direction of travel. The male tests each side of these objects with his forked tongue and recognizes which side the female has pushed against (Ford & Low 1984).

House mice develop complex communication networks based entirely on smell (Hurst 1989). Without even encountering each other directly, mice know the sex and also the social status of others just from the scent left behind in their urine. A pregnant female mouse will abort the litter she is carrying if she is exposed to the urine of a strange male mouse. This is an example of the 'Bruce effect', named after Hilda Bruce who discovered that female mice tended to abort if a strange male mouse (i.e. not the father of her offspring) was around. Although it seems a rather maladaptive response on the part of the female mouse to abort a whole litter, it appears that she is 'making the best of a bad job'. Male mice tend to kill newly born mice that they have not fathered themselves, so a female that resorbs the embryos and starts a new litter with the new male is more likely to have surviving offspring than one that carries the litter to full term and then has it killed.

Even more remarkably, mice have recently been shown to use smell to distinguish mice that are genetically similar or genetically different from themselves in one particular genetic region. The Major Histocompatibility Complex (MHC) is a group of genes that were originally thought to be entirely concerned with recognition of 'foreign' bodies or tissues. It has now been established that odour cues associated with differences in the MHC affect not just cellular recognition but behaviour as well. Which companions a mouse nests with and which mate it chooses are both affected by its genetic similarity to these other mice at this particular genetic region (Lenington 1994). By smelling out mates that differ genetically from themselves, mice have a built-in way of avoiding inbreeding.

It has recently been discovered that similar effects may operate in humans, suggesting that our sense of smell may be more sensitive than we realize. Human females asked to rate the smell of T-shirts that had been worn by six different men reported that the smells were pleasanter from men most different from them in the MHC region and least pleasant from men who were similar (Wedekind *et al.* 1995). What is more, the smell of the MHC-dissimilar men reminded the women most of their own actual or former mates, implying disassortative mate preference in humans.

Sixth and even seventh senses are a reality for some animals. Duck-billed platypus (Fig. 3.8) are able to detect the electric fields generated by muscle activity in animals such as the crayfish and shrimps on which they feed. (Scheich *et al.* 1986). Even though playtpus close their ears, eyes and nostrils while they are underwater, they have electroreceptors on their bills that enable them to detect their prey and will even show interest in an electrode placed in the water if it shows electrical pulses like an 'artificial shrimp' (Manger & Pettigrew 1995). Sharks and rays are even more sensitive to electrical fields generated by muscle movement and, almost unbelievably, can detect fields as small as $0.005\ \mu V\ cm^{-1}$. Electric fish go one stage further and create their own electrical environment. Using specially modified muscle tissue, they set up an electric field around themselves. Objects in the field that are either good conductors or poor conductors of electricity distort the field in characteristic ways, and the fish use the distortions to find out what the objects are. Electric fish tend to live in very turbid waters where vision is not much use and they use their electric sense to detect obstacles and to find prey. They also carry on a great deal of their social life by communicating through electrical signals (Hopkins & Bass 1981). The Earth's magnetic field can be detected by certain bacteria which contain pieces of magnetite (Blakemore 1975), by some fish such as skates, and even by migrating birds (Wiltschko & Wiltschko 1988).

The problem of pattern recognition

Information about what stimuli animals are capable of detecting is clearly an essential preliminary to understanding their behaviour. But it is only the first stage, and it would be easy to make the mistake of thinking that all we need to know about how animals respond to stimuli is to understand how their sense organs transform energy from the environment (light, sound, magnetism, etc.) into nerve impulses, which can then be interpreted or recognized by the brain. The reason this would be a mistake is that the process of interpretation itself is extremely complex and far from fully understood. In fact, even the most sophisticated computers are surprisingly poor at it. It is relatively easy to make computers sensitive to light or sound (to give them the equivalent of sense organs, in other words) but much more difficult to make them good at recognizing speech, handwriting or even simple objects in the way we do all the time.

Figure 3.8 A duck-billed platypus (*Ornithorhynchus anatinus*) hunts for prey in the bed of a stream which it disturbs with its 'bill'. Around the edge of the bill are electroreceptors capable of detecting the minute electrical activity which results from the contracting muscles of the prey – worms, crustaceans – as they try to escape. The platypus's eyes are closed – all detection is via these sense organs. (Drawing by Nigel Mann.)

Figure 3.9 illustrates dramatically how seeing shapes differs from recognizing objects. Both (a) and (b) have the same corner elements that go to make up a cube, but whereas it is difficult to see the cube (a), the cube is obvious when three opaque stripes are introduced (b). Our brains clearly interpret the same elements in quite different ways.

To understand why the interpretation of information supplied by the sense organs poses such difficulties and in turn to appreciate some of the 'solutions' that different animals have adopted, we can use the very simple example illustrated in Fig. 3.10. A common object like a cup is easily recognizable when seen from different angles and distances even though its apparent shape varies quite dramatically. Thus, even though our retinas are stimulated in different ways, we have no difficulty in saying that we are looking at the same object. However, we can also discriminate the cup from a saucer made from the same material even when, just going by outline, the cup seen from above looks more like the saucer seen from above than it does the cup seen from the side.

The 'problem' of pattern recognition can thus be broken down into two sub-problems: that of **recognizing the similarities** between patterns of stimulation that are very different but in fact come from the same object seen from different viewpoints or under different conditions; and that of **discriminating** patterns that are very similar but in fact come from different

Figure 3.9 The difference between 'seeing' and 'perceiving'. Exactly the same elements of the cube are present in both (a) and (b) but whereas the cube is not obvious in (a), it suddenly becomes so with the addition of three 'opaque' stripes. (After Kanizsa 1979.)

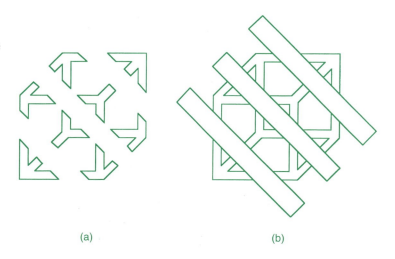

(a)   (b)

(a)

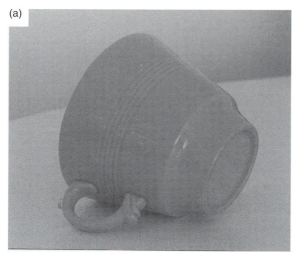

(b)

(c)

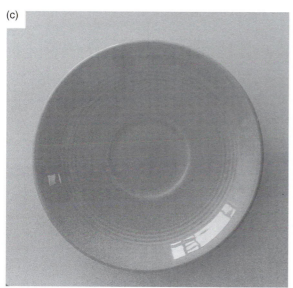

Figure 3.10 (a), (b) Two very different views of the same object – a cup. We have no difficulty in recognizing them as the same object seen from different viewpoints even though the outline of a cup seen from above (b) is much more similar to that of a saucer (c) than it is to (a.)

objects. Exactly the same problem arises in the sound domain as with vision. If someone with a high-pitched voice and someone with a bass voice both say the same word, 'tell', the actual sound pattern reaching our ears will be very different and yet we can recognize them as the same word spoken in two voices. Yet if two words 'tell' and 'dell' are said by the same person, the sound patterns are much more similar but we recognize them as distinct words.

The difficulty is that the two sub-problems of pattern recognition – detecting similarities on the one hand and discriminating differences on the other – tend to have mutually incompatible solutions. Thus, one way of recognizing the similarity between the cup seen from above and the cup seen from the side would be to give the label 'cup' to anything of roughly the right size that had at least one curved edge. This would successfully classify the various views of the cup together but it would also erroneously include the saucer as a cup. In other words, successful detection of similarities between the different views of the cup would lead to complete failure of discrimination between cup and saucer. Pattern recognition involves a variety of different solutions that animals have adopted to circumvent this dilemma. All involve compromises and are prone to some inevitable errors that reveal a great deal about what the worlds of animals are like.

Sign stimuli (key features)

One common solution we find is that animals respond to only one special part of the array of stimuli presented to them. Their pattern recognition is thus accomplished by picking out one or two key features that all instances of whatever they are trying to recognize have in common. Dragonflies attempting to lay eggs on the shiny metal of a car (picking out 'shiny surface' as the key feature of the water they normally lay their eggs in), or the male robin described by Lack (1943) which flew down and attacked a bunch of red feathers on the lawn (robins have red breast feathers), are two examples of animals responding to 'sign stimuli'. The animals are responding to restricted features of their environment, apparently ignoring others. By using a small number of key features, they can recognize as similar patterns that might be much more difficult to recognize as the same if larger numbers of features were being used. Male robins present very different stimulus patterns (at different distances, viewpoints, etc.) if a complex analysis were done, but as long as all male robins have red breasts and isolated bunches of red feathers are a rare occurrence, using this one key feature makes recognition easy, simple and view-invariant. In the case of very important stimuli – such as those denoting the approach of a predator – a few false alarms are a small price to pay for rapid identification.

There are some dramatic examples of the advantages and disadvantages of relying on sign stimuli. Turkey hens which are breeding for the first time will accept as chicks any object which makes the typical cheeping call. On the other hand, they ignore visual stimuli in this situation and deaf turkey hens kill most of their chicks because they never receive the auditory sign-stimulus for parental behaviour (Schleidt *et al.* 1960). Roth (1948) describes how male mosquitoes respond with high selectivity to the sound of their females' wings which beat at a characteristic frequency, different from their own. The males can, in fact, be attracted to a tuning fork vibrating at the correct frequency. Sign stimuli thus provide a rough and ready way of recognizing patterns in the environment that nicely illustrate the problems we discussed in the last section. They give easy recognition despite variation in exactly how a stimulus is seen or heard. But they can lead to a failure of discrimination if unusual circumstances (such as the presence of a human with a tuning fork) appear.

Tinbergen (1951) pointed out that such 'false positives' are one of the most conspicuous characteristics of innate behaviour and described a striking example of male sticklebacks that normally respond aggressively to the red coloration of rival male sticklebacks also displaying to a red mail van visible through the windows of the aquarium. It is important to stress that the sticklebacks have good eyesight and can certainly discriminate between a fish and a mail van (Tinbergen 1951), but in this case they reacted to the sign stimulus of red and neglected other characteristics.

This is not always true, however. Sometimes we find that the sense organs themselves are responsible for the selective response to a particular sign stimulus through being relatively insensitive to anything else. The calls of male green tree frogs, for example, have two peaks of sound energy – a low one at 900 Hz and a high one at about 3000 Hz (Gerhardt 1974). The ears of the females are tuned to pick up these two frequencies in particular: in the female's auditory system, the so-called 'amphibian papilla' is most sensitive to frequencies between 200 and 1200 Hz while the 'basillar papilla' responds best to sounds of about 3000 Hz. (Capranica & Moffat 1975). Narins and Capranica (1976) showed a remarkable sex difference in the hearing of a related species of tree frog, which accounts for the fact that males and females respond to different sign stimuli in the call of the male – a double 'co-qui'. By playing back tape recordings of the males' call to the frogs, they were able to show that males (who attack other males that approach them) respond only to the 'co' note, and females only to the 'qui'. Neurophysiological recordings made it clear that each sex heard only the note of relevance to itself because the neurons of the inner ear were tuned differently for males and females.

Selective responsiveness to sign stimuli has been a particularly fruitful area for showing how the behaviour of an animal can be related to underlying neurobiological mechanisms. Ewert & Traud (1979) provide a particularly good example in their study of the way in which toads respond to snakes (Fig. 3.11). In response to real snakes, toads fill their lymphatic sacs with air, making their bodies appear much larger than they really are and assume a stiff-legged posture, presenting their flank to the snake. Toads will perform this stiff-legged snake posture to quite crude dummy snakes made of flexible pieces of cable. The 'sign stimulus' for this behaviour seems to be quite easy to reveal, as even a model consisting of just a horizontal piece with a little head attached produced almost as much (93%) response as a real snake.

Ewert and his colleagues have analysed in some detail the sign stimuli that toads use for recognizing both prey and enemies. In both cases, movement and contrast with the background are important, but whereas a toad will turn towards a small, horizontal, worm-like object moving in a horizontal plane (worm stimulus, Fig. 3.12), it will not turn towards the same object orientated vertically with the longitudinal axis perpendicular to the horizontal direction of movement. In the toad's view, worms do not walk on their heads! Large, expanding (looming) objects are sign stimuli for escape and a toad will turn away from or freeze to a very crude cardboard model coming towards it.

It is possible to link the toad's responsiveness to prey and enemy to cells in its retina and even specific parts of its brain. A plan of the amphibian retina is shown in Fig. 3.13, and although it is possible to record the electrical activity of all types of retinal neurons, the most revealing have been made by placing

Figure 3.11 Stiff-legged posture of a toad (a) to a real moving snake and (b) to a moving model made from electric cable. (From Ewert & Traud 1979.)

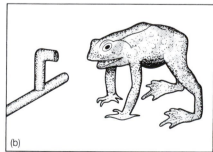

Figure 3.12 A toad's response to three moving models. The toad turns to follow a model and its response is measured by the number of times it turns to follow the model in 1 min. (From Ewert 1980.)

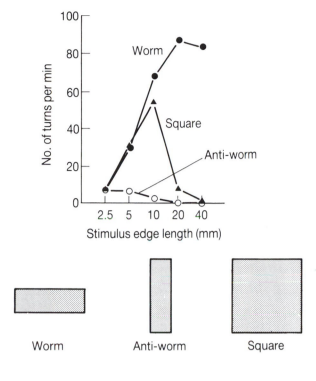

Figure 3.13 Diagrammatic representation of an amphibian retina showing the connections between the various sorts of cells. Excitatory synapses are shown by ——◀, inhibitory synapses by ——●.

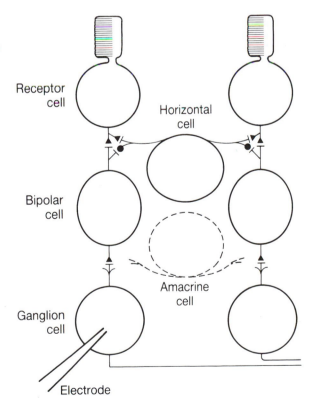

electrodes in the ganglion cells since these turn out to be of several distinct sorts, distinguished by how they respond to different visual stimuli. In a pioneering paper entitled *What the frog's eye tells the frog's brain*, Lettvin *et al.* (1959) showed that there were six different sorts of ganglion cells, with four being most common, and this has subsequently been confirmed for toads as well. Class 1 (rare in toads) and Class 2 retinal ganglion cells show a vigorous response whenever the animal is looking at a small moving object. The retinal ganglion cells have a curious property that explains how they can 'recognize' when an object is small and moving. Each retinal ganglion cell responds to a particular section of the total visual field. Thus, any given cell will not respond every time the toad looks at an object but only if the image of that object falls on a particular part of the retina, known as the 'receptive field' of that cell. The receptive field of each retinal ganglion cell is divided into two parts (Fig 3.14), an inner excitatory centre and an outer inhibitory ring. This means that whenever an object is picked up by the inner part of the receptive field, the retinal ganglion cell is excited and fires, but if the object falls on the outer part of the receptive field, the retinal ganglion cell's activity is **inhibited** (Fig. 3.14). Thus inhibition, far from being a purely negative process, is crucial to the frog's ability to recognize small objects.

Figure 3.14 Diagrammatic representation of a retinal ganglion cell's on-centre receptive field. The ganglion cell receives input from a number of different receptors through the horizontal and bipolar cells. Light falling on receptors at the centre of the field excites the ganglion cells and evokes action potentials. If light also falls on receptors on the surrounding (outer) area, this decreases the activity of the retinal ganglion cell and evokes fewer action potentials.

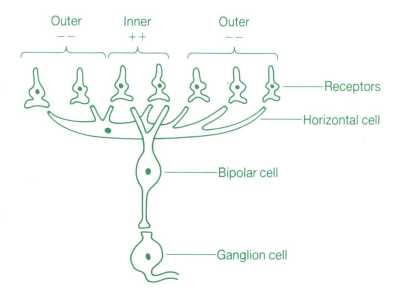

The different classes of retinal ganglion cells have different sizes to their excitatory receptive fields, with the Class 1 and Class 2 cells having smaller such fields (2–5°) than the Class 3 (6–8°) or Class 4 cells, which have the largest fields of all (10–15°) and do not respond at all to a small object unless its visual angle exceeds about 5°. If a square shape is moved in front of a toad's eye, the impulse frequency in the different classes of retinal ganglion cells also varies with the size of the square used. In Class 2 cells, with their small excitatory centres, only quite small shapes give maximum response because larger squares cover both the excitatory and the inhibitory areas of the receptive field and this has the effect of reducing the total activity of the cell. Class 4 cells, with their large excitatory fields, are active when the toad is viewing a much larger square because the inhibitory surround is not stimulated. What this all means is that Class 2 and 3 cells are activated mainly by prey stimuli and Class 3 and 4 cells mainly by 'enemy' stimuli. However, it would not be correct to say that there are specific prey or enemy detectors in the toad's retina and that there we had accounted for the toad's response to 'sign stimuli' in terms of the physiology of its eye. What the retinal ganglion cells do is to extract basic information about the size, angular velocity and contrast of the object the eye is looking at, leaving it to the brain to distinguish between food and enemies. Ewert (1980) describes in some detail how the toad's brain performs this task. In fact, the analysis of sign stimuli and their neurophysiological basis has been one of the most successful areas of 'neuroethology' and excellent accounts are also given by Camhi (1984) and Young (1989).

'Supernormal' stimuli

As we have seen, a common method of investigating sign stimuli is to present animals with dummies or models of a natural situation and then to change these in certain ways to see which are most important to the animals. A very curious phenomenon has emerged from a number of these studies: it is often possible to produce a model which produces a *greater* response from an animal than does the natural object. Examples of such 'supernormal' stimuli are found in all sorts of animals, but some of the oddest have come from the incubation behaviour of birds. Tests have been made with the herring gull, the greylag goose and the oystercatcher and in all three, the larger an egg is, within broad limits, the more it stimulates incubation. Figure 3.15 shows an oyster-catcher trying to incubate a giant egg in preference to its own.

Male sticklebacks preferentially court females which are large and have distended abdomens indicating that they have a lot of eggs. Rowland (1989) showed that when presented with two dummy females, a male would first direct his courtship to the largest and fattest one, even when the 'fat' female had an abdomen distended far beyond the normal range for female stickle-backs.

Herring gull chicks respond to very crude models of their parents (Tinbergen & Perdeck 1950). The adult herring gull has a red beak with a yellow spot on the lower mandible at which the chick pecks. This causes the parent to regurgitate food. However, the chick will not only peck at a headless beak: it will peck even more vigorously at a long thin red knitting needle, par-ticularly one with yellow bands on it. Nor is it only our animal relations that respond in such ways. It is obvious that lipstick makes human lips into a super-normal stimulus (Fig. 3.16) and much of the advertising industry is based on persuading us to respond to exaggerated versions of everyday situa-tions.

Recently, computer modelling has been used to show that supernormal responses may be an almost inevitable outcome of recognition systems based

Figure 3.15 An oystercatcher attempting to brood a giant egg in preference to its own egg (foreground) or a herring gull's egg (left foreground). The bird's original nest was equidistant between the three 'test' eggs. (From Tinbergen 1951.)

on just a small number of cues. Arak and Enquist (1993) used a simple model which possessed an artificial 'retina' of 36 receptor cells and which could recognize simple shapes such as flowers, or birds with different length tails (Fig. 3.17). The model spontaneously responded more strongly to exaggerated versions of the test stimuli than to the test stimuli themselves, a phenomenon that Arak and Enquist describe as a 'hidden preference' and they relate it directly to the phenomenon of supernormal stimuli.

At this point, we can ask why – in an adaptive sense – animals respond to supernormal stimuli at all. It would appear that by responding to exaggerated versions of natural stimuli, animals are behaving 'stupidly', and against their own interests. Even saying that such behaviour appears spontaneously in a computer model does not really solve the problem. Why has natural selection not acted against animals that behave in such ways? It is here that we see clearly the importance of asking different sorts of question about animal behaviour and of understanding its evolution as well as its causation. We have just asked why, in a causal sense, animals respond to sign stimuli and supernormal stimuli and have seen that such responses are ways of solving the difficult problem of pattern recognition. We have come to the conclusion that they are not perfect solutions and so now we have to ask why, in an evolutionary sense, the animals have not done better. Another way of putting this question is to ask whether an animal that did *not* respond to sign stimuli and supernormal stimuli would be better able to survive and reproduce than one that did and sometimes made mistakes. We can consider three different answers.

Figure 3.16 Human lips are made strikingly supernormal by the addition of lipstick.

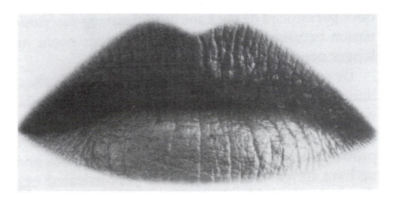

Figure 3.17 Arak and Enquist's (1993) simplified model of how an insect might recognize a flower. (a) The image of a flower falls on the input layer of an artificial neural network. In the model, all cells in the input layer were connected to all cells in the hidden layer which were in turn connected to all cells in the output layer, but for simplicity only one set of connections is illustrated. The network acquired the ability to discriminate between different flowers (input patterns) by adjustment of the connections between cells. The network 'learnt' that one flower type was more profitable than another (recognition task, b). When it was then presented with a variety of test stimuli, it responded to some of them, particularly ones which were exaggerated versions of the profitable flower, even more strongly than to the original. (Test stimuli that were responded to more strongly than to the profitable training stimulus are denoted by ★.)

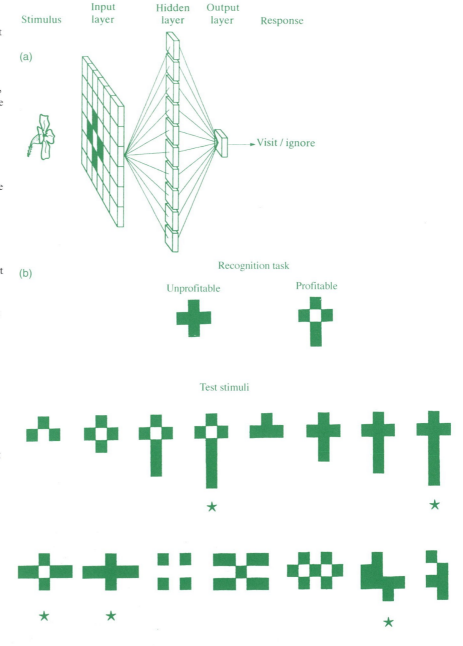

1. Mistakes are so rare that they do not matter

Many of the examples we have discussed of animals responding to sign stimuli and supernormal stimuli have been of animals put into very unnatural situations, i.e. ones that would rarely, if ever, occur in nature. For example, ostrich eggs are not a feature of the natural environment of oystercatchers, nor are knitting needles generally found in herring gull nests. Equally, red mail vans are not a major natural hazard for male sticklebacks. Consequently, simple 'rules of thumb' such as responding to the biggest or brightest or longest object around will usually serve the animal perfectly adequately and may be the easiest to evolve given the structure of its sense organs. False positives will be rare or non-existent and probably confined to interference from ethologists.

2. Mistakes occur, but they are not too costly

Using a simple rule of thumb, such as courting the largest fattest female around for a male stickleback, will normally result in a male having the greatest number of eggs laid in his nest, since he will be directing his courtship towards the most fecund females. However, the fact that the fattest females may sometimes not be full of eggs but full of worms shows that there are dangers in solving a pattern recognition system by relying solely on simple sign stimuli. Where mistakes are relatively rare, these may persist ('the rare enemy' effect as Dawkins (1982) describes it). But where the animal consistently and frequently behaves in maladaptive ways, we expect the evolution of pattern recognition systems that are more discriminating. As we will see in the next section, many animals have done exactly that – evolved pattern recognition systems that are very much more sophisticated than the relatively simple ones we have so far discussed.

3. The mistakes are not mistakes at all

If responding to the biggest, brightest or most colourful male around means that a female mates with a strong healthy male and as a result has strong healthy offspring, then females with a tendency to respond to supernormal males may be strongly selected for. Not only that, but males sporting supernormal plumage may be favoured, leading to an evolutionary exaggeration of both male plumage and female preference (Fisher 1958). In the second part of this chapter, we will discuss the importance of sign stimuli and super-normal stimuli in the evolution of animal signalling, not just in communication between the sexes but in many other kinds of signals too. As we will see, simple rules of thumb, far from being disadvantageous, may be the key to the evolution of the large and exaggerated signals that are typical of much animal communication.

More complex situations

Where mistakes or false positives are common and/or dangerous, we expect natural selection to favour animals that do not make them. It is not difficult to find cases where response to simple sign stimuli has been selected against, sometimes even where selection seems still to be in progress. The habit of brood parasites such as cuckoos of laying their eggs in the nests of other birds provides powerful selection pressure on the potential foster-parents to discriminate between their own eggs and those of other species. The brood parasite benefits if it can 'deceive' a host into caring for its young, but the cost to the host can be very high. A young cuckoo, for example, kills the host's own young by pushing all existing eggs out of the nest (Fig 3.18). So the host – such as a reed warbler – not only loses the eggs it has already laid, but loses again through working hard to feed the rapidly growing cuckoo rather than laying a replacement clutch itself. The host appears to respond to the 'sign stimulus' of an open gape in its nest, but more careful analysis shows that there is an evolutionary 'arms race' between the cuckoo chick to be accepted and the host to recognize it. The selection has been particularly severe at the egg-recognition stage.

Each individual female cuckoo specializes in just one species of host and lays eggs that match, often closely, the eggs of her host (Fig. 3.19). Females from different strains of cuckoos (known as gentes) somehow choose to lay in the nests of species whose eggs resemble their own. Wagtail-cuckoos produce white spotted eggs and lay them in wagtails' nests and pipit-cuckoos produce brown spotted eggs that they lay in pipits' nests. Females deposit a single egg in the nest of their chosen host and take one egg away. The cuckoo egg develops quickly and usually hatches before any of the host's clutch, which gives the young cuckoo the chance to destroy all the other eggs.

Even though cuckoos breed throughout Europe and northern Asia, they

Figure 3.18 Reed warbler feeding a young cuckoo. When the cuckoo hatches, it ejects the eggs or young of the reed warbler by pushing them out of the nest using its back, so that the parents lose their entire brood and feed only the cuckoo. (Photograph by J Horsfall.)

Figure 3.19 Top row, real cuckoo eggs from (left to right) a reed warbler nest, meadow pipit nest and pied wagtail nest. Different cuckoos specialize in different hosts and lay eggs that match the host's own. 2nd row, model cuckoo eggs used in the experiment: (left to right) reed warbler, meadow pipit, pied wagtail, redstart. 3rd row, current favourite British hosts of cuckoos: (left to right) reed warbler, meadow pipit, pied wagtail, dunnock, robin, sedge warbler, wren. 4th row, suitable but rare hosts: (left to right) redstart, spotted flycatcher, reed bunting, chaffinch, blackbird, song thrush. 5th row, unsuitable hosts: (left to right) linnet, greenfinch, bullfinch, great tit, blue tit, pied flycatcher, wheatear, starling, swallow. (From Davies & Brooke 1989a; photograph by M. de L. Brooke.)

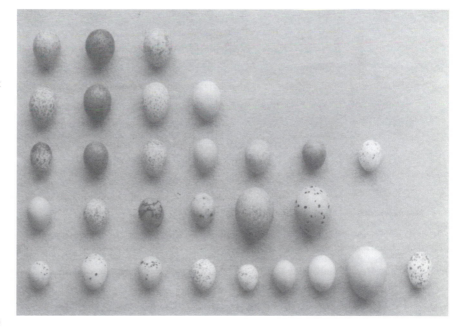

parasitize only about ten host species. It is the difference between the small number of species that are parasitized and the much larger number that are not which provides us with the evidence that egg recognition has evolved under the evolutionary pressure from brood parasites.

Davies and Brooke (1989a, b) investigated the responses of 24 species of song birds to eggs of the European cuckoo by placing model cuckoo eggs in their nests and recording whether they were accepted or thrown out. They found that most of the species that are currently parasitized by cuckoos, such as reed warblers, meadow pipits and pied wagtails, showed some degree of discrimination in that they would accept model eggs if they were similar to their own but not if they mimicked those of other species. However, the other species that are not currently parasitized by cuckoos, such as spotted flycatchers and reed buntings, rejected all model eggs, irrespective of type. These species would seem from their diet and nest locations to be highly suitable as cuckoo hosts, which raises the intriguing possibility that they were once parasitized by cuckoos until their egg discrimination became so good that cuckoo eggs could no longer survive in their nests.

By contrast, other species, such as greenfinches, linnets, great tits and swallows, accepted all the model eggs, whatever they looked like, but these species, because of either diet or nest site, have probably never been suitable hosts for cuckoos and so have never been selected for sophisticated egg discrimination. Meadow pipits in Iceland, where the cuckoo does not occur, were much less discriminating about model eggs in their nest than the same species in Britain, where the cuckoo is a threat. This is a clear illustration that the ability to discriminate is itself under selection pressure.

Generalized feature detection

Our understanding of complex pattern recognition in animals is very far from complete, but one characteristic does seem to be very widespread. This is that sensory systems are adapted to pick out 'information-rich' parts of the environment (Barlow 1972). We all know that a complex scene can often be conveyed by a quick line drawing and this is because lines or edges carry a great deal of information – about where one object ends, for example, and another begins. An ability to detect lines, edges, discontinuities of light and dark and so on would therefore be a great help in analysing many different scenes, because the same general features, although in different combinations, occur in many different situations. Thus the letters 'A' and 'F', for example, are both made up of three straight lines, so line-detectors would be useful in recognizing both. To discriminate between them, on the other hand, it would be necessary to recognize the particular angles, lengths and arrangements of those three lines.

This, in a very simplified version, seems to be the basis of the way complex pattern recognition in animals occurs. At an early stage of sensory processing generalized feature detectors are picking out spots, edges, movement or sudden changes of illumination points. These are then put together to define more complex objects at a later stage of analysis. Generalized feature detectors have now been discovered in a wide variety of animals, both invertebrates and vertebrates.

In the horsehoe crab (*Limulus*) it is possible to see how these general features are extracted from the overall pattern of illumination reaching the eye. Each eye is compound, i.e. it is made up of several hundred units called ommatidia, each of which contains ten or more retinula cells. The axons of the retinula cells come together to form an optic nerve from each ommatidium, but the ommatidia do not operate entirely separately from each other – there is interaction between them in the form of lateral inhibition. Each ommatidium inhibits the actions of its neighbours but this has the important result of accentuating contrasts in illumination.

To understand how this works, consider the case where the whole eye is viewing a uniformly lit scene. All the ommatidia will be equally active and all will be inhibiting each other with roughly equal strength. Now consider what happens when a small bright light falls on just part of the eye. The ommatidia that are stimulated by the light will respond strongly and they will also strongly inhibit their neighbours. Ommatidia just outside the patch of light will not be responding much to the light, but they will be strongly inhibited by nearby active ommatidia. They will therefore have an overall response that is less than ommatidia further away. Conversely, the ommatidia just inside the patch of light will be responding to the light and they will be less inhibited by their inactive neighbours than ommatidia in the centre of the patch of light that are all surrounded by strongly inhibiting neighbours. The result is that the boundary between the light and dark area of the scene will be the most active (Fig. 3.20), and the feature of an 'edge' will be accentuated (Hartline *et al.* 1956).

Lateral inhibition is thus as important to the ability of this compund eye to recognize shapes as it is in the complex retina of vertebrates, where we have already seen that the retinal ganglion cells respond more strongly to a small patch of light than to a large one (p. 142). Lateral inhibition enables the retinal ganglion cells to extract the feature 'spot' from the incoming sensory information because they are excited by activity in some of the photoreceptor cells and inhibited by activity in others. By monitoring which of the lower level photoreceptor cells are active, the higher level retinal ganglion cells can begin to detect patches of light of different shapes.

This idea of different levels of activity was at one time extended to the view that a whole hierarchy of processes might exist with low level feature detectors feeding in to more and more complex feature detectors higher up the visual pathways (Barlow 1972). A row of spot detectors, for example, could all feed into a higher level cell that could then become a line detector, two line detectors could provide the basis for a corner detector and so on. There is considerable evidence that some such a hierarchical arrangement exists, with progressive extraction of information at different points (Hubel & Wiesel 1959), but there are problems with what the highest level of all might

Figure 3.20 (a) The action of lateral inhibition in *Limulus* eye shows that the response of ommatidium A is *reduced* when light is shone on ommatidium B, as B actively suppresses A; (b) shows how this lateral inhibition works to enhance the response to an edge or discontinuity. If light shines on ommatidia A–E but not on F or G, ommatidium E, at the edge of the light, will give the greater response. Light falling on A–D will activate them but they will also be inhibited by the action of their active neighbours on both sides. Ommatidium E, however, will be inhibited on only one side (by D) as F is not active.

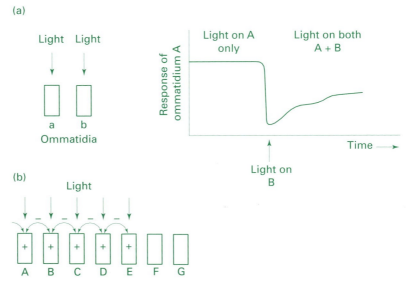

be. At the very top of Barlow's hierarchy was supposed to be one cell that responded to very specific sets of features – sometimes referred to as a 'grandmother cell' (it only responds when you look at your grandmother). But in order for humans and other animals to be able to carry out the discriminations we know they can, there would have to be not just a 'grandmother cell' but a 'grandfather cell' and a 'cousin Evelyn cell' and so on for everything we can recognize. Further analysis of the way the brain analyses incoming information has not only failed to find 'grandmother cells' but has also led to the realization that serial processing of images up separate pathways leading to individual end-point cells may not be the way brains – human or non-human – operate at all.

For example, there is a class of neurons in the temporal lobe of the brains of rhesus monkeys that fire up to 20 times more rapidly when the monkey is looking at a face (monkey or human) than they do in response to lines, gratings or complex objects that are not faces. They can therefore be referred to as 'face-detecting' neurons (Baylis *et al.* 1985) and since many of them respond more to some faces than others, they seem to be involved in the recognition of individuals. However, they do not behave like 'grandmother cells' because a given cell can be involved in the identification of several different faces. The information about which face is seen is therefore not given by the firing of one particular cell, but is spread over a whole ensemble of neurons and it is the pattern of neuronal activity in the whole set that changes when different faces are looked at (Rolls 1994).

Another reason for abandoning the idea of single end-point cells is that we now know that the visual system is not built as a single serial hierarchy but has multiple parallel pathways. Each area of the brain, whether concerned with vision or not, has multiple outputs, with the different areas specializing in different functions. In mammals, there are separate pathways concerned with movement and with the form of objects (Zeki 1993). Quite how information from these different parthways is eventually put together we still do not fully understand. How animals recognize patterns, although an exciting and active area of research for ethologists, physiologists and psychologists, still leaves us with a great deal to discover.

Communication

Many of the most important external stimuli to which animals respond come from other animals of their own species – from their parents or from their own offspring, for example, or from rivals or potential mates. Having looked at some of the discoveries that have been made about the sensory worlds of animals, we are now in a position see how a knowledge of the reception and processing of sensory information helps us to understand how animals respond to each other, and in particular how they communicate using signals. Signals are conspicuous behaviour patterns, often combined with structures like plumes or crests that have been specially evolved to affect the behaviour of another animal. The single large claw of the male fiddler crab (*Uca*) is an example of a structure evoled for a signal function (Fig. 3.21). The claw is far bigger than is needed for feeding or other non-signal purposes. It is brightly coloured and waved rhythmically in the air at females and rival males as the crab stands by its burrow, suggesting that it has been evolved to be a conspicuous stimulus to other crabs.

In this section, we will first discuss what communication means and then we will look at the evolutionary pressures that operate on animal signals, including the importance of arresting the attention of the receiver and the

Figure 3.21 A male fiddler crab (*Uca*) whose hugely enlarged right cheliped is purely for display – a classic example of a 'releaser' in Konrad Lorenz's sense. The feeding cheliped can be seen on the crab's left; in females both chelipeds are this size. (Drawing by Nigel Mann.)

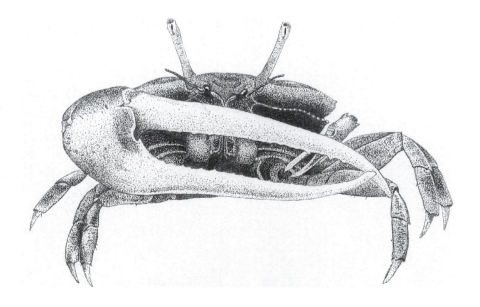

controversial question of whether animal signals are 'honest'. Finally, we will look in detail at two remarkable but very different animal communication systems – those of vervet monkeys and of honey-bees.

What is communication?

'Communication' is said to occur when one animal responds to the signals sent out by another animal. Although this sounds a reasonable definition, there are cases where it is not so straightforward. Burkhardt (1970) cites the example of a foraging ant which lays a scent trail when it has found food. Its nest mates follow the scent trail, but so does a small snake (*Leptotyphlops*) which uses the scent trail to find the ants' nest, where it devours the brood. Most of us would probably refer to the scent trail as a signal, evolved by the ants for communicating with each other, but although the snake undoubtedly obtains information from the ants, we would probably not want to say that the ants communicated with it.

The reason we would not call this communication has nothing to do with the fact that ants and snakes are of different species because there are plenty of examples of interspecific communication. The nectar guide patterns on flowers (p. 129) communicate with bees to the mutual advantage of both and can be seen as a specially evolved signal to help the bees find the centre of the flower. Warning or startle coloration of prey animals – for example, eye-spot patterns of moths (p. 320) – are other examples of interspecific signalling.

The reason we can describe flowers and moths' eye-spot patterns as communicating but not the ants with the snake, is that the ants derive no benefit whatsoever from the response of the snake (they are of course harmed by it), so that there is no suggestion that natural selection favoured ants that attracted snakes whereas the flowers and the moths both benefit from the response they evoke. The ants have evolved to communicate with each other by their scent trails because it is of mutual advantage to recruit other workers to a food source. One of the penalties of this signalling system is that a predatory snake can use the same information but, provided snakes are relatively rare, the advantages the ants derive from communicating with each other will outweigh the occasional disadvantages.

Even this mutual advantage proviso to the definition of communication is not without its difficulties, however, and we must recognize that not every-body uses it. For example, both Altmann (1962) and Hinde & Rowell (1962) studied social communication in rhesus monkeys and attempted to describe their usual signals, but the former study lists about 50, the latter only 22. The reason for this discrepancy lies in their different uses of the term communica-tion. Altmann's definition of social communication is a 'process by which the behaviour of an individual affects the behaviour of others'. All kinds of move-ments and postures may do this – the sight of a monkey feeding, for example. Hinde and Rowell, on the other hand, use a definition similar to that given above and they classified as visual signals only those that were like to have evolved for this purpose. Even though we shall concentrate on specially evolved social signals in this brief discussion of communication, it is impor-tant to recognize that animals do receive a great deal of information in the way Altmann describes. Members of a rhesus monkey troop will respond to a range of diverse stimuli from the dominant male. His general body posture and manner of moving will convey information quite apart from any signals, such as threat movements, that he may make. Almost every feature of his body when moving or at rest is in strong contrast to that of a subordinate male (see Fig. 3.23). In such cases it may be difficult to decide whether a 'specially evolved' signal is involved or not.

We should also be aware that even movements or postures that are scarcely distinguishable to us may have considerable significance for the animals themselves. In blue tits (*Parus caeruleus*), for example, quite different messages are conveyed to other birds depending on whether a bird raises the feathers on the top of its head or the ones just a little way down the nape of the neck (Fig. 3.23). In the next section, we will try and explain why some signals are loud or conspicuous and others so small that humans have difficulty knowing they are there at all.

A further complication is that not all communication takes the form of a clear signal followed reliably by a clear response. In many bird species, domi-nance is associated with special patterns of feather coloration known as 'status badges'. The most dominant male house sparrows (*Passer domesticus*) have the biggest black bibs (Møller 1987) whilst in Harris' sparrows (*Zonotrichia querula*) the most dominant have the blackest heads (Fig. 3.24). But whilst subordinate birds may sometimes respond to a status badge by giving way

Figure 3.22 Typical body postures assumed by (a) subordinate and (b) dominant rhesus monkeys. (Drawing by Priscilla Barrett from Hinde 1974. Used with permission of McGraw-Hill Book Company.)

Figure 3.23 Two blue tits dispute over a food item on the ground. The bird on the left is giving way, about to retreat, and shows the raised crest feathers, indicative of fearfulness. Its opponent has a flattened crest, but the nape feathers are conspicuously elevated giving its neck a thickened appearance. Nape raising indicates high levels of aggressiveness. (Drawing by John Busby.)

over a food source or avoiding the dominant altogether, they do not always do so. Sometimes the subordinate bird appears to take no notice of the status badge, raising doubts as to whether it is really a signal at all. However, further analysis suggests that status badges are best regarded as signals but whether or not other birds respond to them depends on factors such as where the encounter takes place (Wilson 1992) and how hungry the two animals are. Møller (1987) showed that when food was plentiful, subordinate house sparrows responded to the status badge of dominants by giving way to them, but when food was scarce, the high status of a dominant counted for less and fighting was much more likely to occur. This of course makes evolutionary sense since the cost of giving up a piece of food when there is not much available elsewhere can be very great and even fighting a dominant animal for possession of it may become worthwhile (see p. 412).

Well-established dominance relationships in a group of animals that know each other may cause yet more problems for an observer trying to decide whether they are communicating or not. Bossema and Burger (1980) and van Rhijn (1980) studied a small group of captive jays in which the younger birds tended to give way to the more powerful older members of the group. But once the rank order was established, the older birds gave only a minimal 'signal'. The older birds could sometimes displace younger ones

Figure 3.24 (a) In house sparrows, a male's dominance correlates with the size of the black 'bib'. (Photograph by Mike Amphlett.) (b) Møller (1987) found that birds with the biggest bibs (grade 5) were the most dominant. (Drawn from Møller 1987.)

from, say, a piece of food simply by looking at them. Only if this glance was ineffective would they escalate and perform a recognizable threat display. Similarly, most people who have studied primate groups over a long period can recount episodes where a severe fight between two individuals has had repercussions that last for months or years. This may result in the loser fleeing or showing responses to the animal that defeated it while other animals with different past experiences show little or no response. An observer unfamiliar with the history of interactions between the animals might be quite unable to detect that any communication was occurring.

We should not conclude this discussion about definitions of communication without pointing out a rather curious type of signal whose function seems not to communicate information itself, but to qualify other signals that follow it. This phenomenon has been called metacommunication and is best known from play situations in carnivores and monkeys. Figure 3.25

Figure 3.25
Metacommunication in lions. The male's posture with lowered forequarters is only seen as a preliminary to play. He invites the cub to play and cuffs it gently as the cub joins him. (From Schaller 1972.)

shows an adult male lion inviting a cub to play. His posture with forequarters lowered – a 'bow' – is not seen in any other context and its message is that all aggressive movements that follow are play. Dogs, wolves and coyotes use almost exactly the same posture and may also wag their tails during play fights (as Darwin (1872) noted in his book *The Expression of the Emotions in Man and Animals*). Bekoff (1995) showed that the bow is used to convey the message 'I want to play despite what I am going to do or have just done' and occurs when the animal is using actions borrowed from another context such as biting and shaking movements of the head, which are particularly likely to be misunderstood. Monkeys adopt a 'play face' in similar situations. Playful aggression is a conspicuous part of behavioural development in carnivores and it has obviously been necessary to evolve some convention that allows stalking and attacking to take place without the damage of real fights.

The evolution of animal signals

The diversity of ways in which animals communicate with one another – songs, visual displays, scent trails or slight changes of posture, for example – prompts us to ask why different animals use such different signals. We also have to explain why it is that some signals are large and conspicuous while others are so subtle that we are left wondering whether we should define them as signals at all. We now know that there are at least three different selection pressures operating on animal signals.

(1) What travels best in the environment between signaller and receiver.

(2) What best stimulates the sense organs and brain of the receiver.

(3) How far the interests of the sender and receiver coincide.

What travels best through the environment between signaller and receiver?

Looking across the animal kingdom as a whole, it is clear that much of the diversity of animal signals comes from the diversity of their ways of responding to external stimuli that we have already discussed. Animals that rely on vision for finding their way about and detecting prey will also tend to communicate through visual signals, and so on. On the other hand, animals that never

Figure 3.26 A pond skater or water strider (*Gerris*) resting on the water surface by feet (tarsi) which repel water and so allow its tiny weight to be supported by the surface tension. These same tarsi are the means whereby it responds to the ripples set up in the surface film by struggling insects which are potential prey. They use the same stimuli to communicate between one another.

Figure 3.27 Tail-twining – tactile communication by adult dusky titi monkeys (*Callicebus moloch*). (From Moynihan 1967.)

use vision, such as the blind workers of some termite colonies which never leave their subterranean tunnels, rely on tactile communication often enhanced by scents that are detected by chemoreceptors. Water striders (Fig. 3.26) use the ripples on the surface of a pond to detect both predators and prey and they also use the ripples in communicating with each other. Both males and females produce surface ripples by vertical oscillations of their legs, but whereas females produce only low frequency (3–10 Hz) waves, males produce both low and high frequency (80–90 Hz) waves, to which other males respond (Wilcox 1979). Sometimes, of course, animals use several different modalities depending on how close they are to each other. Mammals use hearing, sight and smell at a distance but for many of the more social ones that spend a good deal of time in physical contact, tactile communication is also important (Fig. 3.27). Chimpanzees touch and kiss each other's hands in gestures of reconciliation (de Waal 1996).

Within a sensory modality, we can often explain the diversity of signals by understanding the constraints imposed by the medium through which the signal must travel. Most visual signals depend on natural daylight, available only to diurnal animals. Visual signalling at night is obviously limited as even bright moonlight is only one-tenth as strong as daylight. Some nocturnal animals such as fireflies overcome this limitation by manufacturing their own light and signalling at night. In fish living in deep water, where light is scarce, we also find animals making their own light.

Visual signals are influenced not just by the amount of light available but also by the fact that some wavelengths are scattered or absorbed more than others in their journey from signaller to receiver. This effect is particularly striking in seawater, where blue light of about 475 nm travels particularly well, but red light (longer wavelength) or ultraviolet (shorter) are simply filtered out by the water. This is the reason why so many reef fish are blue or yellow, or have striking black and white stripes (Fig. 3.28) that are also visible in seawater (Lythgoe 1979).

Even on land, some wavelengths of light travel better than others and so some colours and patterns may be more conspicuous at long distances. In warblers of the genus *Phylloscopus,* there is particularly clear correlation with habitat and the brighter species live in the darker habitats (Marchetti 1993).

Figure 3.28 The intense black and white markings in the tropical blue-headed wrasse (*Thalassoma bifasciatum*) make it clearly visible through seawater. (Photograph by M.S. Dawkins.)

Birds such as the cock-of-the-rock (*Rupicola rupicola*) actually choose to display in places in the forest and at times of day when their signals maximize their colour and brightness (Endler 1993). It is also striking that visual signals that operate over long distances, such as the white tails of rabbits and some deer (Fig. 3.29) or the eyebrows of mangabey monkeys, have a broad range of frequencies. White is very conspicuous at all distances.

Sound is even more subject to degradation as it travels than light. This is true of all sound frequencies, but high frequencies attenuate and become scattered by obstacles much more than low frequencies in all types of habitat. The most extreme case of this is the very high frequency (50–100 kHz) sound used by bats for echolocation (p. 131). Sound of this frequency attenuates so much with distance that a 100 kHz sound bat call drops to 1/50 000 of its initial

Figure 3.29 A white-tailed deer (*Odocoileus virginianus*) shows alarm. The tail is raised revealing its white underside and the white fur of the rump flared out, the whole movement producing a vivid flash of white as the deer leaps off. It is perhaps as much a signal to a predator that it has been seen, as a warning to other deer nearby. (Drawing by Nigel Mann.)

intensity only 4–5 cm away. Such sounds would therefore be quite useless for communicating except for animals that were very close and low frequencies are generally better for communication over long distances. As mentioned on p. 139, Gerhardt (1974) showed that the male green tree frog uses a call with two different frequencies. A female at a distance will pick up the lower frequency and head towards a male. As she gets closer, and the sound gets louder, she responds to both the low frequencies and also the higher ones that she can now hear.

Animals can also signal over longer distances by elevating themselves above ground. Crickets that sing from trees or shrubs can spread their signal over 14 times the area of those that sing from the ground, and consequently attract more females (Paul & Walker 1979). The territorial songs of birds are usually delivered from a raised song post which also increases their effective range, while grassland birds such as meadowlarks and pipits sing while they fly.

Because there is less attenuation of sound in water than in air, sound travels further and many aquatic animals use sound extensively for communication. It is ironic to recall that Jacques Cousteau's pioneering book on free-swimming scuba diving was entitled *The Silent World*; the development of the underwater microphone has now allowed us to discover the amazing range of sounds produced by fish and whales. Payne and McVay's remarkable study (1971) of humpback whales suggests that their 'song' could be picked up by other whales several hundred miles away! This is certainly a long-distance record for animal communication.

Chemical signals are particularly well developed in insects and mammals. (There are good reviews by Payne *et al.* (1986) for insects, Ralls (1971) and Brown & Macdonald (1985) for mammals and Duvall *et al.* (1986) for vertebrates generally.) Some chemical signals are designed to last only a short time: they are very volatile and tend to have low molecular weights. The chemicals used by some ants to signal alarm fall into this category of being very short-lived – if they did not, then it would be impossible for the ants to signal the precise localization of a new source of danger. Ant alarm pheromones disperse well over a short range of 3–5 m, but they have usually faded below detection levels within a minute or even less. It is even possible to make use of the volatility of scent to achieve patterning in time. For example, some moths release their sex-attractant pheromones in pulses at approximately 1 second intervals (Dusenberry 1992).

On the other hand, other chemical signals are designed to last for a long time, allowing a message to persist in the absence of the signaller. Territory markers, for example, need to be persistent and therefore their constituents must have a fairly high molecular weight (Wilson 1965). The molecular weight cannot be too high or the substance will be difficult to secrete and may not disperse well. Spotted hyaenas (*Crocuta crocuta*) mark their clan territories both by smearing grass stems with paste from their sub-caudal scent glands and by depositing faeces at latrines (Fig. 30). Where the territory is relatively small, scent is placed strictly along the territorial boundary, but where a small clan occupies a large territory and it is impossible to keep the boundary fully scented, both pheromones and faeces are deposited at strategic places within the territory (Mills & Gorman 1987). Either way, other hyaenas are informed that a territory is occupied even when the owners are not around. Brown hyaenas (*Hyaena brunnaea*) have two separate anal glands for different messages. They leave secretions as two separate pastes (one brown, one white)

Figure 3.30 A spotted hyaena scent-marks the clan territory. It carefully locates a suitable grass stem in the cleft of its extruded scent gland so that the paste-like exudate is placed conspicuously and at the right height above the ground. (Photograph by H. Kruuk.)

next to each other on stalks of grass. Moth pheromones strike a compromise between persistence and good dispersal. In favourable wind conditions males can detect the scent of females from 4–5 km downwind.

As yet another example of an animal using for communication a system it also uses in another context, the electric fish *Eigenmannia virescens* has an unusual short-range signal: the male discharges its electric organ in a series of 'chirps' which Hagedorn and Heiligenberg (1985), having transposed them into sound for our ears, describe as 'short and abrupt during aggressive encounters' and with a 'softer and more raspy quality during courtship'.

What stimulates the sense organs and brain of the receiver

Once a signal has travelled through the medium, it still has to be detected by the receiver and discriminated by it from other signals. A major selection pressure on signals is that they must each be maximally effective at stimulating the receiver to which they are aimed. Here we find that a knowledge of sensory processing in receivers is particularly helpful in understanding why signals have evolved in the way they have, since it is often possible to see that signals evolve out of responses to quite different situations.

For example, female jumping spiders (*Maevia*) choose which male to mate with not by what he looks like but, at least initially, by whether he moves in a way that resembles food. Using videotape sequences of male courtship so that they had precise control of the stimuli presented to females, Clark and Uetz (1992) showed that a female spider orientated to the movement of the male's legs by turning towards him in just the same way that she would turn towards a moving prey item. It thus seems that successful males (provided they can turn off the female's prey-catching behaviour in time) can attract females by mimicking prey, thereby tapping into a pre-existing feeding response in the females.

Female water mites (*Neumania papillator*) also initially respond to males as if they were food (Proctor 1991). Water mites are ambush predators and both sexes hunt prey by lying in wait with their front legs in the air and then grabbing passing crustaceans (Fig. 3.31). A male approaches a female vibrating his legs at a frequency of about 15 Hz, which is within the range of vibrations made by copepod crustaceans. The female grabs him with her front legs just as she would prey, but fortunately this does not injure the male as he is so

Figure 3.31 Courtship in the water mite (*Neumania papillator*). The male vibrates his foreleg in front of the female. The female raises her first two pairs of legs into a 'net-stance' which is the posture used to catch prey. The female will often respond to leg-trembling in the male by orientation and even clutching, but this does not injure or even deter the male, who continues with courtship and eventually deposits spermatophores.

much bigger than the female's normal prey, nor does it even deter him from further courtship. He starts to deposit spermatophores (packets of sperm) near the female and it is only at this point that she behaves any differently from the way she responds to food. She switches to sexual behaviour and picks up a spermatophore. Further evidence that the male courtship is effective because the female at first confuses him with food comes from the fact that, when female water mites are deprived of food, they grab courting males far more often than when they are well fed.

The male oriental fruit moth (*Grapholita nolecta*) also uses the female's response to a food stimulus to entice her to mate, but here a chemical signal is involved. The male initially finds the female by the scent she emits, but when he is very close, he wafts his own scent at her. The most effective part of his pheromone is a chemical ethyl *trans*-cinnamate, which is also found in fermented fruit juices, a favourite food of both sexes (Lofstedt *et al.* 1989).

The effectiveness of the signal in each of these cases depends on mechanisms that are already present in the receiver for non-signal functions. The signaller 'exploits' the responsiveness to the receiver by mimicking food in a quite specific way and so increases his chances of mating. In other cases, the signal may gain its effectiveness by attracting the receiver's attention in a more general way. We have already seen that the visual systems of many animals have evolved to pick up 'information-rich' features of the environment such as movement, edges or points of high contrast because these are important in the recognition of many different patterns, such as predators, prey, mates or simply for orientation through the environment. These general features are also used in signals. Male lizards (*Anolis*), for instance, display by bobbing their heads up and down and extending their dewlaps (flaps of skin under the chin) (Fig. 3.32). The 'assertion display' which is given both to attract females and to deter males starts with a series of very distinct head-bobs, which are larger and faster than the rest of the assertion display. Fleishman (1992) measured the typical distance at which the assertion display is shown and found that at this distance, the initial head-bobbing maximally stimulates the lizard's motion sensitivity in the peripheral retinal field. The result is to trigger a 'visual grasp response', mediated by the optic tectum, which causes the receiver to shift its gaze and bring the image onto the high acuity fovea. Once the receiver's attention has been grasped in this way, the rest of the display follows with lower amplitude bobs. Here we have an example of a signal having its effect by tapping in to a very general property of visual systems – sensitivity to motion in the peripheral part of the retina – rather than by mimicking any specific object.

To be effective, then, a signal has to be stimulating, either through mimicking something a receiver already responds to in another context, or through

Figure 3.32 A male *Anolis* lizard in full 'assertion display'. He bobs his head repeatedly with the brightly coloured dewlap fully extended.

being 'conspicuous' in a broader sense by possessing those properties of movement, loudness, colour and so on that sensory systems in general have been evolved to pick up. In our discussion of the way in which animals respond to stimuli in their environment, we saw that they sometimes respond even more strongly to exaggerated or supernormal versions of natural stimuli than they do to natural stimuli themselves and it is therefore not surprising to find that many signals seem to create their effect through being supernormal injust this way.

A good example is the signal given by the male long-tailed widow-bird (Fig. 3.33). The male has extraordinarily lengthened tail feathers, probably

Figure 3.33 A male long-tailed widow-bird (*Euplectes progne*) whose display flight is observed by a female. (Drawing by Nigel Mann.)

larger in proportion to his body size than in any other bird. He flies in conspicuous sweeps up and down over the vegetation, displaying his ornaments to the
female. We may regard such a tail as supernormal already and Andersson
(1982) showed that the females are indeed attracted by them, but they are
even more attracted to males whose tails have been artificially lengthened still
further. He captured wild male widow-birds that had already set up breeding
territories, and altered the length of their tails. He cut the tail feathers off and
then glued feathers back on, making the tail shorter than before, or longer, or
the same length (to control for the effects of tail manipulation). Male with
artifically lengthened tails (supernormal versions, longer than any naturally
occurring bird) attracted more females than any others. Møller (1988) carred
out a similar experiment with swallows (*Hirundo rustica*) and showed that in
this species, too, females were attracted to males with supernormal tails (Fig.
3.34). Males with artificially lengthened tail 'streamers' attracted mates earlier
in the breeding season, had more second broods and produced more
fledglings.

Why should females prefer supernormal male ornaments? In Chapter
6 we discuss some of the explanations that have been put forward and hotly
debated since Charles Darwin published his book *Sexual Selection and the
Descent of Man* in 1871. But whatever advantage females now gain from such
apparently surprising mate preferences, it is clear that we already have an
explanation for their evolutionary origins. We need look no further than the
structure of the female's sense organs, with their tendency in many modalities to respond to sign stimuli and extract relatively simple information. As
we have seen, simple 'rules of thumb' such as 'respond to the biggest round
object around', or respond to vibrations of a particular frequency or to
moving colourful objects, seem to be very common in the way animals recognize important patterns in the world around them. Such rules of thumb
may also lead animals to respond to supernormal versions of the original
stimulus. Of course, if it were not to the net advantage of females to respond
to supernormal males, we would not expect them to persist in doing so; and
so just identifying the phenomena of sign stimuli and supernormal stimuli
may allow us to link causal explanations of behaviour with one of
Tinbergen's other questions about behaviour (phylogeny and evolutionary
origins), but still leaves us with a problem over present adaptive significance

Figure 3.34 In the monogamous swallow, a male normally takes about 8 days after arriving in the breeding area to find a mate (Control I and Control II). With artificially shortened tails it takes 12–13 days to find a mate but with an elongated tail only about 3 days. (From Moller 1978. Drawing by Nigel Mann.)

(the question of current function). That problem we will take up in more detail in Chapter 6.

How far the interests of the sender and the receiver coincide

The third selection pressure operating on animal signals that we will consider is whether there is conflict between the interests of the signaller and those of the receiver. For example, if two males are fighting over a female, each would benefit if the other fled without a fight, so potentially each would gain if it could convince the other that it was the biggest and strongest and that fighting would not be worthwhile. There will also be selection on each animal not to be taken in by the bluff of the other, so that the conflict (over the female) gives rise to a complex scenario of potential cheating on one side and resistance to being cheated on the other. On the other hand, where the interests of the signaller and receiver coincide – for example, in the contact calls between members of a group of birds or primates – there is little scope for cheating. If it is in the interests of all members of the flock or troop to keep together, none of them gain from cheating about its location. As we will see, recognition of the degree of conflict that exists between signaller and receiver offers an explanation to a very puzzling aspect of animal communication which we have already mentioned: why some signals are flamboyant and conspicuous and others so inconspicuous or *sotto voce* that ethologists have difficulty in deciding whether communication has taken place at all (p. 156).

Dawkins and Krebs (1978) identified conflict between two animals as a major determinant of what signals are used. Sometimes two animals are in direct conflict – as when two males fight over a female or when a predator chases a prey – and we often find that the signals used under such circumstances are very large and costly. For example, red deer fighting over groups of females signal to each other with loud roars (Clutton-Brock & Albon 1979) and gazelles running away from wild dogs signal with conspicuous jumps or 'stotts' (Fitzgibbon & Fanshawe 1988). On the other hand, when animals are cooperating and there is little conflict between them, the signals tend to be much weaker and less strident and have even been described as 'conspiratorial whispers' (Krebs & Dawkins 1984). Examples of this would be the soft sound signals made by honey-bees in their hive and the barely audible 'grunts' made by vervet monkeys to other members of their troop, both of which we shall discuss in more detail at the end of this chapter.

Conflict between signaller and receiver clearly leads to selection pressure for signals to be 'honest' – in other words, proof against cheating or bluffing (Zahavi 1987). But how can signals be made honest? What guarantee could a receiver have that its opponent is not cheating? In the case of the red deer (*Cervus elaphus*), for instance, the stags spend several weeks of each year either defending a group of females or, if they have none to defend, challenging one of the owners for possession of theirs. A conspicuous feature of their challenges is the loud bellows or roars that the stags deliver before and sometimes even instead of overt fighting. Sometime a stag will withdraw without any fighting having taken place at all, just a 'roaring match'. This means that, potentially, a stag would gain if he just out-roared his opponent and chased him away without a fight. But the challenger would only gain from such a retreat if the roaring itself conveyed honest information about real fighting ability. The stags do indeed seem to encode honest information about their fighting ability into their roars.

When a challenger approaches another stag, he does not immediately start a fight. He starts by giving a small number of roars. The owner replies by roaring at a slightly higher rate. The challenger then gives a larger number of roars per minute and the two stags escalate this roaring match, roaring at

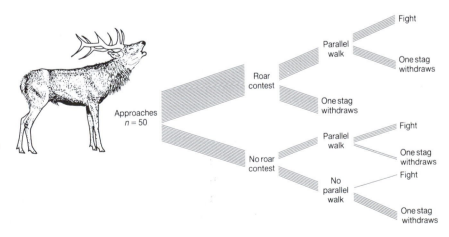

Figure 3.35 Contests between red deer stags. The males challenge each other by roaring as an indication of fighting ability. The figure shows the number of approaches to within 100 m involving two mature stags that lead to roaring contests, parallel walks and fights. Each parallel line represents one encounter. (From Clutton-Brock & Albon 1979.)

Approaches
n = 50

Roar contest

No roar contest

Parallel walk

One stag withdraws

Parallel walk

No parallel walk

Fight

One stag withdraws

Fight

One stag withdraws

Fight

One stag withdraws

higher and higher rates. If the challenger is 'out-roared' by the owner, he may give up altogether and go away, but if both roar at the same rate, fighting is likely to occur (Fig. 3.35). Clutton-Brock and Albon (1979) argued that the roaring signal is one of the ways the stags have of assessing each other's true fighting ability before letting themselves in for a damaging fight and that the ability to roar at a high rate will be a reliable indicator of a good fighter, because fighting and roaring are both exhausting and both make use of the same set of thoracic muscles. Roaring at a high rate would therefore be impossible except for a stag in very good condition with powerful muscles. A challenger that backed off when confronted by a stag able to roar at a higher rate than he could would therefore be likely to be saving himself from being injured by a stag able to fight harder than he could. The point is that reliable information about fighting ability cannot be conveyed by signals that are so small and easy to produce that they can be given by all stags, regardless of their fighting ability. It has to be given by a signal that only good fighters can manage. Receivers that respond to signals that even weak stags can give will lose out compared with receivers that only retreat when confronted with an honest signal of fighting ability.

Similarly, predators that avoid prey giving signals that could be given equally well by individuals that are easy to catch as by those that are hard to catch, will lose out to predators that use 'honest' signals of escape capacity. The conspicuous 'stotting' or leaping in the air shown by gazelles and springbok when chased is probably just such an honest signal (Fig. 3.36). Fitzgibbon and Fanshawe (1988) found that stotting is done most by animals that are in good physical condition and so are likely to be able to outrun wild dogs in a long chase. Moreover, the dogs discriminate between gazelles stotting at different rates, concentrating their efforts on the animals that stott least and are most easily caught and ignoring those stotting at a high rate. Singing by skylarks (*Alauda arvensis*) when chased by merlins (*Falco columbarius*) appears to be another example of a potential prey animal signalling honestly to its predator. Cresswell (1994) found that merlins were much less likely to catch a skylark that gave a continuous full territorial song when it was chased than a skylark that did not sing or one that just gave a brief song, despite the fact that singing loudly and continuously when there is a falcon in active pursuit would seem to be a risky strategy. Singing, however, appears to be done by male skylarks in good condition and so by learning to avoid a singing prey, the merlin avoids trying to catch an uncatchable prey.

Another less obvious area of conflict is that between a parent and its various offspring. Although parents of many species care for their young and make a considerable effort for them, there is also conflict within a family because offspring may demand more food than parents are selected to give (Trivers 1974). The parents will be selected to care for all their offspring and to raise as many as they can. Each offspring, however, will be selected to demand extra for itself at the expense of the parent's other offspring. Godfray (1995) argued that this conflict results in parents giving food to those of their offspring whose need is greatest, but of course the parents have to decide which of their offspring is in fact in greatest need. If they do this through some sort of 'begging signal' there is always the danger that an offspring will give a

Figure 3.36 Thompson's gazelles (*Gazella thomsoni*) flee from a group of wild dogs which they have just detected. Some do not run directly but intersperse their flight with conspicuous leaps – 'stotting'. (Drawing by Nigel Mann.)

dishonest signal of need, signalling that it is hungrier than it really is. Young canaries, however, have a begging signal that appears to be both conspicuous and honest. Their mouth gapes have a red lining, the coloration being due to suffusion of blood. A red mouth is extremely stimulating to the parent which put food into the reddest mouth it can see (Kilner & Johnstone 1997). When the chick has been fed, its mouth becomes paler as the blood is withdrawn to the gut to aid digestion and so, inevitably, is less stimulating to the parent. Redness of mouth in a chick thus seems to be an honest signal of hunger, and parents that allocate food to their chicks according to which ones have the reddest mouths will have the greatest chance of rearing the most chicks.

In summary, then, animal signals are affected by at least three selection pressures – the medium through which the signal must travel, the characteristics of the animal on the receiving end of the signal, and the degree of conflict between signaller and receiver that affects the selection for 'honesty'. These selection pressures are not, however, mutually exclusive and, in the design of any one signal, any or all of them may be at work. Indeed, some sort of compromise is usually inevitable. For example, although cooperative signals may be 'conspiratorial whispers' if the animals are very close to each other, a whisper would be completely ineffective over long distances. The constraints of having to make a signal detectable at a distance might thus turn a cooperative signal such as an alarm call into a 'conspiratorial shout'.

A further compromise may come about because of conflicting selection pressures from different receivers: signals that are conspicuous to intended receivers such as mates may also be conspicuous to predators. Counter-selection from predators may, indeed, be an important reason why naturally occurring signals are sometimes found to be less stimulating to a receiver than supernormal versions. Signallers may have to 'tone' down' their signals or limit the size of their handicaps if the danger from predators becomes too great.

Mimicry, deception and honesty

As we have already seen, an unwelcome by-product of communication is that receivers may sometimes be duped into responding against their own interests. Small fish that snap at the worm-like lure of an angler fish and promptly get eaten, or the hedge sparrow that feeds a baby cuckoo, are being deceived into performing actions that are not to their own benefit. There are also a number of insects, including beetles and butterfly larvae (so-called myrme-cophiles), that live parasitically in the nests of ants and which trick the ants into carrying them into the nest by mimicking the signals given by ant larvae (Hölldobler 1971). Some ground-nesting birds such as plovers lure predators away from their nests by feigning a broken wing (Fig. 3.37). Again female fireflies of the genus *Photurus* can flash in two ways. They can either flash in the pattern of their own species to attract males, or they can mimic the flash pattern of another smaller species of the genus *Photinus*. When *Photinus* males

Figure 3.37 The 'broken wing trick'. A killdeer plover (*Charadrius vociferus*) trails its wing and hobbles away from a predator approaching its nest. Once the bird has been followed far enough, it flies off! (Drawing by Nigel Mann.)

approach in response, the *Photurus* females kill and eat them. Lloyd (1965, 1975), who discovered this extraordinary piece of deception, aptly called the *Photurus* females 'firefly femmes fatales'!

Such examples are sufficiently numerous to show that signals are certainly not always honest in the information they convey, despite the fact that selection for honesty is a powerful one in the design of animal signals. Sometimes, as in the case of cuckoos that deceive their hosts into caring for them, such behaviour is part of an ongoing coevolutionary arms race in which the deceived organisms are in process of evolving the ability not to be deceived. In other cases the receiver is only rarely deceived, so that on average it is beneficial to respond to the signal. As long as the deception is relatively uncommon, it can persist. Many cases of mimicry seem to fall into this category. The predator that avoids black and yellow wasps will sometimes lose out through being deceived into avoiding a perfectly palatable hoverfly, but on average it gains from avoiding black and yellow prey because so many of them are dangerous or distasteful. A predator that avoids all black and yellow prey will in the long run be better off than a predator that tries them all and constantly ends up with a mouthful of distasteful insect. Conversely, the mimics gain an advantage only for as long as they are rare. Once they become common, predators that ignore the signal and attempt to eat them will be at an advantage, (see Fig. 5.3, p. 266).

None of the examples of deception we have discussed so far necessarily implies that the signallers are consciously setting out to deceive other animals. Most of the cases we have discussed so far are plausibly seen as evolutionary deceit, i.e. natural selection having favoured animals that behaved in certain ways and elicited particular responses from other animals with no conscious thought about what they were doing. In describing palatable insects as deceiving their predators through having body coloration that resembles distasteful ones, there is no implication that the insects think about what colour they should be. But in other cases, it is less easy to be so sure. Ristau (1991) has argued that plovers giving a broken wing display (Fig. 3.37) monitor the behaviour of the predator, such as a fox, that is threatening their nest and, for example, adjust their behaviour according to whether the fox is following them or still heading for the nest. Munn (1986) described two species of insect-eating birds that appeared to deceive their flockmates by 'crying wolf'.

By giving alarm calls when there were no predators present, they induced other birds to take off and so gained unrestricted access to food themselves.

Amongst primates, baboons have been reported to behave in ways that strongly suggest they are deliberately deceiving others. Byrne and Whiten (1988) describe a range of such incidents, including one where a subordinate baboon, on being attacked by a dominant animal, gave every appearance of having spotted a predator in the distance. When the dominant animal responded by looking in the same direction, the subordinate took the opportunity of its attention being distracted and escaped. Such incidents are inevitably anecdotal, but Mitchell & Thompson (1985) and Byrne & Whiten (1988) argue that there are now enough comparable examples to support the idea that at least some animals can be deliberately dishonest in their communication with others.

We conclude our discussion of communication with two remarkable examples from very different groups of animals. Both have changed the way that humans look at animals because both have turned out to have far greater complexity than had previously been expected for communication among non-human animals

The honey-bee dance

Not only have honey-bees evolved one of the most remarkable of all systems of animal communication: their dances also present intriguing problems of measurement and interpretation of the type which we have been discussing.

It has been known for centuries that honey-bees must pass information about flower crops to one to another, but this phenomenon will now always be associated with the name of the German zoologist, Karl von Frisch. He was the first to unravel the nature of their communication and his major book, *The Dance Language and Orientation of Bees* (1967) provides a full survey of the whole dance system. Like others before him, von Frisch had noted that if a source of sugar solution was put out to attract bees it was often many hours before the first one found it, alighted and drank. However, once a single bee had located the source it was usually only a matter of minutes before many other foragers arrived – somehow the information had been passed on. It took von Frisch some twenty years of painstaking observation and experiment before he had worked out the bee's communication system to his own satisfaction. The conclusions that he came to were so extraordinary and unparalleled that he himself declared that no good scientist should accept them without confirmation. Following World War II, other zoologists did confirm von Frisch's results and even worked with him on some final experiments. There is now no doubt about the nature of the bee dance itself.

Von Frisch marked foragers as they drank at a dish of sugar syrup and then watched their behaviour when they returned to the hive, using glass-sided observation hives for this purpose. The forager usually contacts a number of other bees on the vertical surface of the comb and gives up her cropful of sugar solution to them. She then begins to dance, and we may first consider the case where the food dish which she has just visited is close to the hive, within 50 metres. Her dance then is rapid in tempo and forms a roughly circular path just over her body's length in diameter. The bee moves in circles alternately to the left and to the right. She stays approximately in the same place on the comb and may dance for up to 30 seconds before moving on. Other foragers face the dancer, often with their antennae in contact with her body, and follow her movements closely, being themselves carried through her circular path (Fig. 3.38). The 'round dance' as this is called, stimulates other workers to leave the hive and search nearby. It appears to convey the information, 'search within 50 metres'. It may also convey some olfactory cues, because if the food source is scented the dancer will carry this scent on her body and perhaps in the sugar solution itself. If the sugar dish is not scented the forager may mark it to some degree by opening the Nasanoff scent gland on her abdomen as she drinks.

Thus far the bee dance is not very exceptional, because many ants and termites have similar alerting displays and pheromones which help to organize foraging activity when a new food source has been found near the nest (Wilson 1965). The extent of the honey-bees' communication system is not revealed until the food source discovered by the forager is further from the hive, beyond 100 metres.

Von Frisch observed that as his food dishes were moved beyond 50 m the forager's round dances gradually changed in form. A short straight run became incorporated between the turns and on this run the dancer wagged its abdomen rapidly from side to side. At about 100 m distant the dance had become the typical 'waggle dance' illustrated in Fig. 3.39 and this form

Figure 3.38 The 'round dance' of the honey-bee worker on the vertical face of the comb: her path is indicated by dotted lines. Note how she is closely attended by other workers who follow her round the path of the dance.

Figure 3.39 The 'waggle dance' of the honey-bee: further details are given in the text. (a) The dance is occasionally performed on the horizontal entrance board to the hive; if so the waggle run 'points' directly towards the food source. (b) On the vertical comb the angle of the waggle run to the vertical is equal to the angle the sun makes with the food source. (c) Shows the honey-bee's convention that directly downwards on the comb represents directly away from the sun. In the same way, upwards represents towards the sun.

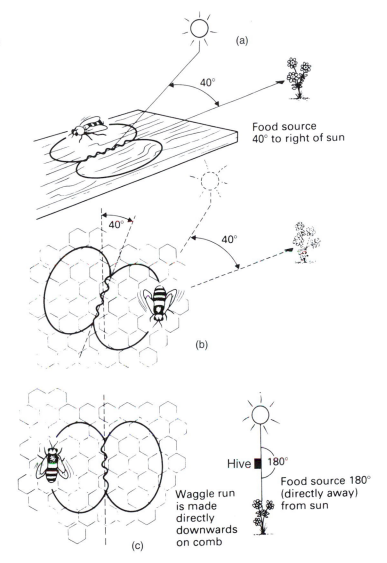

remained the same as the dish was moved further, to 5 km or even beyond. More recent work has revealed that during the waggle run the bee produces bursts of high-pitched sound (Esch *et al.* 1965). It is this waggle dance which, von Frisch claimed, transmits so much more information and is 'read back' by the dance followers as they follow every move the dancer makes.

The waggle dance certainly contains information about both the distance and the direction of the food source. Distance is correlated with several features of the dance. Von Frisch concentrated on measuring its tempo, which falls off with distance, steeply at first and then more gradually. Thus there are 9–10 complete cycles per 15 s with the food at 100 m, but only two when the food is 6 km away. The duration of the waggle run, the number of waggles it contains and the duration of the sound pulses also correlate with distance, all three features *increasing* with it. Von Frisch had no evidence that enabled him to identify which of these distance cues was important for the dance followers. Bees fly at a very constant speed in still air and they interpret distance information in terms of flight time and effort. Von Frisch found that dancers indicated a greater distance for food sources upwind than for those down-wind. (It is always the *outward* flight path which is indicated.)

It is its relation to the *direction* of the food source which is perhaps the most remarkable feature of the waggle dance. The little south Asian bee, *Apis florea,* is a close relative of the honey-bee. This species builds a single vertical comb on a tree branch in the open with a flattish platform on top and the bees perform their waggle dances on this platform. The waggle run is made *towards* the direction of the food source, i.e. it operates like a pointer (Lindauer 1961; Gould *et al.* 1985). Now, occasionally honey-bees will dance on the flat landing board provided at the entrance to most types of hive. If they do so, then like *A. florea*, their dances also point directly to the food source (Fig. 3.39a). However, this observation did not help von Frisch because he watched the dances at their normal site on the vertical face of the combs inside the hive. There he noted that the average direction of the waggle run was consistent within a dance and that it was the same for all the foragers who danced after feeding at the same dish. Given that other bees foraging at different dishes made waggle runs orientated at other angles even when the distances were comparable, this was strong circumstantial evidence that the angle related to direction in some way. Now came the crucial observation. Von Frisch recorded dance after dance throughout the day as foragers repeatedly returned from the same food source and he found that the direction of the waggle run gradually changed. Its mean direction shifted by about 15° per hour and this could mean only one thing: that it relates to the apparent move-ment of the sun. Figure 3.39b and c show the way it does so. The foraging bee, like many other insects, uses the sun as a compass and records the position of the food source with respect to it. To get to the food it steers, say, 40° to the right of the sun. When dancing on the vertical comb the sun is not visible but the bee transposes the angle to the sun into the same angle with respect to gravity. The honey-bees' 'convention' takes vertically upwards to represent the present position of the sun. Thus the forager dances with her waggle run 40° to the right of vertical. As her course with respect to the sun will have to change as the day progresses, so she changes the angle with respect to gravity on the comb to match the sun's apparent movement across the sky. Von Frisch had at last understood the honey-bees' dance language and the world was forced to accept that another animal apart from ourselves – and a humble insect at that – could convey information in a symbolic fashion.

Von Frisch and his co-workers had no doubt that the dance did communicate. Other foragers picked up the dance's rhythm and orientation as they followed through the dancing bee's movements on the comb and they then transcribed back from an angle with respect to gravity to an angle with respect to the sun. This assertion was based on numerous experiments in which an array of food dishes was offered, so arranged as to test the accuracy with which foragers recruited by the dance interpreted its information on distance and direction.

Figure 3.40 shows the two commonest types of configuration: a line of dishes at different distances on the same bearing from the hive tested distance communication; a fan pattern of dishes in an arc equidistant from the hive tested for direction. The figures also indicate the results of typical experiments showing that the number of recruits is highest at the dish closest to that at which the original forager fed and recruiting falls off to each side. (Further details of the experiments are given in the captions.)

This type of evidence was very generally accepted as convincing proof that the dance was the effective communication system, although it was clearly not perfect and there is some spread of recruits around the direction and distance indicated by the dancer. This 'noise' is not surprising in view of the crowded, jostling bees on the comb. Commonly, dancers do not have adequate space to keep a consistent line or rhythm. Nor, with natural food sources, do minor inaccuracies matter at all. For the most part the dancers will have been foraging at large sources of food – a grove of lime trees in blossom or a field of clover. Provided the recruits get some idea of distance and direction this will suffice, the more so because the foragers are very responsive to the scent which the dancers carry into the hive on their body. As mentioned earlier, they may also mark the food source with scent themselves. Indeed, it was doubts over the importance of scent which led to the later controversy

Figure 3.40 Experiments carried out by von Frisch and his co-workers to test the communication of distance and direction by the honey-bee waggle dance. (a) Distance test. Foragers were trained to a dish of dilute scented food 1050 m east from the hive. Then a series of scented plates without food were put out at distances varying in the same direction from 100 m to 2000 m. Dancing was induced by suddenly increasing the sugar concentration at the feeding dish. Recruits were counted, but not captured, as they approached the different scent plates. The numbers above them record the number of visits made to each scent plate during the test. Most recruits appeared at plates close to the feeding station. (b) Direction test. The procedure is much as for the distance test, but the scent plates are put in an array at the same distance but different directions from the hive. The majority of visits were made to dishes close to the bearing of the training station. (From von Frisch 1967.)

(a)

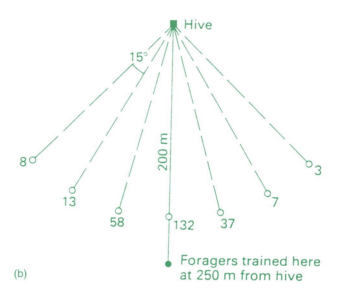

(b)

○ Scent plates at which recruits were recorded, the number of visits is given by each.

over the interpretation of von Frisch's results. Scent and wind direction tended to be ignored in some of the original tests. It could just be possible that the dance merely alerted and random search did the rest.

This would imply that the information contained in the waggle dance relating to distance and direction existed but was not communicated. One obvious riposte is to ask, why then has such a remarkable relationship evolved? However intuitively reasonable this reply seems, it does not really supply a secure argument for it assumes that every biological phenomenon must be functional. We have other examples of behaviour which certainly contains information but which just as certainly is not used by conspecifics. Dethier (1957) has described the searching movements made by flies after they have located and then exhausted a small source of food, such as a drop of sugar solution. Their 'dance' can convey information to a human observer about the 'shape' and concentration of the source, but other flies do not respond. Certainly other bees do respond to the waggle dance, but the 'olfactory' hypothesis also requires that it arouses them to go out and search, just as ant foragers do. Von Frisch has already shown that recruits pick up olfactory cues about the food source from the dancer – its scent will adhere to her body.

Gould (1975) summarizes the evidence and the controversy very attractively. He himself did some elegant and better controlled experiments which effectively proved that the information in the dance *was* communicated as Von Frisch originally described. However, we still lacked the most perfect demonstration, i.e. an artificial bee model to which recruits would respond. There had been a number of attempts to make a model bee but none were successful until recently. Now Michelsen, Lindauer and their collaborators (see Michelsen *et al.* 1989) have succeeded in constructing a relatively crude model of brass covered with bees wax and left in the hive for some hours in order to acquire the colony odour (Fig. 3.41). The model can be moved through a waggle dance path on the comb and, most importantly, it has an artificial 'wing' attached to it which can be vibrated electrically so as to produce an acoustic field around the model similar to that of a real dancer.

The model works! It recruits foragers to visit food dishes which have not been visited at all previously, and the proportion of bees turning up at different directions and distances follows that of the dance pattern. Sound turned out to be crucial: without it, no bees were recruited and this may be the reason for earlier failures with silent models. Not only does the fact that the model works at all provide final conclusive proof of true communication in the bee dance, it also gives us the opportunity to investigate the communication process in detail. Michelsen *et al.* (1992) have begun to experiment with the model, trying to separate the effects of sound and wagging for conveying information. They have confirmed that both are necessary if bees are to be recruited, and that it is the *duration* of the waggle run which is the key factor in distance communication. However, the two components – wagging and sound – do not have to be contiguous throughout the run; if either is 'on' for only part of the run the bees interpret the length of the run as that of the

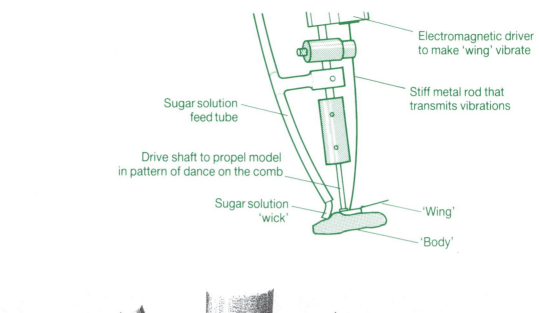

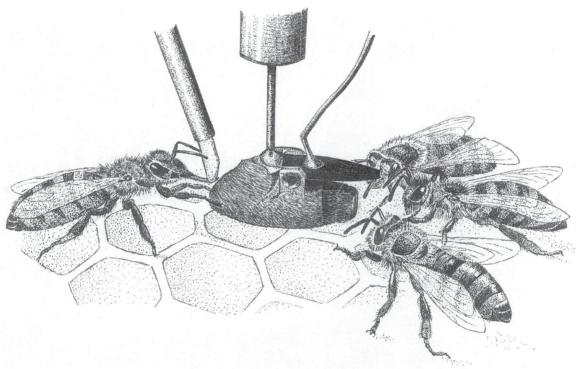

Figure 3.41 The dancing bee model of Michelsen, Lindauer and colleagues. Its construction is shown above. The body can be made to move on the comb in the form of a dance, 'waggling' during the straight part of the run, whilst the wing (a single piece of razor blade) is vibrated to produce sounds mimicking those of the dancing bee. Below, the model is shown on the comb. Workers fan out to its rear and follow its dance. One is accepting sugar solution which the model offers at its head end. (Drawing by Nigel Mann from photographs in Michelsen 1989.)

longer of the two components. It doesn't seem to matter which, and to this extent, they are redundant to each other.

So far, it has proved more difficult to pin down how sound and wagging operate for the transmission of information about direction to the food source, although there is no question that the overall orientation on the comb is crucial. Different experiments to separate sound and wagging directions have not yielded consistent results. It is certainly not easy to run experiments of this type because the model never recruits workers to the same extent as real dancers and this makes it necessary to run large numbers of repetitions. Nevertheless, the advent of the model is a major advance in the investigation of this most fascinating communication system and we should soon obtain answers to the remaining questions.

The ability to communicate in this fashion enormously increases the efficiency with which honey-bees can exploit the food resources in their environment. In a large colony, the foraging behaviour of tens of thousands of workers can be coordinated so that they spend the maximum amount of time working the most productive flower crops. As one crop fades, dancing dies away and foragers tend to stay in the hive until they are attracted by the dances of workers who have found other crops which are now productive. The duration and frequency of dancing by a returning forager is directly related to the concentration or abundance of the nectar or pollen she has found and, accordingly, the number of recruits will come to match the available supplies.

When they return to the hive, the foragers give up their nectar to house workers, who are engaged in brood-rearing and food storage. These latter also play a key role in matching supplies and demand. When the colony is short of food, a returning forager is instantly solicited to regurgitate and her response is to set out again, probably dancing first to attract more recruits if her food source is a good one. Conversely, there are times when there has been a great flow of nectar or pollen and supplies in the hive are superabundant. If so, the returning forager will have great difficulty in finding a house bee to accept the nectar she proffers or a pollen storage cell with space enough in it to receive her load. Foragers can be seen running rapidly around the combs, those with nectar offering food to a sucession of their sisters with no success. At such a point there is commonly a shift in their behaviour and they can be seen moving about, often in the brood-rearing regions of the combs, in a characteristic jerky fashion, shaking their bodies as they do so, with frequent brief pauses followed by an abrupt change in direction. This 'tremble dance,' as it is called, may continue for a long time – 30 minutes in some cases. It has been recognized for many years, but only recently has it become understood. It is a kind of *anti*-recruitment signal. Seeley (1992) has shown that it is induced by the delay in having food accepted and it reduces quite sharply the amount of dancing by foragers who encounter a tremble dancing bee. It may also cause bees to switch their behaviour towards brood care. The immediate effect is to reduce the incoming surplus of food. It is one further adaptive element in the honey-bee's remarkable communication repertoire.

The calls of vervet monkeys

A very different communication system, but one that has also been referred to as a 'language' is found in the vervet monkeys of Africa (*Cercopithecus aethiops*). Much of our knowledge of the meaning of the vocalizations of these elegant, group-living primates (Fig. 3.42) comes from detailed field studies carried out by Dorothy Cheney and Robert Seyfarth in Amboseli National Park in Kenya (1990). They were able to get very close to the monkeys, document the interactions between different members of the group and, most importantly of all, experimentally manipulate their communication system by playing sound recordings back to them from loudspeakers hidden in the grass. It was already known that the monkeys gave different alarm calls when they saw different predators. While this is not unique in the animal kingdom, as both chickens and ground squirrels also have different alarm calls for aerial and ground predators, the predator-specific alarm calls are particularly well developed in vervets.

When a vervet monkey sees a leopard (*Panthera pardus*) or other large cat, it gives a loud barking alarm call (Fig. 3.43). When the other vervets hear this, they run up into trees, a manoeuvre that appears to give them some protection from the leopards because, being so small and light, they can escape to branches where the leopard cannot catch them.

On the other hand, if the vervets catch sight of either of the two large species of eagle that prey on them (the martial eagle, *Polelaetus bellicosus* or the crowned eagle, *Stephanoaetus coronatus*), they give a completely different alarm call – a sort of double-syllable cough (Fig. 3.43). Since both these eagles are highly skilled at taking monkeys both from the ground and from the trees, running up into tree-tops, which is highly adaptive against leopards, would be disastrous against the high speed aerial attack of the eagles. A monkey on the ground hearing the eagle alarm call immediately looks up into the air and then runs into the thickest bush available. A monkey in a tree will actually come down from the tree and hide in a bush. Yet another type of alarm call is given when the monkeys encounter snakes such as pythons, mambas or cobras which hunt for monkeys by hiding in the grass. The snake alarm call is a sort of 'chutter' (Fig. 3.43). When a vervet hears a snake alarm call, it stands up on its hind legs and looks down into the grass. Once several monkeys have spotted a snake, they will often approach and mob it.

Figure 3.42 Part of a troop of vervet monkeys foraging on the ground in the East African savannah. (Photograph from Oxford Scientific films.)

Vervets thus have three different alarm calls for three different sorts of predators. Using the technique of hiding loudspeakers in the grass near where the monkeys were feeding and then playing different calls to them, Cheney and Seyfarth were able to show that the monkeys responded quite differently to the three alarm calls. In each case, their initial response was to look towards the loudspeaker and scan the surrounding area. But then their behaviour was quite different. If they were on the ground, the leopard alarm call would cause many of them to run into trees. The eagle alarm call resulted in them looking

Figure 3.43 Sound spectrograms of typical alarm calls given by adult male and female vervet monkeys to leopards, martial eagles and pythons. (From Seyfarth & Cheney, 1986.)

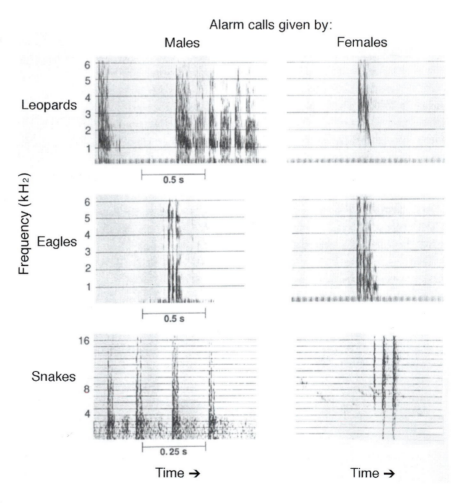

up into the air and often hiding in bushes, whereas the snake alarm call caused them to stand bipedally (Seyfarth *et al.* 1980a, b). The interesting thing about these playback experiments is that, while they do not rule out the possibility that during the approach of a real predator, the monkeys could simply be responding to the predator itself, they do show that the alarm calls themselves carry information about which predator is around. Even without seeing the predator, the vervets 'know' what action to take.

Cheney and Seyfarth (1980, 1986) were also able to show that young vervets are less specific than adults in what they give alarm calls to. Infants often make 'mistakes' and give alarm calls to warthogs, pigeons and other animals that are not dangerous to them. Their calls are not, however, given at random. An infant may give an eagle alarm call to a harmless bird or even a falling leaf, but at least it attaches the eagle alarm call to 'something overhead' and restricts the leopard alarm call to mammals, even harmless ones, on the ground.

There is no evidence that the adult monkeys actively teach the younger ones to be more accurate and discriminating in making their alarm calls. Rather, what seems to happen is that, on hearing an alarm call from a juvenile, adult monkeys will look up, just as they would to a call from an adult. If they see no danger, they will return to what they were doing before. If, however, they see a real predator nearby, the adults will give an alarm call themselves, followed by the appropriate behaviour. In this way, by watching the response and hearing the calls of adults, the young vervets learn what the various alarm calls should be attached to (Seyfarth & Cheney 1986).

As well as their dramatic and obvious alarm calls, vervet monkeys have other, more subtle ways of communicating with each other. One of these is through their grunts which are harsh, raspy noises that vervets make in a variety of social situations, usually when the animals are alert but relaxed. For example, vervets often grunt when they approach a socially dominant individual, when they approach a social inferior, when the group is about to cross an open space and when they have just seen another group of monkeys. To the human ear, all these vervet grunts sound much the same. Cheney and Seyfarth (1982) played grunts recorded in these various situations (approaching a superior, approaching a subordinate, starting to move into an open space and having seen another group of monkeys) through loudspeakers hidden in the grass. They found a very consistent difference between the responses to at least two of these context-related calls. Grunts that had been recorded when the monkeys were about to cross an open space resulted in the monkeys looking away from the speaker towards the horizon, regardless of what they happened to be doing at the time. By contrast, grunts that had been recorded when a monkey had been approaching a dominant animal caused the monkeys to look towards the loudspeaker.

Although humans could detect no differences between the grunts, the monkeys seemed to be able to. When the calls were analysed with a sound spectrograph, physical differences between the calls were seen to be present and it was apparently these that the monkeys were picking up. Such subtle distinctions in animal communication are, of course, the opposite of what we described as ritualization, where signals become large and distinctive under the selection pressure of being clearly understood by the reciever animal. However, we also saw that when animals are cooperating and their interests coincide, they may adopt 'conspiratorial whispers' or very inconspicuous signals. Vervet grunts appear to be a very good example of this. The monkeys all have a common interest in staying together as a group, partly because they may be related, but also because they derive mutual benefits such as protection from predators. It is as much in the interest of one monkey to understand and respond correctly to the grunt of another as it is to the giver of the grunt to be responded to: if the group is to cross an open space or to respond to the presence of a strange group of monkeys, they all benefit if they all respond together. As a result, even the subtlest nuances in the vocalizations of other animals are responded to. The fact that it takes sophisticated sound analysis to demonstrate to humans that there are physical differences in the monkey sounds, which the monkeys are already sensitive to, should alert us to the intriguing possibility that we have as yet barely scratched the surface of how animals communicate with each other.

4   Motivation and decision-making

In the last chapter, we saw that an important part of answering Tinbergen's causal question about how behaviour 'works' is understanding how animals recognize significant patterns of stimulation in their environments. But we also hinted that this was only part of the answer. An animal may be able to recognize a food item, but whether it will actually respond to the food at any given time will depend on a whole range of other factors such as whether there are predators around and how depleted its own body food reserves are. Understanding the causal basis of behaviour, in other words, also means understanding the constantly shifting priorities that lead the animal to 'decide' to do one behaviour at one time and then, a few seconds, minutes or hours later, to do something quite different.

The first thing to be said about such decision-making processes in animals is that they are disconcertingly complex. For any behaviour to occur, nerves, muscles and hormones all have to operate in particular ways. The brain has to direct and coordinate the behaviour in ways we do not fully understand. Faced with this baffling complexity and yet still wanting to give some sort of causal explanation for what they observe, ethologists have often used the term 'motivation' to refer to the various internal factors that give rise to behaviour, even though they may have been ignorant of exactly what these internal factors might be in physiological terms. The term 'motivation' has caused some confusion in the past but, as we will see, it does have its uses in describing how animals 'decide' what to do at any one time. It is the first step in unravelling the complexity of the internal workings of their bodies, a step that becomes redundant as our knowledge of physiology increases.

We will therefore continue to use the word 'motivation' but will be careful to use it to refer to factors inside, as opposed to outside, the animal's body. This is not because there is a clear-cut distinction between external stimuli and stimuli internal to the body – we saw in the last chapter how blurred this distinction can become. Rather, it is to emphasize that when an animal 'decides' what to do, it has to weigh up all sorts of stimuli, internal and external, through some sort of internalized decision-making process and give priority to one behaviour over another. The internal state of the animal, which is the net result of stimuli arising both inside and outside its body, constitutes its 'motivation'. What we are concerned with in this chapter is how the internal decisions are made about what behaviour to perform.

To give an illustration of the complexity of the motivational changes that may occur and the subtlety of the internal decision-making processes that animals have, we may take an example that may be observed throughout the world where this ubiquitous species has spread: a house sparrow (*Passer domesticus*) 'deciding' whether or not to approach a patch of food. When a sparrow discovers some food, it will often not approach immediately but instead sit on a nearby perch close to cover and give a distinctive call that has been described as a 'chirrup'. The call has the effect of recruiting other sparrows to come and join it on the perch (Elgar 1986a; Fig. 4.1a, b). All the birds then fly down to the food and feed together, each gaining the advantage of being surrounded by other individuals which are all watching out for danger. Often the bird that has spotted the food in the first place will not fly down to the food at all unless it is joined by several other birds. This is particularly true if the food is near a potential source of danger such as a human being. The birds are evidently 'deciding' between approaching the food and avoiding the predator but using the chirrup call to shift the balance of factors between the two behavioural options. If they can recruit other sparrows to the food source and so achieve 'safety in numbers', they will feed, but if no other sparrows join them, predator avoidance takes precedence.

But even the decision to give the chirrup call in the first place is

Figure 4.1 (a) The time taken for a first-arriving (pioneer) house sparrow to be joined by another sparrow is inversely proportional to the pioneer's chirrup rate (from Elgar 1986b). (b) The rate at which the 'chirrup' call is given is higher when the ambient temperature is lower. The pioneer is likely to give higher priority to feeding when it is cold and is therefore more prepared to share the food source with others who will 'pay' by the increased vigilance a group will provide. (From Elgar 1986b.)

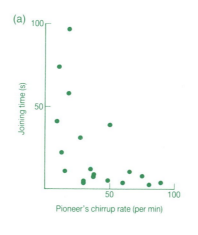

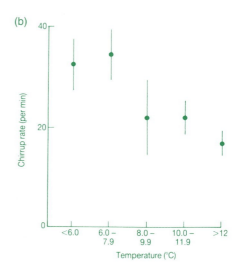

complex. In cold weather, sparrows chirrup at a much higher rate than when it is warmer (Elgar 1986b; Fig. 4.1c). This is probably because their food requirements are greater in cold weather and so the birds give higher priority to feeding even when predators are around. Another factor which affects chirruping is whether or not the food source itself is divisible. If the food is divided up into portions, so that several different birds can feed without interfering with each other, a sparrow will chirrup vigorously, stimulating others to come and feed with it. But if the same amount of food is in one indivisible lump, it will chirrup much less and be more likely to feed alone. Under these circumstances the sparrow 'decides' not to gather a flock around it, apparently because of the competition that results when several birds try to eat the same piece of food (Elgar 1986a).

From this example, we can see something of how decision-making processes must work. A sparrow evidently responds to the sight of food but whether or not it 'decides' to feed at any one time depends on many different stimuli from the environment – food location, food type, temperature, presence of predators, and so on – as well as factors within its own body, such as degree of food deprivation. Understanding the causal basis of such decision-making would mean, for example, understanding what it is about the body of a sparrow that changes when the temperature drops that causes it to increase its chirruping rate. Sparrows at low temperatures can be reliably observed to give chirrup calls; but have they undergone a hormonal change, or a change in level of fat reserves, and what exactly are the pathways by which a change in the internal state actually leads to an increase in vocalization?

Such questions cannot, at present, be answered. Some of the difficulties arise because ethologists, who study the behaviour of whole animals, have concentrated on decision-making in animals with complex nervous systems and have been most impressed with cases, such as the house sparrow, where the behaviour is evidently influenced by a great many factors. This in turn makes it difficult to give a causal explanation of what is observed. Neurobiologists, on the other hand, have looked at the causal mechanisms of behaviour either in animals with simpler nervous systems such as molluscs and insects or, when studying vertebrates, have concentrated on manageable components such as the control of feeding and what makes an animal start or stop drinking. What we know little about is the control of 'higher-level' decisions, such as how an animal that is stimulated to both feed and flee produces behaviour that results in its first giving a recruitment call and then feeding in a flock.

As we shall see throughout this chapter, there are many different opinions as to how best to go about trying to proceed from this point. Some people believe progress can be made only through the techniques of physiology – by looking inside the animal and recording nerve impulses, muscle movements and hormone levels. This would involve building on our knowledge of simple behaviour in invertebrates with relatively simple nervous systems, and eventually achieving an understanding of more complex behaviour in more complex nervous systems. Others believe that such a complete explanation is still so far off that we do best to invoke intervening variables such as 'motivation', at least as a temporary measure, in our search for causal explanations. We will look at these different ways of explaining behaviour, how the concept of 'motivation' is used and how controversies over its use have arisen. We will do this in the way we stress throughout this book – by looking at different levels of explanation as well as constantly keeping our eyes on the interplay between them. We start by looking at decision-making on a behavioural level and then go on to discuss what is known of its physiological basis.

Decision-making on different time scales
Time budgets and daily routines

Almost all animals show different behaviour patterns depending on the time of day and many show great regularity in their habits. Male red jungle fowl crow just before dawn, then descend from the trees where they have spent the night to forage on the ground in the company of hens. They rest in the heat of the day, forage again in the late afternoon and then go to roost at dusk. In general, the daily routines of wild animals and the amount of time they spend on different activities (their time budgets) can be seen as strategies for meeting their needs whilst coping with changes in the environment such as variations in temperature, activities of predators and availability of food.

To some extent, these daily changes in behaviour depend on animals having an 'internal clock', which times their behaviour more or less independently of events in the external world. Figure 4.2 records the activity of a flying squirrel living in a rotating cage. Each row represents 24 hours and the dark bands show when the animal was actively moving around. The squirrel is a nocturnal animal and it was normally active from just after 'dusk' until the lights came on again at 'dawn'. At the beginning of the records shown in Fig. 4.2 the lights were switched off permanently. Now, even though it was living in

Figure 4.2 The spontaneous wheel running activity of a flying squirrel in total darkness. Each line is the trace from an event recorder over 24 h. Active periods appear as broad dark lines as the pen of the event recorder moves frequently up and down on the paper. Before these records were taken the squirrel had been kept on a regular cycle of 12 h light/12 h dark – indicated at the top. In total darkness it retains its rhythm and is active only in the period that had previously been dark. In the absence of external cues, however, its natural circadian rhythm, which is slightly less than 24 h, asserts itself and the period of activity begins a little earlier each night. (From De Coursey 1960.)

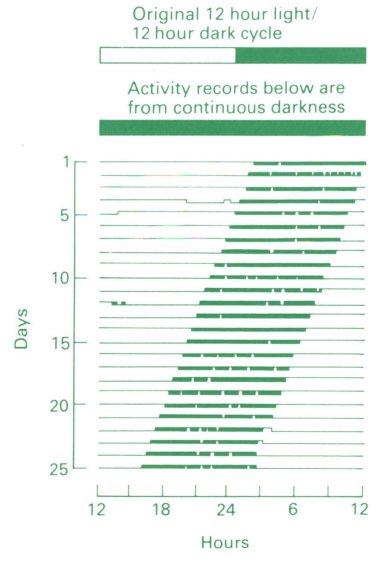

complete darkness, the squirrel nevertheless maintained fixed periods of activity. These lasted almost exactly the same time as the original dark periods and occurred at extremely regular intervals. Interestingly enough, these intervals are not 24 hours, but just over 23 hours, so that in real time, the squirrel begins its activity a little earlier in each 24-hour period. A rhythm of activity of this type is called circadian from *circa diem* 'about a day'. Circadian rhythms are widespread amongst plants and animals and affect not just behaviour but all aspects of metabolism (Aschoff 1981, 1989).

Wild animals spend a surprisingly high proportion of the 'active' part of their days resting and apparently do nothing (Herbers 1981), but this too can be seen as adaptive. Periods of inactivity may be essential for digestion (Diamond *et al.* 1986), energy conservation or simply keeping out of the way of predators. Indeed, spending certain parts of each day sleeping in a hidden place may be one way that animals have of removing themselves from potentially dangerous situations (Meddis 1983). The internal clock is not, however, the only factor determining what animals do at different times of the day, because animals can also adjust their time budgets according to circumstances. We can see quite major changes in how animals fill their days, depending on what demands are being made on them. Female baboons spend more and more time feeding as their infants get older and make bigger lactational demands on them (Dunbar & Dunbar 1988) and they give up some of their rest periods in order to feed more. Then when most of that available rest time is used up, they will feed at times that would otherwise be devoted to social interactions. Ruddy turnstones (*Arenaria interpres*) are migratory shorebirds, many of which overwinter on rocky shores in Western Europe and then fly up to 3500 km to breed in the arctic tundra of northeastern Canada and Greenland. In the 3 weeks just prior to migrating to their northen breeding areas, the birds increase body weight by over 40% and do so by a major adjustment of their time budgets: they spend a greatly increased amount of time feeding, even though as a result they allocate less time to looking out for predators and become markedly less vigilant (Metcalfe & Furness 1984). In this case, feeding (to give them enough fuel for their long journeys) is given a higher priority than usual. Like many other migratory birds, turnstones eat much more than they normally do just before migration and lay down extensive body fat reserves. Hormones such as prolactin appear to influence this fat deposition, but the physiological basis for the seasonal changes in feeding behaviour involves a complex interaction between different hormones, changing day length and endogenous factors (Wingfield *et al.* 1990; Berthold 1993).

Decision-making from minute to minute

Most studies of animals have involved studying their activities over periods rather shorter than that of a whole day. For example, if we watch a bird foraging through the trees or an animal building a nest, the behaviour may change from one minute to the next. If we look closely, we can see that the animal is not just feeding, in the sense that it is continuously taking in food, but that its feeding is divided into searching, capturing food, preparing the food, eating it, flying to another tree and so on. The animal is constantly making decisions about the next stage in its sequence of behaviour – when to stop doing one thing and go on to the next.

An important idea for understanding such sequences is to explore the possibility that animals make decisions that 'optimize' some factor such as the net rate of energy intake. So, as an animal eats up the food available in one area, it will be getting less and less food for more and more effort if it continues to search for food there. However, if it moves to another area where the food has not been depleted it may gain energy at a higher rate, despite the fact that it would have to use up some energy in flying or walking to the new place. Optimal Foraging Theory (OFT), which in its original form assumed that the well-adapted animal would optimize its energy intake, predicted that the decision to stop feeding in one area and move to the next could be calculated by knowing how much energy was involved in both staying and moving away. The point at which a bird, say, would gain more by flying to another tree than by staying to look for food in an already depleted one would be the point at which it should 'decide' to move. Clearly flying to very distant trees, which would use up a lot of energy en route, will become the optimal behaviour only when the present tree is severely depleted (Fig. 4.3). Optimal Foraging Theory predicts that an animal should stay in one area, even though it is gradually depleting the food there, until the net rate of energy intake drops to the average for the environment. At this point, the animal should decide to move on to seek out something different.

It will be obvious that Optimal Foraging Theory assumes that the animal

Figure 4.3 Optimal foraging theory (OFT) applied to the behaviour of a great tit feeding in a patch of food. (a) The curve represents the average net energy gain as a function of time it spends in a patch. The longer it feeds in one patch, the more depleted that patch becomes and so the gain (food intake) begins to go down. Food intake is plotted cumulatively over time, so this results in a 'levelling off' of the gain curve. (b) Calculation of the optimum time the bird should stay in a patch that is being depleted depending on whether there is a long or a short travel time to another (undepleted) patch. The optimal predator should stay in the first patch just long enough to maximize the slope of AB (representing the average food intake for the habitat as a whole). So, the optimum time to remain in the patch is given by the tangent of the line AB to the gain curve. The slope is steeper the nearer the next undepleted patch. (c) Cowie's (1977) test of OFT using great tits searching for food hidden in pots. The solid line is the result predicted taking into account the cost of travelling and searching in a patch. The 12 dots are the means and standard error of six birds tested with both long and short 'travel times'. Each bird had six trials in each environment, and so was able to learn the optimal time allocations in each one.

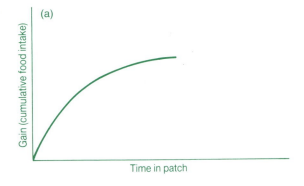

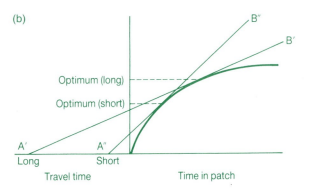

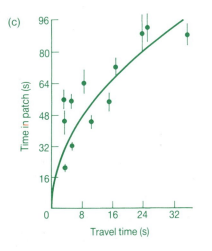

can estimate what conditions are like elsewhere in the environment and that it has some way of comparing the net rate of energy gain it is experiencing now with what it would experience if it flew to another patch and started eating there. The fact that, at least in some cases, OFT does describe behaviour remarkably well, suggests that even if animals do not consciously know these things, they are behaving 'as if' they do. Cowie (1977) studied great tits foraging for pieces of mealworm hidden in 'patches' consisting of sawdust-filled pots in different parts of an aviary. He made moving on to the next patch more or less difficult by putting lids on the pots that were easy or hard to remove. So a difficult lid, by taking more time and effort to remove, was taken to be the equivalent of a long distance between patches. The fit between the time the birds spent in a patch depending on 'distance' to the next one and the time predicted by OFT was remarkably good (Fig. 4.3c). The birds were behaving as though they knew how much food was in each pot and how difficult each sort of lid was to prise off, even though they had no previous opportunity to learn the characteristics of the 'habitat'. The sequence of their decisions – to fly to an area where food is likely to be found, to prise off a lid, to push away sawdust, to eat, to continue searching in the same spot or to fly off to another one – certainly suggested that they acquired this knowledge; otherwise the real decisions of the real birds and the predicted decisions of the model would not have coincided so well.

But do birds – or any animals – 'know' about their environments? Perhaps all that is happening is that a bird moves to another patch whenever it has failed to find a food item for a certain length of time, that length of time varying with its past experience of hunting in that environment. It may be following such a relatively simple 'rule of thumb' and not even unconsciously working out 'travel times' or 'energy gains' at all (and certainly not understanding the equations of OFT!). We do not consciously calculate the trajectories of balls thrown into the air, but we often catch them all the same. An automatic process takes over and we behave 'as if' we knew all about the physics of flight and the aerodynamics of moving objects. Some possible 'rules of thumb' that foraging birds follow might be 'leave after catching n prey', 'leave after x seconds', 'leave after y seconds of unsuccessful search', etc., depending on the exact conditions a bird finds itself in (Stephens & Krebs, 1986).

Not all cases of animal decision-making fit the predictions of OFT as well as did Cowie's great tits. A possible reason for this discrepancy arises from one of the basic assumptions of the theory we encountered earlier – namely, that the animal is doing all it can to maximize its net rate of energy gain. Suppose that this is not always the case. In winter, when food is scarce and the survival of a small bird may depend critically on its obtaining enough energy and not expending too much in the process, it might be reasonable to assume that energy gains are paramount: energy and survival become practically synonymous. But when food is plentiful or when an even greater danger is posed by dehydration or being eaten by a predator, why should any animal go all out for maximizing its net rate of energy gain? There are many other threats to survival and reproduction and we might therefore expect natural selection to have favoured animals with wider horizons and more flexible 'rules of thumb'. In other words, we might expect that their decisions about what to do next would be influenced by factors other than just those to do with feeding. Natural selection leads animals to optimize their fitness (Chapter 6). Although Optimal Foraging Theory helps us to understand some important evolutionary reasons why animals change their behaviour, we should positively expect animal decision-making to be influenced by more than just energy intake.

It is not difficult to find examples. Even great tits which, as we have seen, have been observed to optimize the time spent in a patch and travelling between patches, do so most obviously when they are very hungry. When their hunger is reduced, great tits become less than optimal foragers (Ydenberg & Houston 1986). They compromise their feeding and spend more time on the look out for other birds intruding into their territory. Another example is the house sparrows we have already discussed, sitting on the fence and making different decisions about whether to fly down to feed depending on the nearness of danger, the divisibility of the food source and so on. Fish, too, balance the demands of feeding with other activities. When sticklebacks are very hungry, they prefer to feed in high density swarms of water fleas because in this way they get the maximum amount of food with the least effort (Milinski & Heller 1978). However, when a model kingfisher was flown over a tank containing hungry fish, the sticklebacks preferred to feed in a much lower density swarm (Fig. 4.4). A plausible explanation of this effect is that kingfishers are predators of sticklebacks and if a stickleback feeds in a high density swarm, it may get a lot of food, but finds the many, jerking water fleas so distracting that it is less able to keep a look out for predators. The low density swarm of fleas, while giving less food, allows a stickleback to keep an eye out for danger so the decision to feed on the swarm of lower density paves the way, in turn, for a series of even shorter term decisions, to snap at a water flea or to look upwards to see if a predator is about to strike (Milinski 1984). In all of these cases, the internal state of the animal – such as its state of hunger or its state caused by the presence of a predator – has to be taken into account.

Decision-making from second to second

We have so far looked at decision-making at the level of gross categories of behaviour such as 'feeding' or 'vigilance' or 'flying to a tree', each of which takes at least a few seconds to perform and each of which involves very large numbers of muscles. But we can be even more specific and detailed than this. When an animal is feeding, what movements does it actually make? Does it hammer open a mussel like an oystercatcher, graze like a sheep or stalk its prey like a lion? If we are ever to understand how the action of muscle systems leads to behaviour, we need to look in more detail at these movements themselves – the movement of limbs and heads and bodies. We should not underestimate the complexity of the machinery underlying animal movement. Even an apparently simple behaviour such as a water snail opening its mouth and

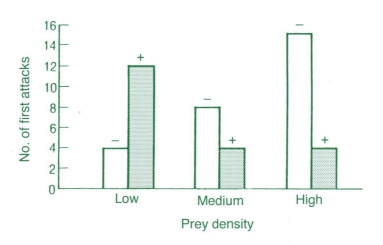

Figure 4.4 Number of first attacks by sticklebacks when presented with prey (water fleas) at different densities either with (+) or without (−) a model kingfisher overhead.

scraping its radula along the surface of a plant – an action that takes only 1–2 seconds – even this involves the contraction, in the correct sequence, of 25 different pairs of muscles controlling the mouth (Kater & Rowell 1973). There are, of course, many other muscles controlling the head and the snail's foot as it moves along. There is a veritable 'symphony beneath the skin' involving hundreds of muscles, each coming into action at a particular time. And if this symphony is complex for water snails, it is even more so for a male mallard duck courting a female, the oystercatcher, the sheep and the lion, all feeding in their idiosyncratic ways. We should remember that no man-made robot comes anywhere near the complexity, the finely tuned adjustment and the flexibility of the behaviour of an animal.

When confronted with all the different possible movements animal can make, it is often useful to focus on easily recognized and distinct sequences of behaviour lasting only a few seconds or so. These short sequences are the behaviour patterns such as the courtship strut display of the sage grouse that we discussed in Chapter 1. They have descriptive names like 'pecking', 'drinking', 'wing-flapping' and together they constitute the *ethogram* or behavioural repertoire of a species.

Early ethologists laid great stress on these short sequences of behaviour, which they called Fixed Action Patterns, implying that they are immutable and always performed in exactly the same way. Certainly the courtship patterns of some ducks studied by Lorenz (1941) and the 'strut display' of the male sage grouse we discussed in Chapter 1 do have this rigid 'clockwork' quality (Dane *et al.* 1959). However, not all behaviour patterns are so stereotyped. There is enough similarity between different instances of 'drinking' in chicks or tail-wagging in dogs, that we can recognize these behaviour patterns on different occasions, but they are not exactly the same from individual to individual or even from occasion to occasion in the same individual. For this reason, the term Fixed Action Pattern gives an inaccurate impression and a more neutral description – such as simply 'behaviour patterns' – is preferable.

Most 'behaviour patterns' are recognized intuitively, but film and video-tape analysis allows us to be a bit more objective. Dawkins and Dawkins (1974) used slowed-down videotape to show that the behaviour pattern of 'drinking' in chicks was much the same whether or not the chick was thirsty or even whether it was drinking a substance that turned out to be distasteful (Fig.

4.5). There was much more variation in the intervals between drinks than in the drinks themselves, suggesting that a decision about whether or not to continue to drink was made between drinks, not during the drink itself.

Sometimes behaviour patterns that themselves last only a few seconds can be strung together into sequences with a high degree of predictability that lasts much longer. The 'song' of the humpback whale, mentioned in Chapter 3, is a long series of sounds which is then repeated in precisely the same way. Each repetition lasts about 40 minutes – about the same length of time as Beethoven's Fifth Symphony!

Mechanisms of decision-making

The theory of natural selection leads us to expect that animals should make their various decisions – on all the different time scales we have discussed – in such a way that they make the best or optimal decisions. This means the ones that maximize their chances of successful reproduction. As we have seen, animals may not always 'optimize' feeding efficiency because of other conflicting demands, such as having to look out for predators or care for young. But in a well-adapted animal, the choice of what to do next should be an optimal *compromise* between all the different options open to it. This should be as true whether we are considering daily time budgets or second-to-second switches in behaviour.

Our discussion of the adaptiveness of behaviour sequences has also shown the importance of considering the internal state of the animal. Animals

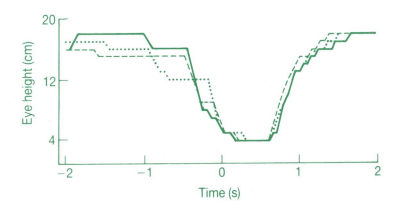

Figure 4.5 Superimposed graphs of the height above the ground of a chick's eye during three successive drinks, lined up on the moment when the bird's bill strikes the water (time 0). (From Dawkins & Dawkins 1974.)

do not always forage optimally – it often depends on how depleted their food reserves are, for example. To understand the mechanisms by which adaptive sequences of behaviour come about we therefore need to understand what is meant by 'internal state' and ultimately discover what the difference in 'state' between a hungry and a less hungry animal is. In this section, we will look at what is known about how an animal's state affects its decision-making.

The first problem we come up against is the complexity of the mechanisms involved. An insect has about a million nerve cells, a cat's brain about 10^9. Sense organs, muscle cells and hormones add to the huge number of elements that may be involved in an 'explanation' of behaviour. Many workers have dealt with this complexity problem by studying mechanisms of behaviour without referring directly to neurons or hormones at all. Instead, they invoke concepts such as 'motivation' or 'state' to explain the as yet unknown causal factors that are at work inside the animal. Of course, these 'unknown causal factors' include neurons, hormones and other physiological elements and the ultimate hope would be to explain motivation in these terms, but some people believe that such explanations are so far in the future that much useful groundwork can be done by not invoking physiology directly. We will therefore start our discussion of the causal basis of decision-making by looking at what such an approach can tell us.

Motivational state

If an animal changes its behaviour over a time scale of minutes or hours when everything in its external environment has remained constant, then we can deduce that something inside the animal must have changed. It is this something – however little we yet know of its nature – which we call 'motivation'. For example, if we gave an animal some food and initially it took no notice but some hours later it began to feed, this would suggest that the animal's internal state had now changed. We should note, however, that not all changes in behaviour are attributable to changes in motivation. A particularly important distinction is between the long-term, near-permanent changes due to learning or maturation, where an animal changes its behaviour as it grows from infancy to adulthood on the one hand, and short-term reversible changes on the other. It is the short-term reversible changes – for example, as an animal changes from feeding to drinking and then back again to feeding all within the course of a few hours – that we refer to as being due to changes in motivation.

A characteristic feature of motivational changes is that it is often not just the animal's response to one specific stimulus that fluctuates, but a whole range of responses that are functionally related to one another. Thus an animal's threshold of response to all stimuli connected with food and feeding behaviour will rise and fall together, so will those connected with sexual behaviour, and so on. It is because of this association of effects with different sets of stimuli that ethologists and other workers have talked about 'specific motivational states', implying the existence of a set of internal causal factors that affect not just one behaviour but a whole functional group. Instead of invoking a different motivational state for every single element of its behav-ioural repertoire, such as a 'chasing motivation' and a 'swallowing motivation' the animal is said to have 'feeding motivation', implying that most or all of the behaviours functionally related to obtaining food (stalking, chasing, biting, etc.) are affected by many of the same causal factors. Feeding motivation thus refers to factors that, whatever they are, affect the whole group of feeding-related behaviour patterns.

Evidence for the operation of common causal factors comes from work such as that of Baerends and his collaborators (Baerends 1976), who studied the changes in various behaviour patterns that occurred when herring gulls were disturbed at the nest while incubating an egg. In the 90-minute period following the interruption, three behaviours (building, preening and re-settling) all decreased at first and then increased at very much the same time, suggesting that all three shared at least one common causal factor, the level of which was falling and then rising again. Baerends (1976) also used an analysis of what behaviour precedes or follows another one to identify which ones might have common causal factors. He found that there were two sets of behaviour patterns. One set involved re-settling, pecking, picking up material and sideways building. The other set was composed of turning, head-shaking, scratching, shaking, yawning and looking around or looking down. If a gull had just performed one of the behaviour patterns in the first set, its next action was most likely to be one from the same set, not one from the second set. If it had just performed a second set behaviour, the converse was true. This sug-gested that, within one set, the various behaviour patterns shared a common causal factor and that this was different (or at least different in effect) from the other set. Possibly the two causal factors acted antagonistically.

With some motivational systems we have to be careful that categories that we define in human terms, such as 'aggression', correspond to those we observe in the behaviour of the animals. Aggression is an interesting case in point. Aggression towards members of the same species – often called social aggression – is reasonably distinguished from the interspecific fighting shown in catching and killing prey and also from the defensive behaviour shown by an animal faced by a predator. Such a distinction can be made both on the basis that the behaviour patterns used in these latter situations are often different (although both may show some overlap with social aggressive behaviour) and also on the grounds that the external and internal factors giving rise to them seem to be different. For instance, in mice, electric shock-induced fighting often takes the form of upright postures with bites directed at the opponent's nose, which more resembles the behaviour of a defensive 'cornered' mouse than that of a mouse attacking an opponent (Blanchard & Blanchard 1981). However, we should not assume that the motivational systems of attack and defence are entirely unrelated. Huntingford (1976) has shown that in sticklebacks there must be some common causal factors between the aggression shown to rival males and the aggression – or perhaps we should call it defensive behaviour – shown towards predators such as small pike. Fish which behaved in a 'bolder' fashion towards pike were also those which showed the highest levels of fighting in territorial disputes.

Note that we are already beginning to say something about what appears to be going on inside an animal, using the temporal correlations between its different behaviour patterns, even though the exact nature of the various causal factors has not been determined. It would, however, be misleading to think of animals as though they had fluctuating sets of causal factors for 'feeding', 'drinking', 'sex', etc. all operating independently of one another. Many stimuli produce non-specific 'arousing' effects which render animals more responsive to a wide range of stimuli and this could be described as a rise in 'general motivation' or, as it is called by some psychologists, 'general drive'. Although it might seem a straightforward question, it is in fact very difficult to collect really conclusive evidence as to whether motivation is general or specific; see good discussions of this point in Toates (1986) and Colgan (1989). In what follows, we shall assume that there is a considerable degree of specificity, although it would certainly be wrong to think of motivation as rigidly compartmentalized. Indeed, we have already seen that different motivational systems interact with one another. These interactions are one of the main ways in which sequences of behaviour are produced.

Behavioural analysis of sequences

Goals

An important idea in the study of sequences has been that of a 'goal' which can be defined objectively as that situation which brings a whole sequence to an end. A dog shown his bowl of food on the other side of a fence will go through all sorts of different behaviour – running, jumping, scratching the fence and so on – and these would all cease once the dog had achieved the goal of getting its food. In fact, it would be difficult to describe the dog's behaviour concisely without calling it goal-directed. The implication would be that all of these different behaviours shared common causal factors such as the sight of a food dish and food deprivation, and would all be activated together and then lowered together when the goal had been achieved.

The nest-building behaviour of the female great golden digger wasp (*Sphex*) fits the idea of goals beautifully. The wasp digs a burrow in the soil and then provisions it with katydids (large grasshoppers) for her offspring to eat when they hatch. Once she has achieved the goal of one completed burrow, she moves on to build another one, but she also has secondary goals in the excavation of each individual burrow. The wasp first digs a downward-sloping main burrow and then a side tunnel, at the end of which is the nesting chamber (Fig. 4.6). Brockmann (1980) altered the depths of main tunnels artificially and showed that the wasps adjusted their behaviour accordingly. If a wasp was confronted with an artificially lengthened main tunnel, she switched to making the side tunnel much sooner than normal, apparently satisfied with the construction even though she had not made it herself. The wasp evidently has a goal of a main tunnel of a particular length and moves on to the next part of the building sequence when this is present.

A similar cessation of behaviour once a goal has been achieved is shown by the reduction in sexual behaviour of the male three-spined stickleback when he has achieved the usual result of his courtship of a female – a new clutch of eggs in his nest. Male sticklebacks of various species build quite elab-orate nests out of plant material (Fig. 4.7). A male will then court passing females and then, having persuaded one to enter his nest and lay eggs, he follows into the nest after her and fertilizes them. After he has done this, his sexual behaviour declines for a while, not because he is exhausted from his activities, but because the sight of eggs in the nest reduces his sexual motiva-tion. Sevenster-Bol (1962) placed eggs in the nests of male sticklebacks and found that their sexual behaviour was reduced whether or not the males had been through the act of fertilization.

In the case of both the stickleback and the digger wasp, the goal that brings the behaviour to an end and causes the animal to start doing something else consists of the usual *result* of the behaviour, demonstrated in each case by an experimenter artificially mimicking the normal outcome and showing that this was what switched off the behaviour. If there had been no interference, the normal result of the behaviour would have acted as a **negative feedback mechanism** (Chapter 1) to end the sequence. The essential feature of a nega-tive feedback mechanism is that some of the results of the behaviour are mon-itored and these results are fed back into the control system of the behaviour, reducing or sometimes increasing the animal's motivation to keep doing the same thing, depending on whether the goal has been achieved or not. It would be a mistake, however, to think that all animal behaviour can be neatly divided into separate sequences each brought to an end by the achievement of a separ-

Figure 4.6 A female great golden digger wasp digs a burrow consisting of a main tunnel sloping downwards with a side tunnel ending in a nest chamber. (From Brockman 1980.)

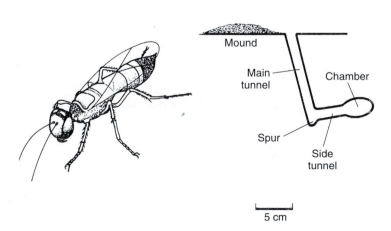

ate goal. In most cases, the situation is a great deal more complex. For example, domestic hens go through a sequence of nest-building activities that involves scraping a shallow depression in the ground and placing nest material around the edges. It might be thought that the goal would be a completed nest. However, hens will continue to show nest-building behaviour even when they are given ready-made nests (Hughes *et al.* 1989). Unlike the digger wasp and the stickleback, therefore, the hens are still motivated to perform the behaviour even though their apparent goal has been achieved.

The idea that it could be the actual performance of behaviour that was important in bringing a sequence to an end, rather than the attainment of a particular goal, was behind a very influential model of animal motivation put forward by Konrad Lorenz in the 1950's. Lorenz believed that all behaviour could be divided into a striving or searching phase which he called 'appetitive', followed by 'consummatory behaviour' which brought the sequence

Figure 4.7 A male ten-spined stickleback 'fanning' at the entrance to its nest. This behaviour plays a key role following courtship and egg-laying by ventilating the eggs. The male drives a water current through the nest by vigorous beating of his pectoral fins. This produces a backward thrust on the body, for which he compensates – and hence hangs still in the water – by driving forward with his tail. (From Morris 1958.)

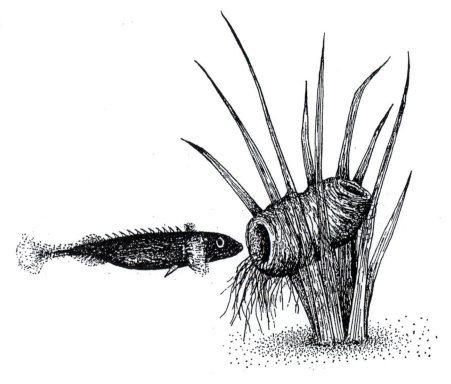

to an end. He stressed that it was not so much the achievement of a goal that was important, but the performance of the behaviour itself. Thus, once the animal had performed a behaviour, it would go through a period of quiescence in which the behaviour would be difficult to elicit. His model would thus fit the nest-building behaviour of the hen but not the tunnelling of the digger wasp nor the sexual behaviour of the stickleback. It certainly does not fit aggressive behaviour, where aggression in various species shows the opposite of quiescence after performance. Sevenster (1961) measured levels of aggression in male three-spined sticklebacks by using the number of bites delivered to a test-tube containing another male. He found that after 10 minutes of such aggressive biting, the tendency for further aggression went up, not down. Wilz (1970) showed that even if the rival male is removed after being attacked, the remaining male continues in such a high state of aggression that he attacks even females and seems unable to respond to them sexually for some time. Similar increases in aggressiveness following a spell of fighting have been reported for other species, including spiders (Riechert 1984) and lizards (Rand & Rand 1978).

It was the fact that Lorenz's model simply did not fit so many examples of animal behaviour that in the end led to its being abandoned. Although you may still sometimes see the terms 'appetitive' and 'consummatory' used in a somewhat loose way, it is probably best to avoid them because of their historical ties to a particular model of motivation which turned out not to fit the facts.

The idea of 'negative feedback', on the other hand, is much more widely applicable, but we now know of many different ways in which the results of behaviour can be fed back into the system. For example, when an animal eats, there are changes in its external world (food is depleted) but there are also resultant changes in the animal's body both in the short term (food in the mouth and stomach) and also longer term (nutritional balance of the body), which can also act as goals and result in changes in the motivational state of the animal. Drinking, nest-building and courtship are just some of the behaviours that are also drastically affected by what the animal itself has done earlier, through the animal's own action on its environment. Feedback effects have now been incorporated into a number of motivational models under the general heading of 'homeostatic models' (McFarland 1971; Toates 1986).

Homeostasis

'Homeostasis' was the term coined by W.B. Cannon in his book, *The Wisdom of the Body* (1974) to describe the relative stability of the body despite the changes that go on in the outside world. Our internal body temperature does not depart much from 37°C even though the external temperature may be much hotter or much colder. In this sense, the goal of the body is to maintain a constant body temperature. We constantly lose water by evaporation, urination and so on, and yet the body's fluid volume remains roughly constant. Homeostasis implies that the body has some means of correcting deviation, so that if temperature or fluid volume falls, steps are taken to restore the balance. The starting and stopping of drinking behaviour can be seen as part of this homeostatic mechanism. Body fluids decrease, this stimulates the animal to drink and this in turn helps to correct fluid loss, returning the body fluids to some 'ideal' or normal value. Figure 4.8 shows a simple homeostatic model of motivation in which this process is represented diagrammatically.

Homeostatic models assume that there is an ideal state or set point for the animal. If there is a difference between this set point, say, for body fluids and the actual state in the body, this so-called 'error' or 'discrepancy' is said to provide the motivation for drinking. You may see such a system described as a 'negative feedback' loop (see Chapter 1) because error is 'fed back' into the system and stimulates the behaviour or physiological response to operate until the error itself is reduced. The system is thus self-correcting.

In many instances, such a negative feedback model of motivation seems to provide a reasonably good analogy to the behaviour of a real animal. Fitzsimmons (1972) showed that rats injected with salt, which dehydrates

Figure 4.8 Simple homeostatic model of drinking behaviour. The actual state of the body fluids is compared to an ideal 'set-point' and any difference activates drinking. Water reaches the body fluids via the mouth and gut (part of the 'body fluid' system) and corrects the deficit, switching off drinking.

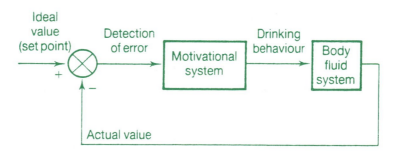

them, drink just enough water to restore their fluid balance. The effects of placing food directly into the stomach are also understandable on a homeostatic model: loading the stomach with food is enough to 'turn off' eating because a full stomach is normally part of the negative feedback loop of the homeostatic feeding system.

It would be quite wrong, however, to think of animals like simple thermostats, switching behaviour on and off whenever they are in particular states. Homeostasis is much more complicated than this in practice (see Rolls & Rolls 1982), and for good functional reasons. Consider the problem of an animal having to maintain a constant fluid volume and composition in its body despite losses of water due to urination, perspiration, etc. and despite variations in the external temperature, the composition of its food and the availability of water in its environment. Its first difficulty is the time-lag which exists between behaviour (drinking) and the eventual effect on fluids in the body. When rats and other mammals are deprived of water for a long time, the cells of their bodies become dehydrated and their extracellular fluid (such as plasma) also diminishes in volume. One of the main reasons why we become thirsty when we take in salt is that the salt stays outside cells (since it cannot get through the cell membranes) and draws water out of the cells by osmosis. Water loss from cells is then detected by specialized osmoreceptors in the brain, which are spread out over quite a wide area of the lateral hypothalamus. (For a diagram of the brain, see Fig. 4.9.) Blass and Epstein (1971) injected

Figure 4.9 The basic divisions of the vertebrate brain. The brains of all vertebrates pass through a stage rather like this during development and the brains of adult sharks and rays are little changed from this plan. However, in mammals and birds the adult brain is dominated by the enormous growth of the cerebral hemispheres and cerebellum. These come to overlie all the rest and obscure the original layout. (From Romer 1962.)

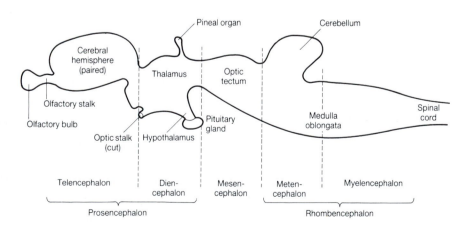

small amounts of saline or sucrose solutions into the lateral hypothalamic areas of rats and found that the rats started drinking, but they did not do so following injection with urea. Since urea is, unlike salt or sucrose, able to cross cell boundaries, it does not draw out water from the cells and the stimulation of drinking therefore seems specifically related to such cellular dehydration.

Now on the very simplest sort of homeostatic model, drinking would occur when cells become dehydrated and cease when they become rehydrated again. But since most animals drink much more rapidly than the fluid can be restored to their cells, the animal would drink far too much if it drank until its tissues and plasma were fully rehydrated. If human beings are deprived of water for 24 hours and then allowed to drink, they will drink almost all that is needed to restore fluid balance within 2.5 minutes even though changes in plasma dilution cannot be detected for 7.5 minutes and are not back to normal for about 12.5 minutes (Rolls & Rolls 1982).

There must, therefore, be a means of detecting that water has been taken into the body before the full physiological consequences have made themselves felt. Miller et al. (1957) showed that at least some of these water detectors are in the mouth and oesophagus and that activation of these causes termination of drinking even before fluid balance is restored. They allowed three groups of previously water-deprived rats to drink for 18 minutes. The first group just had 14 ml of water loaded directly into their stomachs (i.e. by-passing the mouth and oesophagus altogether), the second were allowed to drink 14 ml in the normal way and the third group of rats had no water at all. Putting water into the stomach had relatively little effect on the amount drunk immediately after: the stomach-loaded rats drank almost as much (average 16 ml) as the completely deprived rats (average 21 ml), whereas the rats that had been allowed to drink for themselves drank only 6.7 ml during the 18 minutes of the test (Fig. 4.10). Water is thus detected in the mouth and throat during the course of normal drinking and reduces the subsequent tendency to drink. Distension of the stomach and stimulation of the intestine (together with possible hepatic-portal factors) may also play a part in terminating drinking, the effects being different in different species. In dogs, for example, gastric distension is relatively unimportant for drinking, but it is very important in monkeys. Furthermore, some drinking occurs not because of a fluid deficit at the present moment, but because of an anticipated fluid deficit in the

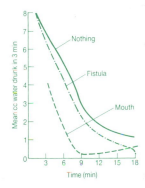

Figure 4.10 The effect of giving thirsty rats 14 ml of water by mouth or by stomach fistula on the amount of water they subsequently drank. Different groups of rats were each given *ab lib.* water to drink at successive 3 min intervals after the initial administration of water. The graphs show how much they drank at each of these intervals compared with thirsty controls that had not been administered water initially. (From Miller *et al.* 1957.)

future. Fitzsimmons and Le Magnen (1969) showed that when feeding, rats drink in anticipation of a water deficit (dry food will make them thirsty), a phenomenon which McFarland (1971) called 'feed-forward'. The state of the body fluids is certainly detected but it constitutes only one among many other factors in determining whether an animal will drink at a given moment. But whatever the precise mechanism, it is quite clear that a simple homeostatic model, involving only the detection of fluid imbalance to initiate and terminate drinking, would be a poor predictor of the drinking behaviour of real animals.

Simple homeostatic models similarly fail to explain all aspects of feeding behaviour. Animals fed to satiety on one kind of food will often resume feeding if given a wider range of foods. Wirtshafter and Davis (1977) showed that rats used to a normal laboratory 'rat food' diet and then given a 'super-market' diet (a variety of sweet and palatable foods) will eat a great deal more than usual and may even become grossly obese. Such observations have led researchers away from a simple view of a particular brain site as the place where feeding is switched on or off (the so-called feeding centre) towards a view of the control of feeding in which many different parts of the brain are involved. As we have already seen, feeding behaviour is also affected by the availability of food, the presence of predators, whether other conspecifics are present or not, and so on.

All models are simplifications. Nevertheless, the attempt to identify certain principles, such as 'negative feedback' can be an important step in understanding some aspects of animal behaviour, even though we may be fully aware that it is not a complete explanation. However, while it is often the case that functionally related groups of behaviour – for example, those related to feeding or sexual behaviour – rise and fall together, the reason for the changes in the level of causal factors for these behaviours may not relate in any simple way to the achievement of a goal or to the maintenance of a homeo-static state. In any case, when we look for explanations of behavioural sequences on a somewhat larger time scale, involving interactions between different motivational systems, we need to invoke some different principles.

Competition

Throughout this book, we have stressed the idea that animals are often stimu-lated to do different things simultaneously and are thus in a state of conflict about what to do. Their different motivational systems compete for control of what McFarland and Sibly (1975) called the behavioural final common path, an extension of Sherrington's idea that rival reflex arcs compete for control over the muscles and, therefore, what the body does. Given this idea of constant stimulation to do many different things, an obvious way to look at sequences of behaviour is in terms of changes in which motivational system has control over the behavioural final common path at any one time. We can expect that this will shift over time as animals are exposed to different hazards, use up their food supplies, and so on. One way in which sequences of behaviour could result from such shifting motivational priorities would be if an animal always started by doing the behaviour for which it had the highest level of causal factors (the highest motivation) and then, when the causal factors for that behaviour declined, to switch to the behaviour with the next highest motivation, and so on. The sequence of different behaviours would thus come about by competition between the different motivational systems (McFarland 1971). For example, suppose an animal was very hungry but only moderately thirsty. It would start by feeding but then as it took in food and became satiated, its motivation to feed would decline until it dropped below the level of causal factors for drinking, at which point it would change its behaviour and start drinking. This might be one way of explaining the sequence of feeding and drinking in doves we discussed in Chapter 1 (Fig. 1.12, p. 29).

Sometimes, we have direct evidence for animals changing their behaviour as a result of competition between different motivational systems. For example, in the smooth newt (*Triturus vulgaris*) courtship takes place under water and there is a sequence of three distinct phases that end with the male depositing a spermatophore and the female picking this up with her cloaca (Fig. 4.11). During the first phase, the male orientates himself in front of the female. The female frequently turns away at this point, but if she remains stationary, he then displays to her with a series of vigorous tail movements, which is called a static display because he remains in one place. Eventually, the female begins to move towards the male and he retreats from her. His transition from a static display to a retreat display appears to be through competition as the time when he does it depends entirely on the female, her response providing a sudden increase of the level of causal factors for the retreat display (Halliday & Sweatman 1976).

Figure 4.11 The courtship sequence of the smooth newt (*Triturus vulgaris*). The male is shown in black and the female in white. The male begins by approaching the female and following her in an orientation phase. During the static display, he fans and whips his tail at her. This leads to the retreat display phase in which the male backs, still whipping his tail and only reverting to static display if she fails to follow. The retreat phase is preliminary to the male, now moving forwards with the female following, depositing a spermatophore. The final phase of spermatophore transfer functions to ensure that the female, in following the male, is so positioned that her cloaca touches and takes in the spermatophore. If this crucial phase fails, the male may switch back to a repetition of the retreat display. Further details are given by Halliday (1975) from which this diagram is taken.

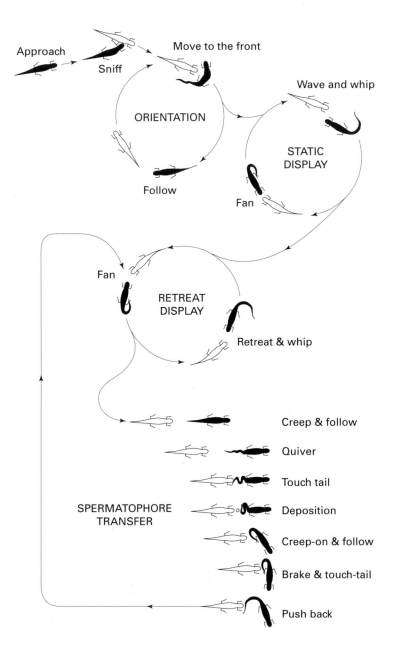

A very dramatic example of competition between motivational systems comes from the incubation behaviour of jungle fowl. Jungle fowl hens undergo a major change in behaviour before and after incubation starts. Normally, the hens will spend some 60–70% of the daylight hours scratching in leaf litter and feeding. When they are incubating their eggs, however, they spend almost all their time on the nest, only leaving it for 5–10 minutes a day to feed and drink. As a result, they may lose up to 17% of their body weight during the 21 days it takes for the eggs to hatch (Sherry *et al.* 1980). Clearly their motivation to sit on the eggs is very strong (a broody hen does little else all day), but what has happened to the motivation to do other behaviour such as feeding? Is it still high and inhibited by the strong motivation to brood? Or has it fallen and is simply outcompeted? The evidence suggests that during incubation, motivation to feed drops dramatically. Even if broody jungle fowl hens are offered food on the nest, they refuse to eat and still lose weight. Their whole metabolism changes so that they are effectively anorexic and their motivation to incubate easily outcompetes their motivation to feed for most of the day. They are certainly not desperately hungry and inhibited from feeding by the motivation to incubate.

We can then ask, what causes the change from incubation to the few minutes of feeding that occurs every day. What makes the hen get off the nest and feed? It seems to be a shift in the competition between feeding and incubating. If hens are deprived of food, thus increasing the level of causal factors for feeding, they leave the nest to feed earlier than usual, stay away longer and eat more. The timing of the change from incubating to feeding thus seems to be determined by the level of the causal factors for the 'second-in-priority' behaviour of feeding (Hogan 1989). However, the causal factors for incubation also seem to fluctuate during the day. An incubating hen is much more likely to leave her eggs and eat at the beginning or middle of the day than she is in the afternoon, even though her state of hunger presumably increases throughout the day. One hen in Hogan's study chose to go without food for 5 days running when food was only available after 1700 h, but the same hen ate normally when food was available in the morning. These results suggest competition between incubation and feeding. The causal factors for incubation fluctuate during the day and are lowest in the late morning, whilst the causal factors for feeding rise steadily throughout the day. The change from incubation to feeding occurs when feeding temporarily wins the competition.

While the shifting balance of competition between motivational systems undoubtedly accounts for some sequences of behaviour, competition is unlikely to be the only mechanism involved. The trouble with 'pure' competition is that it could lead to maladaptive 'dithering' between different behaviours, as the motivation for first one and then the other gets the upper hand. It could even lead to prolonged conflict in which neither emerges as the winner and the animal is literally immobilized by its motivational conflict. Suppose an animal's food and water are located in different places some distance away from each other, so that there is a cost (in energy, time, conspicuousness to predators etc.) to changing from feeding and drinking. If there was a straight competition between the motivation to feed and the motivation to drink, the animal would dither between the two – taking a drink and then, when the motivation to drink had dropped slightly, immediately running to feed until feeding motivation had dropped and then going back to the water, and so on. Such rapid alternation would carry a high cost of changing from one behaviour to the other, with the animal constantly exposing itself to the risk of predation and not properly satisfying either its hunger or its thirst. A much more adaptive strategy for deciding what to do when there is a conflict of motivation is to have a degree of 'authority' with inhibition of competing responses, at least for short periods of time before the next behaviour occurs, just as with reflex organization at the spinal cord level (Chapter 1). In fact, the patterns of feeding and drinking in doves (Fig. 1.12) do indeed show a bout of one behaviour followed by a bout of the other, rather than constant dithering between the two.

Inhibition/disinhibition

Inhibition is of critical importance in animal behaviour, at all levels. In Chapter 1, we saw how the most basic movements of limbs depend on inhibition. Sherrington recognized that flexor muscles must inhibit extensors and vice versa, or their contractions would cancel each other out and no movement would occur. In Chapter 3 we saw that lateral inhibition is very important in enhancing edges and points of contrast. Inhibition at a neuronal level means that the firing of one nerve cell is actually suppressed because of the action of another. Inhibition at a behavioural level means that behaviour that would otherwise have occurred is prevented from occurring by the action of another motivational system.

As an example, we can again look at the courtship of the smooth newt (Fig. 4.11). As we have seen, newts breathe air but perform their courtship under water and this poses a particular problem for male newts, since their courtship consists of vigorous side-to-side movements of the tail which demand a lot of oxygen, while the female's role is much more passive. During courtship, the male rapidly finds himself in a motivational conflict: he is stimulated both to go up to the surface to breathe and to continue with a sequence of behaviour that may lead him to a successful fertilization. Resolving such a conflict by competition would be fatal to his reproductive chances since if he if he took time off to breathe when the causal factors for breathing reached the level that normally stimulated him to go to the surface, the female might lose interest or even pick up the spermatophore of another male. Temporarily, therefore, breathing is inhibited by courtship. The male stays near the female despite rapidly depleting oxygen levels. Halliday and Sweatman (1976) showed how powerful this inhibition can be by altering the behaviour of a female in such a way that courtship was artificially prolonged. They encased a female newt in a 'straight-jacket' (Fig. 4.12). This immobilized female provided enough stimulation for the male to go through the initial parts of his courtship display, but since the female could not move towards him, he did not receive the stimuli necessary to move on to the next stages. He was thus trapped, repeatedly displaying to a female who did not respond.

Figure 4.12 Female smooth newt in a 'strait jacket'. By controlling the behaviour of the female, the male's courtship can be prolonged. (Drawing by Tim Halliday.)

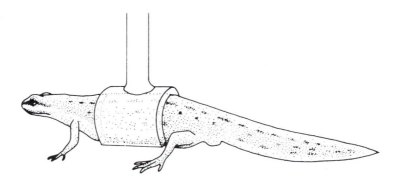

Under these circumstances, the male held his breath for much longer than he would normally do under non-sexual (i.e. no female present) circumstances. Only when the female was removed would he go up for air. Sex inhibited breathing very effectively and breathing occurred only when 'disinhibited' by the removal of the sexual stimulus.

It will be obvious that this cannot continue indefinitely. Clearly, the need to take in air imposes an upper limit on the amount of time that a male can devote to sexual behaviour and so eventually the inhibition of breathing by sexual behaviour breaks down, and he goes up to breathe, leaving the female behind. The sequence of courtship and breathing is thus a complex mixture of inhibition and disinhibition of breathing by sexual behaviour but with an element of competition if the need for oxygen becomes too great.

These two basic elements of interaction between motivational systems – competition and inhibition – are thus used by different animals, and even by the same animal in different circumstances, to give adaptive outcomes to the motivational conflicts that occur when they are stimulated to perform more than one behaviour at the same time. A resolution of motivational conflicts based entirely on competition would result in dithering or even in complete immobility. A resolution based entirely on inhibition would result in an animal never doing second-in-priority behaviour until top priority behaviour had no causal factors left. A male newt that never left a female however long courtship took would die of lack of oxygen. An adaptive mixture of shifting motivational priorities with a degree of 'authority' between them is what we find in most animals most of the time.

The feeding behaviour of 5th-instar locust nymphs (*Schistocerca gregaria*) gives a clear idea of what this authority can look like in practice. These rapidly growing nymphs have a need for both carbohydrate and protein in their diets but they cannot always find exactly what they need in one single food. If the carbohydrate-rich and the protein-rich food are separated in space, the insects are able to balance their diets remarkably accurately by eating both in the correct proportions (Simpson & Raubenheimer 1993). The sequence with which they choose one food and then the other indicates that their behaviour is governed neither by pure competition nor by pure inhibition. Figure 4.13 shows what the locusts actually do, plotted as a graph of their intake of protein against their intake of carbohydrate. A competition-only mechanism would make them 'dither' and run backwards and forwards from one food dish to the other, leading to a shallow 'staircase' effect. An inhibition-only mechanism would lead them to completely satisfy their requirements for one nutrient before switching to the other, giving one big step. As can be seen, most insects fell somewhere in between but generally managed to balance their nutritional requirements very evenly.

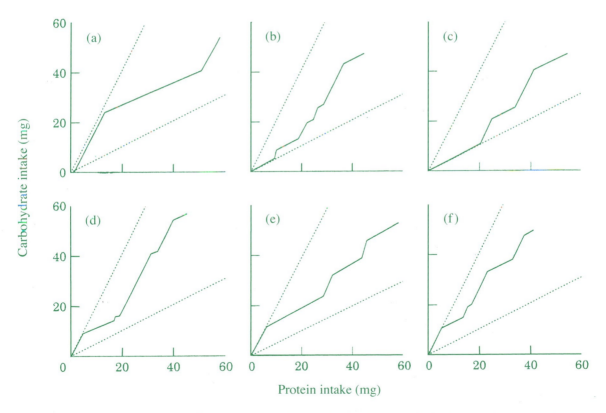

Figure 4.13 Cumulative intake of carbohydrate (vertical axis) and protein (horizontal axis) by six individual locusts given a free choice of both types of food. The dotted lines represent the intake of an animal which eats only carbohydrate (upper dotted line) or only protein (lower line). The solid line shows what each insect actually ate, showing that they fed in 'bouts' of one food or the other before switching over. (From Simpson & Raubenheimer 1993.)

Physiological explanations of sequences
The neuroethology of sequences

We can now begin to see how these ideas of feedback, competition and inhibition that apply to decision-making at a behavioural level have their counterparts in neurophysiological explanations of behaviour too. One particularly good example where we can go so far as to locate the cells responsible for decision-making is the sequence of movements made by an escaping crayfish (Fig. 4.14). When a crayfish is touched, it has a sudden and dramatic way of escaping: it flexes its abdomen so that the body is drawn into a tight curve. This has the effect of propelling the crayfish rapidly backwards. This behaviour is brought about initially by the activation of receptor cells that are sensitive to high frequency water movements and to touch. These are in turn connected to Sensory Interneurons (SIs) which in turn connect to a Lateral Giant Interneuron (LGI) (Fig. 4.14). There are over 1000 Tactile Sensory Cells and over 20 Sensory Interneurons. The Lateral Giant Interneuron is a very distinctive nerve cell and is unusually large – about 100 μm in diameter. There is one LGI per abdominal segment and each one connects to the LGI in the next segment in front, all the way up the nerve cord. In each segment, there is also a very large motor neuron (Motor Giant or MoG) which inner-

Figure 4.14 Diagrammatic representation of the ventral view (dissected) of the abdominal nervous system of the crayfish showing the excitatory pathway from receptor cells to the flexor muscles. The cells are not drawn to scale and only one segment of the lateral giant is included. (Modified after Wine & Krasne 1982.)

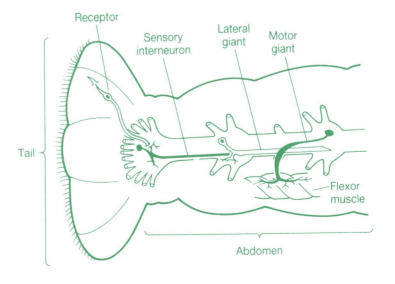

vates a Fast Flexor Muscle (FFM). The contraction by FFMs of all the segments together gives rise to the escape movement of the whole crayfish.

The LGIs in each segment obviously have a key role in making the escape response. Many years ago, Wiersma showed that electrical stimulation of any single LGI gives rise to a fast abdominal flexion very similar to the complete escape response of the whole animal. But without being prompted by the LGIs, the motor nerves do not stimulate the muscles to contract.

It also seems that it is the LGIs which make the decision whether or not to show the escape behaviour at all. Olson and Krasne (1981) showed that when shocks of low voltage were applied to the sensory nerve, these were not large enough to stimulate the LGI to produce an action potential (see Chapter 1). No escape behaviour occurred. Then, as the voltage applied to this nerve was increased, the excitation of the LGI grew until suddenly it 'decided' to respond with an action potential and a full abdominal flexion occurred. An all-or-nothing response by the LGI (strictly, by any LGI in any of the abdominal segments) decides whether the crayfish shows the escape response or not. In this sense, we can see that with a rising level of causal factors (here identified as a voltage), the probability that the response will occur suddenly rises as a threshold is exceeded and the escape behaviour outcompetes everything else.

But we can also see the importance of inhibition. A single action potential in an LGI also inhibits other behaviour that might interfere with escape. In the crayfish, the circuits that extend the abdomen are inhibited when the escape response (which of course involves flexion) is in progress. A single action potential in an LGI is enough to bring about this inhibition. This evokes Inhibitory Post Synaptic Potentials (IPSP) in the motor neurons of the extension and prevents any of the muscles acting. It also inhibits slow postural movements (both flexion and extension) which would also get in the way of rapid escape and which have their own slow extensor and flexor muscles. Each slow flexor and extensor muscle has five excitatory motor neurons and one inhibitory one. The LGI uses the single inhibitory neuron to suppress postural movements and make sure that the crayfish's whole body gives top priority to escaping as rapidly as possible.

In the giant sea slug *Pleurobranchaea*, we find another dramatic example of inhibition, this time mediated not through neurons but through a hormone.

When the sea slug is laying its eggs, feeding behaviour is strongly inhibited by egg-laying. *Pleurobranchaea* is a carnivore and normally eats whatever animal matter it can find, including other sea-slugs and their eggs. Inhibition of feeding during egg-laying prevents it from eating its own eggs. The inhibition is brought about by a hormone released when the animal lays eggs. The hormone works directly on the buccal ganglia that control the muscles of the mouth and simply inhibits movements of the mouth (Davis *et al.* 1977). Even egg-laying, important though it is, can in turn be inhibited by escape. The sea slug's motivational priorities are thus clear and understandable: escape is given top priority because of the serious and immediate consequences of not getting away from immediate danger but second comes egg-laying because it is not advantageous for a sea slug to eat its own eggs (Fig. 4.15).

Even sea slugs, however, have a greater degree of flexibility in their behaviour than might appear from this 'inhibition-only' priority system of escape and egg-laying. When the oral veil of the sea slug is touched, the animal usually withdraws defensively. If the animal is feeding, however, this withdrawal response is not shown and the animal continues feeding. This occurs because neurons in the buccal ganglia that are active during feeding inhibit output to the oral veil (Davis *et al.* 1977). While this looks like a straightforward case of inhibition, the inhibitory action can be two-way. If the animal is satiated with food, the inhibition of withdrawal normally brought about by the presence of food does not happen. When its head is touched, the animal now does withdraw, but this time gives priority to defence over feeding. The low level of causal factors for feeding in a satiated animal means that withdrawal outcompetes feeding, but inhibition still characterizes the interaction between the two systems.

As well as competition and inhibition, we also find that the concepts of feedback and homeostasis are used as much at the physiological level as at the behavioural level of explanation. This is shown particularly well through the study of one area of the vertebrate brain that has proved to be of major importance in motivation, the hypothalamus. The hypothalamus is a relatively minute piece of brain tissue – in the human brain it is smaller than the last joint of the little finger – but is of primary importance in a whole host of reactions. Its position in the brain is shown in Fig. 4.9, where it can be seen that it has a very intimate connection with the pituitary gland. This key endocrine gland

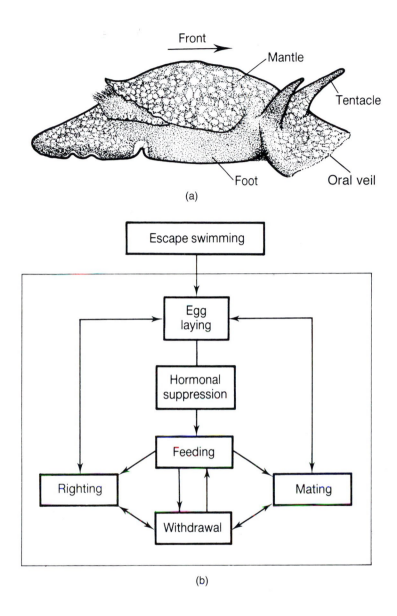

Figure 4.15 (a) The carnivorous sea slug *Pleurobranchaea*. (b) The behavioural hierarchy of *Pleurobranchaea*.

controls the whole hormonal system of the body and is connected to the hypothalamus by the pituitary stalk, which is rich in both nerves and blood vessels.

From its connections with other parts of the brain, its rich blood supply and its links with the pituitary gland, the hypothalamus is well adapted both to measure changes in the metabolism of the body and to set in motion activities which will rectify them. It is, in other words, well suited to serve as part of the homeostatic control systems we have already discussed. We saw (p. 214) that the hypothalamus has 'osmoreceptors', cells that detect water loss by responding to increased concentration of circulating body fluids. They can set in action two compensating systems. The first, acting via links with the pituitary

227

gland, causes the secretion of anti-diuretic hormone (ADH) which increases the resorption of water by the kidneys. The second system, as we have already seen, causes the animal to seek out and drink water. If the link between the hypothalamus and the posterior lobe of the pituitary gland is damaged, ADH may never be secreted. In such a situation, the kidney continues to excrete copious quantities of urine and, to compensate, the animal drinks large quantities of water – a condition known as *diabetes insipidus.*

Another homeostatic system that involves the hypothalamus is the control of body temperature. There are areas of the hypothalamus which are highly sensitive to changes in the temperature of the blood. If these areas are heated artificially by implanted wires, a rat will start sweating and panting. Sweating is controlled peripherally by the autonomic nervous system and the hypothalamus can initiate its activity. The reverse effect is produced when temperature-sensitive areas are cooled; now the animal shivers, another autonomic response. If a rat is cooled in this way for an even longer time, shivering alone is inadequate and, if given the right material, the rat begins to build a nest, or enlarge the one it has, in order to insulate itself. Both the reflex response of shivering and the more complex behavioural one of nest-building are initiated by the hypothalamus.

The hypothalamus is also implicated in the control of feeding. Mammals and birds normally keep their weight very constant and adjust the amount of food they eat accordingly. Rats with damage to the central areas of the hypothalamus (the ventromedial nucleus) were known to lose this sensitive control of their eating and develop the hyperphagia we mentioned earlier. Figure 4.16 shows the food intake of rats whose ventromedial nucleus has been damaged compared with that of sham-operated control animals. After a few days' post-operative depression of food intake, the brain-damaged rat begins to eat huge amounts of food – at least four times the normal amount. This so-called 'dynamic phase' of hyperphagia lasts for 3 weeks or so. Beyond this, food intake declines and eventually settles down, with the animal eating about double the normal amount. Needless to say, hyperphagic rats become grossly fat and are very inactive. By contrast, rats with damage to a different part of the hypothalamus, the lateral area, show completely opposite symptoms. Initially they stop eating and drinking altogether and would die unless some intervention were made to stop them. Eventually these rats would start taking in food and water again and regain normal body weight, even though they failed to respond appropriately to food or water deprivation – that is, they failed to eat or drink enough to make up for their deficit (Teitelbaum & Epstein 1962).

It used to be thought that these results together consituted evidence for two feeding 'centres' in the hypothalamus, the lateral hypothalamus being the hunger centre and the ventral hypothalamus being the satiety centre. However, more recent work has shown that the role of the hypothalamus is more complex than this. New techniques enable damage to the brain to be monitored much more precisely, and these have shown that the failure to eat and drink that had been thought to be characteristic of rats with damage to the lateral hypothalamus could be induced by damaging only the fibres containing the neurotransmitter dopamine that pass through the lateral hypothalamus. Such rats would stop feeding and drinking even though there had been no damage to the lateral hypothalamus itself. Indeed, damaging the lateral hypothalamus without affecting these fibres resulted in rats that fed and drank normally (Winn *et al.* 1984). However, such rats did behave oddly in some respects: they did not drink as much as a normal rat when injected with saline or with angiotensin, a substance that is formed naturally in the body during dehydration. In normal rats, both saline and angiotensin make them start drinking. Similarly, the rats with damage to the lateral hypothalamus did not start eating when injected with 2-deoxyglucose, which normal rats do. This suggested that, although the hypothalamus was clearly important to

Figure 4.16 The daily intake of normal rats and those with bilateral lesions in the ventromedial nucleus of the hypothalamus. Further explanation is given in text. (From Teitelbaum 1955.)

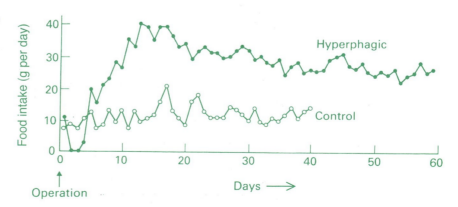

feeding and drinking behaviour, it might not have the major control function that would justify it being called a feeding or drinking 'centre'. Rather, several different brain regions, including the frontal cortex, would be involved in the control of ingestion (Winn 1995). This in turn coincides well with what we have already discussed at a behavioural level. Animals do not eat and drink only in response to food and water deprivation. On the contrary, it is remarkable how many different factors, ranging from the presence of predators and members of their own species to the palatability of the food itself and even their expectation of where it is to be found, affect what they do at any one time. The complexity of decision-making that we observe in the natural lives of so many animals inevitably means that input from many different parts of the brain will be involved as the animal 'decides' what it will do next.

Hormones and sequences of behaviour

Hormones form a chemical message system within the body which is probably as old as the nervous system itself. The endocrine organs, which produce the hormones, and the nervous system share the common functions of communication and coordination both within the animal and between it and the outside world.

The functions of the two communication systems – endocrine and nervous – remain essentially complementary. The nervous system can pass information from one part of the body to another very rapidly, but it operates on a time scale from milliseconds to minutes and is not particularly well suited to transmitting steady, unchanging messages over long periods of time. The endocrine system cannot respond so rapidly, but its cells can maintain a prolonged steady secretion into the bloodstream, lasting for months if necessary. Moreover, hormones can reach every cell in the body via the bloodstream, whereas the nervous system generally controls only the muscles.

Circulating hormones are one of the major components of the 'motivation' of behaviour: we have already seen how feeding behaviour is suppressed by egg-laying 'motivation' through the mediation of a hormone in *Pleurobranchaea*. In vertebrates, too, we have direct evidence of hormones being one of the major causal factors leading to the occurrence of behaviour. Sometimes hormones apparently 'force' behaviour from an animal even under the most inappropriate circumstances. For example, injections of the pituitary hormone prolactin will cause isolated male cichlid fish to perform the parental fanning behaviour normally associated with keeping eggs aerated in the nest (Blüm & Fiedler 1965). Injected males will begin to fan in a bare tank devoid of all the normal external stimuli for fanning – CO_2 in the water, fertilized eggs, nest, etc. – and the amount of time they spend fanning is dependent on the dose of prolactin they have received.

Such central action is dramatic, but hormones are highly potent substances which can affect 'target organs' in other parts of the body and thus affect behaviour indirectly. For example, they may affect the responsiveness of a sense organ. Under the influence of oestrogens (sex hormones produced by the ovaries), the sensory field of the perineal nerve supplying the genital region of the female rat enlarges and is maximal at the time of oestrus (Komisaruk *et al.* 1972). This means that at the peak of oestrus she is more easily stimulated by the thrusts of the male's penis and can orientate her body so as to help intromission.

Correspondingly, testosterone (the sex hormone secreted by the testes) causes changes in males such as thinning of the skin covering the glans penis in rats. This means that the tactile sense organs are more exposed and, under the influence of the hormone, the male becomes more sensitive to the movements of the female during intromission. In male birds, testosterone affects a wide variety of behaviours including aggressiveness, rate of singing, and even incubation and feeding of the young (Ketterson & Nolan 1994). In some female birds, oestrogen causes shedding of ventral feathers and increased vascularization of the skin on the lower breast to form a brood patch. Hinde and Steel (1966) found that a female canary is more responsive to stimuli from the nest cup after her brood patch has developed and that this affects behaviour during nest-building and incubation.

From the behavioural viewpoint, however, the most important target organs for hormones are regions of the central nervous system itself. Here we often find an elaborate interaction between the way hormones act on the brain and the way the brain's response to an environmental stimulus can, in turn, affect hormones by altering the secretory activity of the endocrine organs. One of the best examples of this interaction comes from studies of reproductive behaviour in the ring dove (Fig. 4.17) initiated by Lehrman and his co-workers (Lehrman 1964). It was known for a long time that female pigeons and doves do not normally lay eggs if kept alone or with other females, but begin to lay soon after a male is introduced. Simple tests show that the sight of a male is enough to cause ovulation, provided he is able to court the female. Thus a male which performs the typical courtship bows, even though separated behind a glass screen, is much more effective than a castrated and therefore non-courting male in the same cage.

Under the warm and well-lit conditions of a laboratory colony, a male dove is usually ready to court the moment he is put with a female, which presumably means that his testes are active and secreting hormones. Visual stimuli from his courtship activate the female's hypothalamic centres which control secretion by the pituitary gland of two hormones: follicle stimulation hormone (FSH) and luteinizing hormone (LH). These stimulate the growth of her ovary which in turn secretes the female hormone oestrogen, under whose influence her reproductive tract begins to grow. After a day or so, both birds select a nest site and begin to build. During nest-building, courtship continues and the birds copulate. When the nest is complete, the female becomes increasingly attached to it and she soon lays her first egg. This follows the release of LH from her pituitary, and for a few days before this her

Figure 4.17 A pair of ring doves. Male and female look very alike and are most easily distinguished by their behaviour. The male is perched on the edge of the dish provided for nesting. He will display there to attract the female, and they will go on to begin nest-building together. (Drawing by Melissa Bateson.)

ovary has been secreting another hormone, progesterone, which induces incubation behaviour. The male, too, secretes progesterone which acts antagonistically to testosterone so that his courtship and aggression die away and he becomes willing to incubate eggs. The birds take turns to sit on the eggs and after a few days' incubation, the cells lining their crops begin to proliferate and slough off, eventually producing 'crop milk' for the young birds. In both sexes, the growth of the crop is controlled by yet another hormone from the pituitary, prolactin. Prolactin not only leads to crop growth but also inhibits the secretion of FSH and LH, which in turn shuts off the secretion of sex hormones. As a result, sexual behaviour between the pair dies down as they incubate the eggs and feed the squabs. After 10–12 days, the young birds leave the nest, by which time parental feeding and prolactin secretion have begun to decline. As prolactin levels in the blood fall so FSH and LH are once again secreted and the male begins to court the female again, ready for the next cycle of reproduction. Both visual and auditory stimulation are important for the female's hormonal response (Barfield 1971), but physical contact is not essential. Castrated males are less stimulating for females unless they are given an injection of testosterone and it seems to be the male's display that has the stimulating effect (Erickson 1985).

Some of the experiments involved in working out this sequence are quite dramatic. Thus the secretion of prolactin normally follows when the doves have been sitting on the eggs for a few days, but the male dove's crop develops even if he does not actually incubate himself – provided he can see his mate incubating! Here, then, a specific visual stimulus leads to the secretion of a specific hormone, but it will only do so if the male is in a particular physiological state. He must previously have participated in nest-building, because if he is separated from his mate earlier in the cycle his crop does not develop, even if he can see her incubating.

These direct effects of hormones can be demonstrated by implanting specific hormones into different parts of the brain. Controlled amounts of hormone can be placed in particular sites through a hollow needle without interfering with an animal's freedom to behave. Hutchinson (1976) and Komisaruk (1967) showed that castrated male doves would show courtship behaviour to females and aggression to other males if testosterone were implanted in the anterior hypothalamus and the preoptic nucleus (so called because it lies just anterior to where the optic chiasma enters the brain). Progesterone implanted in the same places suppresses these behaviour patterns and causes an increased tendency to incubate – a direct demonstration of antagonism between progesterone and testosterone.

Harris and Michael (1964) were among the first to use similar hormone implantation techniques in mammals and studied on the effect of oestrogens on the sexual behaviour of cats. They showed that minute amounts of oestrogen injected into certain parts of the hypothalamus would induce sexual behaviour. Even castrated cats, whose reproductive system remains completely undeveloped, would show full oestrous behaviour. Normally castrated female cats are completely unreceptive to males and will lash out viciously if a male gets too close. Implanted castrated cats will elevate the rump, deflect the tail to one side and make treading movements with the hind legs. They will assume this posture as soon as the male approaches and will submit to being mounted.

The relationship between behaviour, hormones and the nervous system can be very intricate. One dramatic and, at first sight, rather puzzling example is the 'Bruce Effect', named after Hilda Bruce (see Bronson 1979), which we discussed in the last chapter (p. 133). As we saw, pregnant female mice abort their litters and reabsorb the embryos if a strange male mouse (not the father of the litter) comes into contact with them. This happens even if pregnant females are put into cages with material soiled by a strange male and the females merely smell the male without any direct contact with him at all. The pregnancy-blocking effect is brought about by an inhibition of prolactin secretion and consequent reduction in progesterone when the female comes into contact with the smell of a strange male mouse. Females with olfactory bulbs removed remain pregnant even when exposed to strange males.

There are also close ties between hormones and the immune system, some of which are quite unexpected. Testosterone, important though it is for male behaviour and appearance, also has some costly side-effects, the most serious of which is to make the immune system less able to fight infection (Grossman 1985). This immuno-suppressive effect of testosterone appears to be widespread among vertebrates and probably accounts for the fact that sexually mature males are particularly vulnerable to infections from parasites and diseases. The resplendent male, displaying vigorously to females is, paradoxically, least likely to be able to fight infection. This may explain why, when choosing a mate, females appear to pay particular attention to how healthy and disease-free a male is (Zuk 1994).

Measuring motivation

As we have now seen, considerable progress has been made in relating changes in an animal's behaviour to changes going on inside its body, for example, changing levels of hormones. For some invertebrates, we can even relate changes in behaviour to changes in particular neurons or groups of neurons. But giving a complete account of the 'motivation' of a vertebrate, i.e. trying to describe everything that is going on inside its body in as much detail as we can for a crayfish, is still impossible. The nervous systems of vertebrates contain enormous numbers of neurons – to be counted in tens or hundreds of millions. There are many alternative pathways and much parallel processing. Consequently, for the foreseeable future, we will continue to refer to an animal's motivational state to cover our ignorance of the exact details of what is going on. To be useful, though, we must be able to measure quantitatively the level of such motivation. Otherwise, we are in danger of talking about levels of causal factors without having any independent measure of how high or low these are. The whole idea of an animal being 'highly motivated' would thus become at best speculative and at worst circular. But how can we measure how strongly motivated an animal is, when all we have available is what it is doing combined with a few, possibly incomplete physiological correlates of what it is doing? There are two reasons why finding a satisfactory answer to this question is important. The first is that many of the motivational ideas we have discussed so far explain decision-making in animals in terms of the interactions between two or more motivational systems. Unless we have some quantitative idea of the strength of an animal's motivation to perform different behaviours, such models will turn out to be little more than redescriptions of what is observed. The second reason is that measuring the strength of an animal's motivation has practical implications for assessing its welfare. If an animal is kept in a bare cage in which there are no opportunities for it to perform much of its natural behaviour, it is important to know whether the animal is strongly motivated to do something it cannot do or whether the lack of opportunity means that motivation is low too ('out of sight is out of mind'). An animal that is strongly motivated to do something it cannot do is much more likely to suffer than one that is not motivated at all. Unfortunately, different measures of motivation do not always give the same answer, but the following have proved useful and provide us with a starting point.

Amount of behaviour performed

Perhaps the most straightforward way of measuring an animal's motivation is to give it the opportunity to perform a response and then see how much or for how long the behaviour is performed. In the case of feeding behaviour, we might measure the amount of food eaten, and it is usually easier to weigh the amount of food eaten than to count the number of feeding movements. Drinking motivation can be measured by the amount of fluid drunk, and even sexual and aggressive behaviour can be measured with a little ingenuity. Sevenster (1961) put an 'object' stickleback into a glass tube and counted the number of bites directed at the tube by a test fish as a measure of its aggression. Similarly, the number of 'zig-zag' courtship movements directed by a male towards a tube containing a female was used as the measure of sexual motivation. Vestergaard (1980) measured the motivation of domestic hens to dustbathe by how long they dustbathed when they were moved from wire floors (where they could not dustbathe) to a litter floor (where they could). Interestingly, he found an upsurge in the amount of dustbathing the hens showed under these circumstances, suggesting that their motivation to dustbathe rose during the period of deprivation.

In fact, we commonly find that when complex behaviour, such as courtship, has not been elicited for some time, it has a lowered threshold and is performed at a high intensity when it is at last elicited. This phenomenon is sometimes called the 'rebound effect', implying a parallel with the 'reflex rebound' described by Sherrington at the level of the reflex arc; as we discussed in Chapter 1, when inhibition is removed from a reflex, it often returns at a higher intensity than before – in other words, it 'rebounds'. It is possible that the motivation for courtship or dustbathing is inhibited by other activities (p. 220) or by the absence of usual stimuli, and shows something akin to reflex rebound when the inhibition is removed or the right stimulus is provided.

How aversive a stimulus can be made before it is avoided

The aim here is to attempt to prevent the animal from performing a behaviour and see how far it will persist in spite of this. For example, quinine is an intensely bitter substance to us, and other mammals seem to find it equally so. If quinine is put into food pellets or drops of condensed milk at increasing concentrations, a rat will eventually reject usually attractive food as too bitter. The concentration of quinine it will tolerate can be used as a measure of feeding or drinking motivation.

An alternative version of this method is to place a stimulus, such as food, in full view of an animal and then contrive that, in order to reach it, the animal has to overcome some obstacles or 'run the gauntlet' of something it would normally avoid, such as an electric shock or a blast of air. By varying the intensity of the shock and finding how much an animal will accept in order to reach the food, we have a measure of motivation. Female rats will cross an electrified grid to reach a male. The strength of shock needed to stop them reveals cyclical changes corresponding to their oestrous cycle; it is highest during their 'oestrous' or 'heat'. As a variation on this theme, Duncan and Kite (1987) gave cockerels access to hens, but only if they pushed through doors with weights. The weight of the door that a bird was prepared to push was used as a measure of its motivation.

Rate of bar-pressing or key-pecking

The 'Skinner Box' is a useful piece of apparatus for studying both learning and motivation (see Chapter 5). An animal is put into the box when hungry and it is 'taught' that it will receive a small food or water reward when it presses a bar that protrudes into the box or, in the case of a bird, when it pecks at a key. The apparatus is so arranged that rewards do not follow every bar-press or key-peck but come at irregular intervals, averaging out at, say, one reward every 30 seconds. In psychological jargon, this is called a 'variable interval reinforcement schedule' and it means that the animal never knows whether or not a given bar press will give it food. Somewhat surprisingly, perhaps, animals will often press their bar (or peck their key) much more consistently when they get their rewards only irregularly than if a reward comes after every response.

The rate at which the animal presses the bar under a variable interval schedule is so predictable that it can be used as a measure of motivation. The rate at which water-deprived rats will bar-press under these circumstances is reliably related to the length of time for which they have been deprived of water. In some cases, an animal's motivation appears so strong that it will show 'compensation' or 'resilience': thus, if the number of bar-presses required for a given amount of reward is gradually increased, hungry rats may be prepared to work harder and harder to obtain it. Hogan (1967) showed that Siamese fighting fish can be taught in an analogous way, to swim through a loop to obtain a food reward, and they will similarly show 'compensation' if they have to swim the loop many times to get the same amount of food. However, aggressive behaviour does not show the same effect. Although fish will readily learn to swim through a loop for the reward of gaining access to a rival male fighting fish which they then display to, they will not show compensation. They stop swimming through the loop when the number of responses they have to make for each sight of the rival becomes too great.

Vacuum activities

When very highly motivated, animals sometimes carry out behaviour even when the appropriate stimuli are not present. Lorenz (quoted by Tinbergen 1951) describes the case of a starling which went through all the movements of catching and eating an insect even though no insect was present – so the behaviour was performed 'in a vacuum'. Hens kept in wire-floored cages sometimes go through all the movements of dustbathing, even though there is nothing for them to dustbathe in. Vestergaard (1980) calls this 'vacuum' dustbathing and suggests that when behaviour is performed in the absence of suitable stimuli, or at least with very minimal stimuli, the animal is very highly motivated to perform the 'frustrated' behaviour.

These, then are some of the ways in which levels of motivation have been measured. Superficially, they all appear to be measuring the same thing and we would expect that when, say, an animal is deprived of food, all these measures would increase in the same way. However, Miller (1957) describes a number of experiments which show that at least three of the measures of feeding motivation – amount eaten, quinine accepted and rate of bar-pressing – do not all rise together. Over the range of 0–54 hours of food deprivation, the amount of quinine that rats will accept steadily rises and so does their rate of bar-pressing, but their food intake reaches a maximum after only 30 hours and actually declines slightly thereafter. Thus a rat appears to be becoming 'hungrier' in that it will accept food that is more and more bitter and yet it eats less. There are also some situations in which rats will eat abnormally large amounts of food (hyperphagia) and yet do not show any other signs of hunger such as being prepared to overcome a barrier or press a bar for food, or eat quinine-treated food. Hyperphagia can be produced by certain brain lesions and it would seem that what we commonly lump together under the term 'hunger' may in fact be a combination of factors; brain damage elevates some of these factors and depresses others.

There is evidence for a similar multiplicity of factors under the heading of 'thirst'. Very much in parallel with Miller's experiments on feeding mentioned above, experiments by Choy (quoted by Miller 1956) show that different measures of drinking motivation do not correspond with each other very well, either (see Fig. 4.18). Choy had rats with tubes implanted directly into their stomachs so that they could be given liquids without having to drink. He measured drinking in rats previously satiated with water, after 5 ml of concentrated salt solution was put into their stomachs, and he recorded three measures of thirst over time. Up to 15 minutes after giving salt, bar-pressing does not increase, even though the amount of water drunk does. Yet the amount of water drunk levels off 3 hours after the salt is given, even though bar-pressing continues to rise and so does tolerance of quinine in the water. It is obvious that we need a combination of measures to get a reasonable picture of the effect of salt on drinking motivation, which is clearly complex and not a simple, single entity. 'Motivation', as we have seen, covers a range of causal factors inside the animal (hormones, activation of specific parts of the nervous system, etc.) and so it is hardly surprising that its component parts do not always change exactly in step.

Prolonged conflict and stress

We have so far emphasized the idea that animals have decision-making mechanisms that enable them to resolve their motivational conflicts fairly rapidly and choose the option that will optimize their chances of reproductive success. But sometimes the decision-making mechanism fails to give a clear behavioural result for considerable periods of time, either because an animal is physically prevented from carrying out the behaviour it has 'decided' to do, or because the animal remains locked in a motivational conflict from which it is unable to escape.

There are two reasons why animals may fail to resolve their motivational conflicts immediately. The first is that they may need time to collect more information about the various options open to them, so it is highly adaptive for them to delay making a decision until they have had a chance to collect more information.

Conflict – even prolonged conflict – is the natural state for wild animals for much of their lives. While sometimes decisions have to be made at once (for example, delaying escape from an approaching predator would be strongly selected against), there are many other occasions when it is much

Figure 4.18 How three different measures of thirst change in the period following the placing of 5 ml of strong salt solution directly into the stomach of a previously water-satiated rat. The units on the y axis are arbitrary; they are simply the difference between control and experimental rats on the various measures of thirst. (From Miller 1956.)

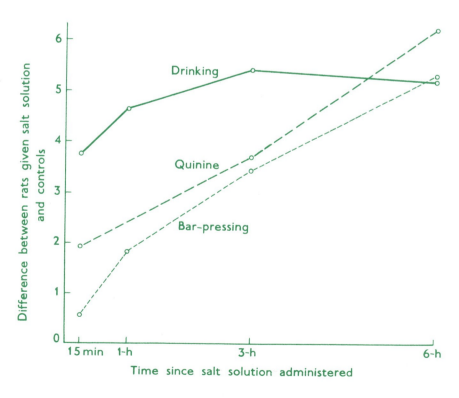

better to wait and collect more information before a decision is made. This is particularly true in social situations where animals may take considerable time to 'assess' each other (Chapter 3). Red deer stags, for example, may take half an hour or more to assess each other's fighting ability before deciding whether to attack or flee (Clutton-Brock & Albon 1979). The roaring matches and parallel walking are used by the stags as indicators of the likely outcome of a real fight and mean that the animals remain in a state of continuous motivational conflict while they gather the relevant information. Even escape from predators may be preceded by a period of lengthy assessment. Prey animals do not flee every time a predator appears but weigh up the risk that a predator will attack them against the loss of feeding time they would incur by moving away (Ydenberg & Dill 1986). Zebras watch hyaenas carefully and are much more likely to flee on some days than others, apparently sensitive to small differences in the hyaenas' behaviour and the size of the hunting group that indicates what prey the hyaenas are after on any one occasion (Kruuk 1972). Such assessments are highly adaptive but take time to make. The motivational conflict is resolved only when, eventually, the new information received tips the balance of causal factors in the direction of one behaviour or another.

The second reason for prolonged motivational conflict, or for animals remaining in a state of high motivation for an appeciable length of time, is that it is for some reason prevented from carrying out the behaviour that it has 'decided' to do. For example, it may be highly stimulated to drink and all set to give top priority to drinking behaviour over anything else, only to find that its water is covered with an immovable glass lid. Such an animal is said to be **thwarted**. Alternatively, the animal may be highly motivated to drink but unable to find any water at all. Such an animal is said to be **frustrated**. Both thwarting and frustration are commonly, but not exclusively, seen where animals are kept in unnatural situations such as cages in which they are unable to perform all the behaviour in their natural repertoire. This has given rise to the view, which we will discuss in more detail in the next section, that thwarted or frustrated animals are 'stressed'. Understanding what happens when an animal remains highly motivated for long periods of time is therefore critical to the study of animal welfare and to the understanding of suffering and stress in animals. However, many of the symptoms of 'stress' that animals exhibit when they are kept in unnatural environments arise as distortions of the decision-making mechanisms animals have evolved for dealing with motivational conflict in quite natural situations. We therefore start our discussion of the effects of prolonged motivational conflict and thwarting as they are seen in natural environments.

Displacement activities

Kortlandt (1940) and Tinbergen (1951) both described a very curious category of behaviour typically observed in conflict situations, which they called **displacement activities**. The most striking common characteristic of displacement activities was their apparent irrelevance to the situation in which they occured. For example, in the middle of a fight, two cockerels will each turn aside briefly and peck at the ground, sometimes picking up stones or grains which they allow to fall again. Fighting gulls may show 'grass-pulling' in the middle of a fight (Fig. 4.19) A male stickleback which has been courting an unreceptive female will suddenly swim to his nest and perform the characteristic parental 'fanning' movements which ventilate the nest with fresh water even though there are no eggs present. A tern which is incubating eggs in its nest makes preening movements just before it takes off at the approach of an intruder. A thirsty dove, which is prevented from getting to its water bowl by a sheet of glass, may preen itself.

Figure 4.19 'Grass-pulling' in the lesser black-backed gull. A male faces his rival across the boundary between their territories. Interspersed with bouts of threat or actual fighting, they seize grass or turf and tug at it flamboyantly, deliberately choosing material that will resist their efforts. (Drawing by Niko Tinbergen.)

In all these examples, the animals seem either to be thwarted or in a conflict between two opposing motivations. The appearance of, say, feeding in the middle of a fight is surprising because it seems irrelevant to the conflict (which would seem to be between attack and escape) that the cockerels appear to be in. Similarly, when a tern is faced with the decision of whether to escape from a predator or stay and incubate its eggs, preening itself seems irrelevant.

There is direct evidence that this irrelevant behaviour is particularly likely to occur when animals are in a state of prolonged conflict and unable to decide what to do. Rowell (1961) put chaffinches in a conflict by flashing a light when they approached their food dish. The birds were hungry but avoided the light and were thus in a conflict between whether to feed or to escape. There were several different perches in the birds' aviaries, so that they could either get as far away as possible from the light by using the most distant perch or approach the food dish by using the nearest perch. Rowell found that the chaffinches did most displacement preening and bill-wiping on intermediate perches. While this suggests that displacement activities are most likely to occur at a balance point between two conflicting tendencies, it is not conclusive. Since the birds paused for longer on the intermediate perches, the same result would be obtained if preening simply occurred at a fairly constant rate, wherever a bird made a pause and was doing nothing else – the longer the pause, the more preening.

In control observations of birds under no conflict, Rowell found that this was indeed the case, but still the amount of preening per length of pause was significantly greater when the birds were in a conflict, which justifies labelling this extra amount 'displacement preening'. The exact causal basis for displacement activities is still controversial (Roper 1984), but displacement activities may not be such an unusual category as was once thought. If they occur in conflict situations and enable animals to gather a bit more information about their opponent or a potential predator, what we may describe as irrelevant may not be irrelevant at all to the animals involved. In the roaring duals of red deer, for example, the roaring and the parallel walking could both, if we did not understand the dynamics of the situation, be described as displacement activities, irrelevant to the serious business of defeating a rival. But they are both important to fighting in the sense that they enable the stags to obtain information about each others' fighting ability.

Conflict and displays

Many other displays can equally be seen as arising from unresolved motivational conflicts unresolved, that is, until the contestants have assessed each other for their relative fighting abilities and other factors we discussed in Chapter 3. These displays were initially called 'threats', although present-day workers emphasize the information about the sender that a signal conveys (status, fighting ability, etc.) rather than its threatening or intimidating qualities. Tinbergen (1952) saw many signals resulting from dual motivation, conflict between simultaneously aroused tendencies to attack and to escape, when neither can find separate expression. Several lines of evidence pointed to this conclusion about the causal mechansims of displays, long before modern ideas about game theory or assessment were conceived of as evolutionary explanations of what was going on.

Firstly, it was noticed that 'threat' often occurred at the boundaries between territories where there is good reason to believe that both tendencies to fight and to flee are aroused simultaneously. This might be called evidence from situation or context. Linked with this conclusion, there was evidence on the mechanism of threat which comes from independent manipulation of the levels of attack and escape tendencies. Blurton-Jones (1959) had a group of completely tame Canada geese which ignored him if he was dressed in familiar old clothes. If he wore a white coat, the geese attacked him uninhibitedly, whilst if he appeared carrying a broom (used to drive the geese into their house for the night) they would flee. The familiar threat postures of the goose (lowered head on outstretched neck, hissing, etc.) appeared only when Blurton-Jones combined wearing a white coat with carrying a broom!

Finally, there was also evidence gained from a close examination of the forms of threat postures which can sometimes be analysed into elements belonging both to attack and to escape behaviour, and for this reason such postures are called 'ambivalent'. For example, Tinbergen (1959) gave a detailed analysis of the threat postures of the lesser black-backed gull (Fig. 4.20a). The bird moves towards its rival with the neck stretched upward and slightly forward, and the head and bill turned down. The wrist joints of the wings are lifted well clear of the body and the plumage is slightly raised. Now, gulls normally launch an attack by beating with the wings and attempting to peck down at their opponent. The raising of the wings and the down-pointing bill look like actual attack. Elements of escape can also be seen, particularly as the two rivals come very close. Now the head moves increasingly back, the bill becomes lifted and the plumage sleeker; the bird may turn sideways to its opponent and more parallel, not towards it. This turning aside looks like an element of escape. Gulls draw back their heads and sleek the plumage preparatory to taking off, so these may also be escape elements. (It is interesting to note that the so-called 'appeasement' displays adopted by submissive birds are almost an exact antithesis of the threat posture; see Fig. 4.20b.)

Not all threat postures can be interpreted as mixtures of attack and escape and indeed, as we have seen in Chapter 3, animals may use quite

Figure 4.20 Threat and appeasement postures. (a) The 'upright threat' of the lesser black-backed gull and (b) the hunched appeasement posture of the same species. This latter represents an almost perfect opposite to threat. The head is held low on a shortened neck and the bill points upwards, whilst the wings are pressed close into the flanks so that the wrist joints – so conspicuous in the threat display – are completely hidden. Now compare the gull postures with those of the domestic dog – an illustration taken from Charles Darwin's *The Expression of the Emotions in Man and Animals* of (1872). He entitled them (c) 'Dog approaching another dog with hostile intentions' and (d) 'The same in a humble and affectionate frame of mind'. The parallels with the gull postures are remarkable and both animals exhibit clearly what Darwin called the principle of antithesis in their communicatory behaviour. (a) and (b) by Niko Tinbergen; (c) and (d) by permission of the Syndics of the Cambridge University Library.

different displays as the only 'honest' way of signalling their fighting ability, rank, and so on. The conflict between attack and escape may still be there but not always manifest in ambivalent 'patchworks' of display (Baerends 1976).

Maynard Smith and Riechert (1984) used the idea of motivational conflict to develop a model of the fighting behaviour of the desert spider, *Aegelopsis*. This spider is very aggressive and fights over access to web sites. In Maynard Smith and Riechert's model there are two variables, fear and aggression, which fluctuate during the course of an encounter between two spiders according to such factors as who owns the disputed web site, relative body weight of the two, territory quality, and so on. There are also some genetic factors involved for it is known that some spider populations are more aggressive than others, even when the external situation is kept constant. On their model, if fear greatly exceeds aggression, the animal simply withdraws. If fear is equal to or less than aggression, then the contest continues, but what behaviour is performed (locate–signal–threat–contact) depends on the absolute levels of the two variables.

The model accounts well for many observed facts about spider fighting behaviour such as the fact that owners of web sites tend to win, larger spiders tend to win, and so on. Maynard Smith and Riechert argue that a model with two variables is the simplest one that will account for all the observed data. It does appear, then, that the traditional idea of agonistic encounters (including threat) as extended conflict has stood the test of time although there has perhaps been a shift of emphasis. Animals may display when they are in a state of conflict not necessarily to signal that they are in a conflict, but, by displaying and seeing the displays of others, to get themselves out of the conflict. They make their decision on whether to attack or to flee from a rival or even a predator on the basis of new information received during the course of the encounter. Their decision to perform one or other behaviour is then made in an adaptive way that would not have been possible without the period of extended conflict and the exchange of information it made possible.

Stress and animal welfare

If conflict or thwarting are prolonged for days or weeks, with the animal allowed no chance of escape, then many different bodily changes are likely to occur. We can lump them under the term 'stress' because the body's responses to a wide range of stressors (thwarted escape, overcrowding, extreme cold and burns, for example) are very similiar. Archer (1979), Broom and Johnson (1993) and Toates (1995) describe the physiological changes involved and point out that most of them are attempts to restore the delicate balance of the body's metabolism when it has been upset. Many people have become concerned that modern intensive farming methods cause stress to the animals kept in them precisely because the animals are thwarted through being unable to perform much of their natural behaviour, and from being unable to escape the confines of their cages or stalls.

Before we can evaluate this idea, however, we have to face up to the major difficulty of giving an adequate definition of 'stress' because many of the changes we observe in confined animals may be part of the normal functioning of the body. For example, when an intruder enters an animal's territory, various physiological changes occur in the body of the defender. There is increased activity in the autonomic nervous system which supplies the viscera and smooth muscles and which is hormonally under volunatry control. The autonomic system also supplies the adrenal glands, endocrine organs close to the kidneys which have a double structure, an internal medulla (supplied by autonomic nerve fibres) and an external cortex. When stimulated via its autonomic nerve supply, the adrenal medulla releases the hormone adrenalin into the body. This causes changes in numerous parts of the body: the sweat glands of the skin begin to secrete, hair becomes erected, the heart beats faster, breathing becomes more rapid and deeper and blood gets diverted to the skeletal muscles from the alimentary canal. These responses are adaptive in the sense that they prepare the animal for the strenuous action of either fighting or fleeing. The increase in heart rate, in breathing rate and the diversion of blood to the muscles from the alimentary canal all ensure that there is plenty of oxygen available for physical activity. Corticosteroid levels are also elevated during coitus and physical exercise in humans and, to judge by its physiological response, a passionate kiss is as much of a stressor as an electric shock (Selye 1973)! The problem in defining stress is thus to decide when normal, and adaptive, preparation for action has become evidence for a pathological state that is likely to be indicative of suffering. Since stress has negative connotations, it seems most appropriate to pay particular attention to cases where stress is (a) prolonged and severe, (b) associated with situations that animals avoid if they possibly can (Dawkins 1990) and (c) where the body's normal motivational processes have been stretched beyond their normal adaptive range (Broom & Johnson 1993; Toates 1995).

In natural territorial defence, for example, there will perhaps be a rapid flush of adrenalin as the intruder is detected, the territory holder will become aggressive, and this will usually result in the intruder leaving. Adrenalin levels will then subside. But if the stressful situation persists – as might happen if two animals were confined in a small cage – then a further reaction begins. This involves the other part of the adrenal glands, the adrenal cortex, which is stimulated to release its hormones, not directly by nerves as is the medulla, but by another hormone, adrenocorticotrophic hormone (ACTH), produced by the pituitary gland. Here, as with the release of adrenalin, the nervous system initiates the response. Stress activates cells in the hypothalamus which itself then stimulates the pituitary to release ACTH. It is not fully clear how the adrenal cortex hormones help the animal to adapt to stress. Some of them are concerned with glucose metabolism and may serve to mobilize the body's long-term food reserves. Corticosteroids both make glucose available and slow down its rate of metabolism, thus keeping glucose reserves available in the bloodstream. Whatever their action, the release of adrenocortical hormones is most dramatic. The cortical cells become drained of their contents and, if stress persists, the adrenals enlarge, sometimes by 25%.

Hunting deer with horses and hounds often involves chases of 20 km or more. With such long chases, the deer show evidence of both behavioural exhaustion (e.g. lying down) and physiological stress (Bateson & Bradshaw 1997). For example, cortisol levels rise but so much glucose is used that reserves of carbohydrates in the blood and muscles become exhausted. In the course of a hunt the muscles show signs of physical damage.

Animals under chronic stress become really ill and may die. Barnett (1964) showed how a wild rat, unable to escape from the territory of a dominant resident male, may die after a few hours of intermittent attack, even though it has no significant wounds. We also know that stress affects the immune system and that corticosteroids can depress immune system function (Maier et al. 1994). The stress which results from a prolonged conflict can even be severe enough to produce physical damage. Animals may develop gastric ulcers or tumours of the pituitary gland. In such cases, it is impossible to regard the responses as being in any way adaptive. They are the result of an animal being kept in artificial conditions and then pushed beyond its adaptive limits. The unfortunate animal finds itself in a situation to which natural selection has provided no solution, so it is hardly surprising that its adaptive control mechansims break down. In the natural world, escape from conflict situations is almost always possible as a last resort; if not, the stress is curtailed by death. It is largely in the artificial conditions in which humans keep animals that prolonged stress and conflict occur.

This association between the symptoms associated with prolonged stress and human treatment of animals has meant that the study of stress has become very important in the study of animal welfare. In particular, there have been efforts to look for behavioural indicators of stress and suffering that might be used as well as or even instead of physiological ones. Perhaps behaviour could give 'advance warning' that something is wrong before pathological physiological symptoms appear. Perhaps behaviour could even be used to help define stressful situations as those that animals are highly motivated to avoid.

Behaviour has been used in the measurement of stress in the following ways.

Stereotypies and other abnormal behaviours

Animals confined over a period of time, may develop 'stereotypies' – fixed sequences of behaviour quite unlike those shown in wild animals, which are performed over and over again in the same way and which have no obvious function. For example, polar bears in zoo enclosures, even if they are quite large, may pace around a set route so often that they wear away the ground, or stand constantly swaying their head and neck from side to side. Sows confined in stalls may repeatedly rub their mouths backwards and forwards over the bars, even making themselves bleed (Fig. 4.21). Different individuals may have individual stereotypies different from others kept in identical conditions. One stalled sow may take two steps forward, move her mouth to left and then right on the bar in front of her stall six times, rub her nose on the side of the stall and then take two steps backwards over and over again, whereas another pig may have a quite different routine.

The causal basis for stereotypies and exactly what they say about the welfare of the animal performing them is a matter of continuing debate. One complicating factor is that the circumstances that lead to the development of a stereotypy in the first place may not be the same as those that keep it going in an adult animal (Mason & Turner 1993). Domestic hens in small cages may develop stereotypic pacing backwards and forwards, apparently out of frustrated attempts to escape (Duncan & Wood-Gush 1972). Even when let out of their cages, however, they may continue to show the same stereotyped behaviour. So the performance of a stereotypy may not necessarily indicate that the animal is stressed just now. Rather, it may indicate that at some time in the past, its welfare was compromised and that it still carries the behavioural scars of its past experience.

Figure 4.21 Stereotypy produced by industrial husbandry. Sows confined in narrow stalls repeatedly bite and worry at the bars. (Photograph by Mike Appleby.)

Another problem with the interpretation of stereotypies is that they may even help the individual doing them to 'cope' with its environment, in the same way that, say, pacing back and forth sometimes relieves tension for a human. There is some suggestion that performing stereotypies is associated with a reduction of some of the physiological symptoms of stress such as corticosteroid levels or heart rate. Cronin *et al.* (1985) even proposed that by performing stereotypies such as bar-biting, pigs induce a state of self-narcotization by releasing endorphins into the bloodstream. However, we should not conclude that stereotypies maintain the physiological balance of the animal and so are no cause for concern. Rather, they are almost certainly indicative of poor welfare at some stage in the animal's life (Dantzer 1991; Mason, 1991; Broom & Johnson 1993).

Other abnormal behaviour has been produced in the laboratory. Masserman (1950) trained cats to open a box for a food reward when a signal light flashed. Later when the food box was opened, the cats sometimes received a strong blast of air. Under these conditions, the animals' behaviour often became severely disturbed. Some of the cats became hyperexcitable, others moped in corners for days on end. They nearly all showed signs of acute stress, with raised blood pressure, hair erection and gastric disorders.

Disruption of normal behaviour

Even when an animal shows no new behaviour patterns, its normal ethogram may be considerably disturbed. Sometimes this may be obvious, for example where cattle slip over on slatted floors and simply cannot maintain their balance. At other times, the disruption may be more difficult to detect. Many zoo and farm animals fail to show behaviour exhibited by animals of the same species in the wild. The problem of correctly interpreting these more subtle behavioural measures of 'stress' is to know where to draw the line. At what point does a difference in behaviour between a 'normal' and a 'stressed' animal become great enough to justify concern for welfare? After all, a domestic dog shows major differences in behaviour from its wild ancestors and, as we have seen, symptoms of so-called 'stress' can be associated with a wide range of activities both painful and pleasurable. To address this fundamental problem, we need to ask how strongly motivated an animal is either to avoid or to approach the apparently stressful situation.

Strength of avoidance and welfare

In our discussion of how to measure the strength of an animal's motivation, we described several different measures that involved the animal being offered the opportunity to *change* its environment in some way. For example, animals can be put into a Skinner Box where moving a pecking key or lever gives them food, water, or access to social companions. We can then ask how hard the animal will work to obtain these things and whether it will still work if, say, it has to press the lever hundreds of times for each piece of food. Animals can also be 'asked' whether they are willing to put up with aversive stimulation such as running through water or even a mild electric shock to get what they want. Such studies give us an idea of not just what animals 'like' but how much they like it in terms of their willingness to 'put up with' aversive stimulation. It can also be arranged that the animal can use the Skinner Box to get away from something it dislikes; for example, pecking the key or pressing the lever might turn off a loud noise or the rocking motions experienced on a cattle truck (it is easier if this experiment is done in a simulator!). In this way, we can break out of the circular definition of stress as something that causes 'stress' symptoms and ask the animals (a) whether they do find something aversive and (b) how aversive they find it (criterion (ii) on p. 237). We also resolve the difficulty that both pleasant and unpleasant experiences can result in stress symptoms by allowing the animals themselves to classify the world into what they find pleasant (and reinforcing) and what they find unpleasant or painful (and aversive). We can then concentrate our concern for their welfare on those situations where physiological symptoms of stress coincide with situations where, behaviourally, the animals have shown they find something strongly aversive and will avoid it if they possibly can. It is animals that are kept for long periods in such conditions that are most likely to suffer (Dawkins 1990; Broom & Johnson 1993).

Sows kept commercially often have to give birth to their piglets in a bare cubicle with a concrete floor. They have no straw to build nests with even though domestic pigs will build quite elaborate nests if given the right materials. Arey (1992) showed that pregnant sows would, if given the opportunity, work to obtain straw for nest-building. He gave the sows access to two chambers, one of which contained straw and one of which contained food (Fig. 4.22). The pig could enter either chamber by pushing with its nose a panel that unlocked the door. All the sows quickly learnt the trick of opening the doors and would choose to go into the straw area almost as often (mean of 17 times per day) as they chose the food chamber (21 times a day). But a somewhat different result was obtained when the 'cost' was increased. Instead of being able to unlock the doors with one push of the panel, the pigs suddenly found that they had to push it 10, 20 or even 300 times before the door opened. In other words, they had to work much harder to get what they wanted. Sows that were 2 days away from giving birth kept pushing the panel to get at the food so that they were now pushing up to 300 times for one entry into the food compartment and would still choose to enter the food chamber 11.4 times a day. On the other hand, they did not seem as willing to work for access to the straw chamber since when they had to push the panel 50–300 times for each entry, they only rarely entered the straw chamber (2.6 times a day). However, the day before they were due to give birth to their piglets, their motivation for access to straw appeared to increase dramatically. They started pressing the panel leading to the straw chamber even though they had to work hard to get in. Even with 300 panel pushes, they entered the straw chamber nearly as often (16.4 times a day) as they entered the food chamber (17 times a day). This suggests that motivation for access to straw is as high as motivation for access to food in the 24 hours before the birth of piglets and therefore, at this time, deprivation of straw would be as serious for the sow as deprivation of food.

Figure 4.22 Apparatus used by Arey (1992) to test the motivation of pregnant sows for nest material and food. The swing doors to either the straw area or the feed area could be unlocked by the sow pushing the panel on the appropriate side with her snout. The 'cost' of each commodity could be adjusted by changing the number of times the panel had to be pushed before it opened. One press only represents a low cost; many presses required means a high cost. The 'price' sows are prepared to pay for straw rises steeply just before they give birth.

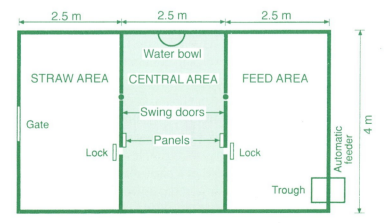

Although measures of stress in animals are increasingly being used to evaluate the welfare of animals kept in zoos, farms and other restricted conditions, it is important to emphasize that any single measure used on its own is likely to give misleading results. An animal's willingness to work for something should not be considered in isolation from measures of what conditions lead to good physical health or physiological measures such as heart rate or corticosteroid levels. There is no simple 'litmus test' of when regulatory mechanisms have moved from adaptive stress to suffering. This is as true of physiological measures of stress as it is of behavioural ones. Indeed, many of the problems of studying stress have arisen because the complexity of the interactions between behaviour, physiology and disease have not been fully appreciated. Only when we take all of these into account can we have a chance of seeing clearly how to improve animal welfare.

Conclusions

Throughout the last two chapters on the causal basis of behaviour, we have tried to strike a balance between trying to explain what animals do in terms of known physiological entities such as hormones and neurons on the one hand and avoiding over-simplication on the other. Animal behaviour, with its shifting motivational priorities and its decision-making on many different time scales, will inevitably have complex explanations and we should not underestimate how difficult it is to 'explain' it. At the same time, major advances have been made in our understanding of behaviour at a physiological level and are likely to continue at an accelerating rate as new techniques such as brain scans enable us to gain access to what was previously inaccessible. We can look forward to a future in which we have a greatly increased understanding of how the brain works to control behaviour. Even with our present knowledge, it is striking that behavioural and physiological explanations have proceeded hand in hand and reinforced each other's findings. Physiological studies have added to and sometimes even superseded vaguer ideas of 'motivation' while behavioural studies have shown what real animals do and guarded against our viewing animals as simple control systems switching on and off when a single variable reaches a particular value. We can no longer think in terms of single 'centres' for feeding, drinking, aggression and so on, but have to look at the integration of information from many different areas of the brain. What an animal decides to do will be affected not only by the stimuli in its present environment but also by what happened to it in the past. For this reason, we now turn to the important topic of learning and memory. Although we dealt with some aspects of development in Chapter 2, our emphasis there was on the role of innate behaviour. In the next chapter we emphasize the role of experience. Equipped with some understanding of the causal basis of behaviour, we are now in a position to see how what animals learn affects the machinery of behaviour itself. The past affects the present state of the animal so that questions about causation and questions about development, although distinct, are intimately related.

5   Learning and memory

Throughout this book we have made frequent reference to learning as a form of behavioural development retained into adult life and as one pathway by which behaviour becomes adapted to an animal's requirements. Now attention must be focused on learning itself. Learning and memory go together because whilst the former involves changing behaviour as a result of experience, its effects cannot be put to use unless the results of the experience can be stored in some way and recalled the next time they are needed.

We want to examine learning in a wide range of very different animals and Thorpe's (1963) book, although published some time ago, remains useful because it does deal with the whole animal kingdom – learning in molluscs and insects as well as in birds and mammals. Thorpe defines learning as '. . . that process which manifests itself by adaptive changes in individual behaviour as a result of experience'. This definition draws attention to two important features. Firstly, learning normally results in adaptive changes and, as we discussed in Chapter 2, learning and instinctive behaviour are both ways for equipping an animal with a set of adaptive responses to its environment. Normally both are found in combination and logically they have much in common. In one case we have the selection of individuals whose genes best operate during development so that, as these animals are most successful, adaptive behaviour evolves in a population of animals. In the case of learning, individuals select and retain the best responses over the course of their own lifetime. The second important point arising from Thorpe's definition is that, strictly speaking, learning is a process which we cannot usually observe directly; we measure what has been remembered as a result of learning.

This ability to benefit from experience links us to our non-human fellow creatures which so obviously share this striking behavioural capacity. Animal learning and memory differ from ours both quantitatively and qualitatively, but there remain many features in common and we can make useful comparisons. In fact, because we can communicate so easily with human subjects, they are in many respects better material for learning studies than animals. For instance, in the laboratory, it is possible to test our memory in two ways: 'recall', i.e. by reciting or writing down a list of nonsense syllables which we have previously learnt; and 'recognition', i.e. presented with a set of nonsense syllables which includes those we have learnt, we record which syllables we recognize. Recognition is always an easier task than recall because the situation provides stimuli which, as we say, 'jog our memory' and help the process of recall. If we have trained a rat to run through a maze we cannot ask it to draw a map of its route on a piece of paper. The only way to test what it has learnt and retained is to put it back in the maze and observe its behaviour. If the rat makes mistakes we have no means of knowing whether it failed to learn adequately or learnt but failed to recall.

This difficulty brings us to the question of learning mechanisms. What happens in the nervous system when an animal learns something? When a young game-bird crouches on the first exposure to the sound of its parent's alarm call, it must utilize pathways already present in its brain whereby a particular auditory input has easy 'access' to the motor system controlling crouching. On the other hand, when a rat learns to press the bar of a Skinner box new pathways must become established, because prior to learning the bar evoked no special response. We know, too, that mere presentation of the bar to the rat is not enough; there must be some sort of reward as a result of doing so. Some part of the learning mechanism records the result of the rat's reaction and if this result is 'good' it increases the probability that the reaction will occur again the next time the situation is presented. Further, we know that somewhere in the nervous system there is stored a more-or-less permanent record of the learning that can be consulted or recalled on future occasions. The investigation of the behavioural and physiological basis of memory has been an active area of research for decades. Recently there have been some real advances and we shall return to this subject after looking at the phenomenon of learning itself.

There was a time during the first half of this century when a huge amount of effort was directed towards the study of animal learning. Psychologists hoped that animals would serve as models for human beings in this respect and the majority of their work concentrated on two convenient laboratory species, the rat and the pigeon. By investigating every variation of learning task, of reinforcement (reward or punishment) and of stimulus presented, they hoped to generate 'laws of learning' which would enable us to predict performance and outcome no matter what subject or what situation. Munn's *Handbook of Psychological Research on the Rat* (1950) is a fascinating record of this 'heroic age' of experimental psychology, the diverse models and the theories of learning which it produced. It has to be said that, for the most part, they attract little attention now. The classification of learning into distinct types, the proscribed learning situations in which they were studied and the rigid models into which all animals and all learning were expected to conform, have been replaced by a much more biologically based approach. This recognizes that an animal's learning abilities must have evolved to suit its own special requirements just as any other aspect of its behaviour; they are learning by instinct. We should not expect learning in pigeons to resemble that of honey-bees except in a very general manner. Furthermore, learning is now seen as a way in which animals attempt to identify key aspects of a fluctuating environment: to detect its regularities and ignore the distracting 'noise' which is not important for them. They will integrate their learning with the inherent biases of perception and response that all animals bring to the world. Encouragingly, these moves towards a more truly comparative and biological approach have come both from psychologists, dissatisfied with the old learning theories, and from ethologists who observe the role of learning in their animals' natural lives. Articles by Mackintosh (1983), Roper (1983) and Rescorla (1988) form a good background to this new synthesis. Books by Dickinson (1980), Hinde and Stevenson-Hinde (1973), Seligman and Hager (1972) and Pearce (1996) provide more examples and a more detailed analysis of some of the changing attitudes towards learning.

Sensitization and habituation

If animals are to learn to change their behaviour to meet a new situation, then clearly anything new appearing in their environment – and we do not mean just a visual stimulus – will have to be taken note of and its importance assessed. The easiest way to get some estimate of importance is to note what happens just before or just after it appears. An alert animal, for instance one which has had its appetite aroused by the smell of food or has just fled from a predator's attack, is particularly sensitive to such stimuli. To take a laboratory example, a rat which has just received a small electric shock to its feet jumps in alarm to any novel stimulus – e.g. a flash of light, a tap on its box – which would normally evoke little or no response. Evans (1968) studied marine ragworms (*Nereis*) which live in tubes they construct in the sandy floor of the sea, emerging to stretch out the front part of their bodies to forage on the surface nearby. Evans kept his *Nereis* in laboratory aquaria in dim light. Under such conditions he found that retracted worms sometimes emerged from their tubes following a flash of light and in one experiment 21% of worms did so. However, if he fed a second group of unexposed worms just once in the dim light conditions, over 60% of them emerged from their tubes to a subsequent light flash. The arousal following feeding had made them much more sensitive.

This phenomenon – a period of high responsiveness following arousal by rewarding or punishing experiences – is, reasonably enough, called **sensitization,** and it is widespread in such situations. Why then, do not animals respond with alacrity to all stimuli, for opportunities to have become sensitized to them must often exist? The solution lies in a second, and in many ways opposite, effect of stimulation. If repeated several times in the absence of any significant accompaniment – no more foot shocks to the rat, no more food for *Nereis* – then animals gradually cease to respond. From being initially sensitized, they calm down and eventually ignore the stimulus. This waning of responsiveness is called **habituation**

If we now think back to the definition of learning – adaptive changes as a result of experience – then perhaps we should regard both sensitization and habituation as learning. Habituation, as we shall see, fits quite well but sensitization really does not. It is too short-lived and too indiscriminate – for a short time the animal is hyper-responsive to a variety of events in the outside world and pays more attention to them. This is probably an essential preliminary to any learning but not the process itself. Soon one of two things must follow. If the novel stimuli which initiated sensitization are not repeated, then the effect wanes quickly. If they are repeated enough times, then the animal goes on to learn. What it learns depends on the results of the situation. Do the stimuli signal pleasant or unpleasant events – food, water, a predator, pain? If so then the stimuli which give the best prediction of these rewards or punishments (together we can conveniently call them **reinforcements**) will be picked out for continuing attention. We say that the stimulus and the reinforcement have become **associated** – and association learning is a major category for discussion later. There is an alternative result of repeated stimulation – that nothing happens, there is no reinforcement. The stimulus then captures less and less of the animal's attention. Birds soon come to ignore the scarecrow which put them to flight when it was first placed in a field. We say that the birds habituate, and habituation can be regarded as a simple form of learning. It is relatively long-lasting and it is 'stimulus-specific', i.e. only the stimuli which are repeated without reinforcement are affected – the animal remains alert to others.

These qualities of habituation will be clearer from an example. Once again the relatively simple behavioural repertoire of *Nereis* has provided good material for studies by Clark (1960a, b). As already mentioned, the worm burrows in mud or sand on the bed of brackish estuaries. Its head and anterior segments protrude from the tube to feed on the surface around the burrow. During its bouts of feeding a variety of sudden stimuli will cause the worm to jerk back into its tube. In the laboratory, Clark could easily get the worms to live in glass tubes in shallow basins of water. He found that jarring the basin (mechanical shock), touching the head of the worm, a sudden shadow passing over and a variety of other stimuli would all cause retraction into the tube, but the majority of worms emerged again within one minute. If these stimuli were repeated at 1-minute intervals the proportion of worms responding fell off until none of them were retracting; they had habituated. Clark found that habituation occurred more rapidly if stimuli were given close together. For example, with a bright flash of light it took fewer than 40 trials at half-minute intervals, but nearly 80 trials if the interval was 5 minutes. The speed of habituation also depended on the nature of the stimulus; mechanical shock, shadow, touch and light flash each produced their characteristic rates of habituation. Further, habituation was to a large extent stimulus-specific; Fig. 5.1 shows how the waning of retraction to repeated mechanical shock is independent of that to a moving shadow.

There are a number of other processes which may be confused with habituation because they also lead to a reduction in responsiveness. Results such as those illustrated in Fig. 5.1 eliminate any possibility that the waning of response is due to motivational changes or to muscular fatigue, but it is often more difficult to eliminate sensory adaptation. Many sense organs eventually stop responding to repeated stimulation. We cease to be aware of our clothes within a few moments of putting them on because the tactile receptors in the skin cease to respond. In *Nereis*, Clark could also eliminate sensory adaptation as an explanation for the waning. For example, the worm soon ceased to retract when touched by a probe, but clearly still detected the stimulus because it then attempted to seize the probe with its jaws. Sensory adaptation is usually a short-lived phenomenon; a few minutes without stimulation is usually sufficient for complete recovery. We sensorily adapt to the feeling of our clothes rather than habituate to them and we should retain the term 'habituation' for a more persistent waning of responsiveness which must be a property of the central nervous system and not the sense organs.

Clark could detect some recovery from habituation within an hour or less and the worm's retraction response was completely recovered and back to full strength within 24 hours. This is not very long and, if we think in terms of ordinary memory span, not impressive, but this is to overlook the role that habituation plays in the life of animals like *Nereis*. It is, in fact, a vital process for adjusting an animal's behaviour to the minute-to-minute events in its environment. Habituation enables it to concentrate on important changes there and ignore the others. Small prey animals cannot spend too long skulking in shelter – they must normally be out feeding. Although at first it is best to retreat rapidly in response to sudden shadows, not every shadow means a bird

Figure 5.1 The rates of habituation of two different stimuli in *Nereis*. The response measured is the sudden retraction of the worm into its tube and trials are given at 1 min intervals. The shaded areas record the responses of a group of 20 worms to a moving shadow. Within 10 trials they have all ceased to respond, but switching to a mechanical shock stimulus (unshaded areas in upper graph) brings back the response in half the worms, and it characteristically habituates more slowly, taking more than 30 trials. The recovery of the response to moving shadow is complete after 40 min, whether the habituation trials to mechanical shock intervened (upper graph), or if the worms were simply left alone (lower graph). Clearly, habituation is quite independent for these two very distinct types of stimulus. (from Clark 1960a.)

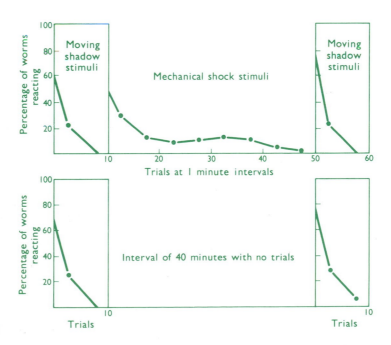

or a fish overhead: it might just as well be a floating piece of seaweed. In general, frequent shadows are more likely to be seaweed than predators and it is certainly adaptive to cease responding when a repeated stimulus has no attendant consequences. Just as certainly, it is dangerous to ignore shadows from then on. By the next incoming tide the seaweed will have moved and a shadow could now signal a predator's approach. The same cautious testing out must begin again.

This is not to imply that all habituation is so short-lived. Newly hatched chicks begin to feed by pecking at any object which contrasts with the background. Responses to inappropriate objects rapidly habituate and do not return – rather, the growing chick begins to learn positively which objects represent food.

In nature we must expect to find sensitization, sensory adaptation and habituation almost inextricably combined, so that the time course and persistence of diminished responsiveness depends on a variety of factors. Habituation processes are shown by all animals, but predators and prey animals, for example, are unlikely to respond in the same ways because the costs and rewards will be so differently balanced.

Associative learning

Habituation is generally regarded as a simple form of learning because it involves the waning of a response that is already there. We tend to think of learning as a process whereby we acquire new responses and new capacities. For this reason, the basic characteristics of associative learning are immediately familiar to us. A previously neutral stimulus or action has sufficiently important consequences to be singled out from other such events. After some repetitions followed by the same consequences, a long-term association is built up between the event and its result and the animal's response changes accordingly. Having found sugar solution there previously, a honey-bee picks out the blue dish from an array of dishes laid out by an experimenter. Rodents exploring new territories quickly learn the shortest routes to shelter for use when a hawk or an owl swoops.

The best known type of associative learning situation is the **conditioned reflex**, inseparably linked with the name of the great Russian physiologist, I.P. Pavlov (see Pavlov 1941) whose school was active around the turn of the century. Everyone has heard of Pavlov's dogs who salivated to the sound of a bell! So far has his influence run that the type of learning he studied is often referred to as **classical conditioning**. Pavlov's influence on behavioural studies and neurophysiology in Russia is still considerable but, perhaps because the reflex theory he developed attracted little favour here, his influence in the West has been less. Pavlov and Sherrington were working at the same time, but from completely different viewpoints. Sherrington studied the organization of reflexes in the isolated spinal cord of dogs and cats, having deliberately cut off influences from the higher centres. Pavlov worked with intact animals and considered that, just as simple reflexes are a property of the spinal cord, so conditioned reflexes are the particular property of the higher centres of the brain, especially the cerebral hemispheres. Pavlov's aim was to study 'the physiology of higher nervous activity', but most of his experiments were, in modern terms, pure experimental psychology. Indeed, Pavlov was really one of the founders of experimental psychology; he was applying objective techniques to the study of learning years before J.B. Watson, whose book, *Behaviorism* (1924) had such a huge influence on American psychology.

Pavlov's classical experiments with dogs often involved the 'salivary reflex'. Dogs salivate when food is put into their mouths and Pavlov could measure the strength of their response by arranging a fistula through the cheek from the salivary duct, so that drops of saliva fell from a funnel and could be counted. A hungry dog was placed on a stand, restrained by a harness and every precaution was taken to exclude disturbances. In this position, it could be given various controlled stimuli such as lights, sounds or touch, and meat powder could be puffed into its mouth through a tube (see Fig. 5.2). A standard quantity of meat powder caused the secretion of a certain amount of saliva. Now Pavlov preceded each ration of powder by, say, the sound of a metronome ticking. At first, this stimulus caused no response, save perhaps that the dog pricked up its ears momentarily. However, after five or six pairings of metronome followed by food, saliva began to drip from the dog's fistula soon after the metronome started and *before* the meat powder arrived. Eventually the amount of saliva produced to the metronome alone was the same as that which was given to the meat powder.

The dog had learnt to respond to a new stimulus, previously neutral, which Pavlov called the **conditioned stimulus** (CS). The salivation response to the CS is the **conditioned response** (CR). Prior to learning, only the meat powder or **unconditioned stimulus** (UCS) produced salivation as an **unconditioned response** (UCR).

Pavlov found that almost any stimulus could act as a CS provided that it did not produce too strong a response of its own. With very hungry dogs even painful stimuli, which initially caused flinching and distress, quite soon evoked salivation if paired with food. The CR is formed by the association of a new stimulus with a reward and in the same way a CR for withdrawal can be formed by associating the CS with punishment. An electric shock to the foot causes a dog to lift its paw; if a metronome is paired with the shock, the dog soon raises its paw to the sound alone.

Pavlov carried out exhaustive tests on the precision with which a particular stimulus was learnt. He found that if a dog was conditioned to salivate when a pure tone of, say, 1000 Hz was sounded, it would also salivate when other tones were given, but to a lesser extent. It **generalized** its responses to include stimuli similar to the conditioned one, and the more similar they were the more the dog salivated. The process opposite to generalization is **discrimination**. Dogs naturally discriminate to some extent, otherwise they would salivate equally to all sounds, but their discrimination becomes refined after repeated trials when only one particular tone is followed by reward. We

Figure 5.2 A typical experimental set-up in Pavlov's laboratory. The dog is gently restrained on a stand facing a panel and, under experimental conditions, is well insulated from external disturbances. Tactile, visual or sound stimuli can be presented in a carefully controlled fashion. Tubes run from the fistula in its cheek, which collects saliva as it is secreted. A simple arrangement of a hinged plate, onto which the saliva drips, and a measuring cylinder where it collects, enables the amount of saliva to be measured together with the intensity and duration of its secretion. Not shown is the device for blowing a controlled amount of powdered meat into the dog's mouth to reward the salivation response.

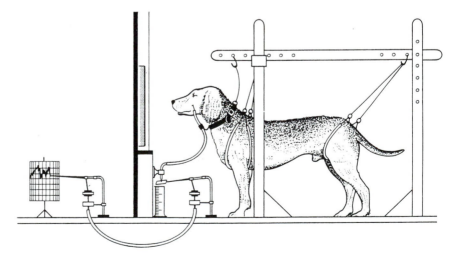

can accelerate discrimination if, as well as rewarding the right tone, we slightly punish the dog when it salivates to others.

This **conditioned discrimination** method has been of enormous value for measuring the sensory capacities of animals. After training to one particular stimulus – it may be a colour, brightness, shape, texture, sound, smell, weight, etc. – we then test to see how far the animal can discriminate this stimulus from others. We present it together with another stimulus of the same type and reward only responses to the former, perhaps giving slight punishment for incorrect responses. The two stimuli are made increasingly similar until there comes a point beyond which the animal can no longer learn to discriminate between them. This marks the limit of its sensory capacities as measured by its behaviour. To give but three examples from many hundreds, this method was used by von Frisch (1967) in his classical studies of the colour vision of bees it was also used to examine the touch sensitivity of the octopus (Wells 1962) and the chemical senses of fish (Bull 1957).

Conditioned reflexes of the type investigated by Pavlov have been observed in many different animals, from arthropods to chimpanzees. For example, birds learn to avoid the black and yellow caterpillars of the cinnabar moth after one or two trials which reveal their evil taste. They associate this with the colour pattern and generalize from cinnabar caterpillars to wasps and other black and yellow patterned insects. Because predators generalize, it is advantageous for different distasteful insects to resemble one another – the phenomenon of Müllerian mimicry (Fig. 5.3).

In nature we rarely observe reflexes in as 'pure' a form as in the laboratory. When foraging naturally, bees do not just learn to associate a colour with the nectar reward: they also learn the position of the group of flowers with respect to their hive, and learn what time of day the nectar secretion is highest, directing their foraging trips accordingly. Pavlov, despite his scrupulously controlled environment, found that his dogs learnt more than one particular response to one particular stimulus. A hungry dog familiar with his laboratory would run ahead of the experimenter into the test room and jump up on to the stand, ready to be hooked up to the apparatus and wagging its tail with every sign of expectancy! Animals, then, don't just hang around waiting for a stimulus to signal reinforcement: they try actively to put themselves into the situations or perform actions which lead to reward or escape from punishment. As we shall discuss later, Rescorla's (1988) account of modern work on classical conditioning shows that animals are not simply making stimulus–response (S–R) connections, but searching for regularities in the sequence of events which face them so as to arrive at reasonable predictions of the best outcome.

Pavlov's dogs running into his testing room introduces us to another

(a) (b)

Figure 5.3 An example of Müllerian mimicry. (a) A wasp, showing its striking black and yellow striped 'warning' pattern. (b) A group of cinnabar moth caterpillars, showing a similar pattern. Two very different sorts of insects have converged on the same pattern to warn potential predators that they are noxious. Each derives a certain degree of protection from bad experiences predators may have had with other similarly coloured prey. (Photographs (a) by Paul Embden and (b) Mike Amphlett.)

familiar form of learning, in which animals come to associate with reinforcement, not a novel stimulus, but some action they have performed. Thorndike, one of the pioneers of American experimental psychology, investigated this type of learning with cats, using a series of 'problem boxes' of the type illustrated in Fig. 5.4. For example, we may have a box with a spring door which can only be opened from the inside by depressing a lever. A cat is shut in and tries hard to escape; it moves around restlessly and after a time – by chance – it steps on the lever and the door opens. The second trial may be a repetition of the first and also the third, but soon the cat concentrates more attention on the lever and eventually it moves swiftly across the box and presses the lever as soon as it is confined. Thorndike gave the descriptive name **trial and error** to this type of learning. The cat learns to eliminate behaviour which led to no reward and increases the frequency of behaviour which is rewarded, but in the early stages there is little system to its activity – the first reward is obtained by pure chance.

The famous 'Skinner Box', named after B.F. Skinner, who used it extensively for the study of learning, is basically a problem box of a convenient form in which an animal learns by trial and error that pressing a bar or pecking a key yields a small reward. Because the animal's own 'spontaneously generated' behaviour is instrumental in its gaining a reward, such learning has been called **instrumental conditioning** (Skinner also used the term **operant conditioning**), but it is no different in principle from trial and error.

Skinner, whose book (1938) was grandly entitled, *The Behavior of*

Figure 5.4 One of Thorndike's problem boxes. A cat is confined inside the cage and must learn to pull the string loop to open the door. (From Maier & Schneirla 1935.)

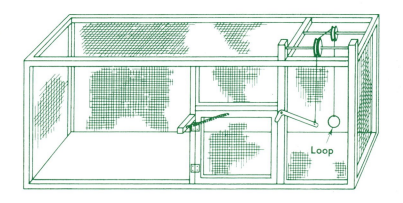

Loop

Organisms has been a very influential figure in learning theory and because his school has concentrated largely on instrumental conditioning whilst others have worked on classical conditioning, there has been a tendency to regard the two situations as leading to two rather distinct types of learning. At first sight they do seem clearly distinguishable. With classical conditioning the animal associates a novel stimulus with a response (the UCR) which was there from the outset. With instrumental conditioning or trial and error it is a novel response, not a novel stimulus, which is learnt.

However, close examination of how the animal behaves in the two situations shows that this distinction may be more apparent than real, at least in some cases. The clearest examples come from pigeons, familiar subjects for instrumental conditioning studies because they can readily learn to peck at a key on the wall of a Skinner box to obtain a food or water reward. If, as is commonly the case, you simply record key presses automatically, then it looks like very conventional trial and error – the pigeon learns this new response and can use it to obtain food or water, depending on circumstances. However, there is real value in bringing an ethological approach into the Skinner box. A pigeon key-pecking for food reward does not behave in exactly the same way as when pecking for water. Close observation of the head and bill reveals that when hungry and pecking for a food reward the pigeons' eyes are partly closed and its bill open (Fig. 5.5a); when pecking for water its eyes are fully open and the bill almost closed (Fig. 5.5b). These are precisely the contrasting features which distinguish a pigeon pecking to pick up food grains when feeding from one which dips its bill into water to drink; (pigeons, unlike most

Figure 5.5 Trained pigeons in a Skinner Box 'pecking' a key to obtain a reward. In (a) the reward is food and the bird's head moves sharply towards the key. As it reaches it, the bill is opened as if to seize a food item and, in a characteristic protective reflex, the eyes are almost closed. In (b) the bird is 'pecking' for water reward and the bill, almost closed, is pressed against the key more slowly whilst the eyes remain fully open. (Drawn from photographs in Moore 1973.)

(a)

(b)

birds, suck up water like mammals). In other words, the pigeon which pecks instrumentally is treating the key 'as if' it were food itself in one case, and water in the other.

Pavlov regarded the conditioned stimulus as coming to replace the unconditioned one, i.e. the dog treats the metronome sound 'as if' it were food. It seems that pigeons in the Skinner box respond in a similar way and such a conclusion certainly forms a bridge between classical and instrumental conditioning and forces us to look at the learning situation more from the animal's point of view. This is not to say that any distinction between classical and instrumental conditioning is meaningless, but we have to recognize that animals will come to any learning situation with some built-in predispositions which will affect the way they respond.

Once it was widely believed that, given the right opportunity, animals could learn to associate any stimulus they could perceive or any response which they could perform with any reinforcement – it was the association which counted. In practice, this assumption has proved quite inadequate. In Chapter 2 we have already discussed some evidence that animals may have inherent biases to learn particular things. In conventional learning situations we usually find that they most readily learn responses which form part of their natural behaviour to the reinforcement. Cats will quickly learn to lick or bite at a lever to get food, but it is very difficult to train them to turn a treadle wheel with their paws for the same reward. Exactly the converse happens if the reinforcement is escape from electric shock. Now the turning of a treadle is easily associated with escape, but licking a lever cannot be. We can understand these biases in terms of the normal way the cat responds to food or runs away from unpleasant situations.

Problems with species-specific biases are well illustrated by avoidance conditioning. One familiar form of laboratory avoidance conditioning is the 'shuttle-box'. A rat is put into a box whose floor is a grid which can be electrified to deliver a mild shock to its feet. The box is divided into two halves by a partition with a gap for easy passage between them. Each half can be electrified separately. On a signal (usually a buzzer or light) the rat has a few seconds to run through into the other half to avoid electric shock. This is not an easy task for rats because during the early trials they are being asked to take refuge in another part of the box where they have also received a shock just previously. It takes them some time to accept this as a refuge, but at least they have the right natural bias – they run in response to shock. Bolles (1970) discusses the varying fortunes of psychologists who have used other species in a shuttle-box situation. For many animals it is a nearly impossible task because their natural response to foot shock as to any alarming stimulus is to crouch or freeze. It is no use expecting a hedgehog to learn to run away from approaching danger – hence the carnage on our roads.

Sometimes the biases animals show are so strong that when placed in unnatural learning situations they will perform patterns which actually delay their getting reward. Breland and Breland (1961), pupils of Skinner, tried to put their techniques to good use by training animals to perform various eye-catching tricks for TV commercials. They knew that it should be possible to train hungry pigs to drop money into piggy banks or get chickens to ring bells, provided these actions were associated with food. It worked – up to a point – but they could not eliminate 'undesirable' side actions. The pigs would root with their snouts on the way to the bank, the chickens scratched and pecked at the ground. Spending time on this behaviour was certainly inefficient in operant conditioning terms. It was never reinforced and it delayed the arrival of food. The Brelands, recognizing the instinctive behaviour patterns which chickens and pigs use in feeding, came to accept that animals do not always simply associate response and reinforcement. They mischievously entitled their paper *The Misbehavior of Organisms* (note Skinner's title); Seligman and Hager (1972) give many other such examples.

The fact that animals do not always learn to respond so as to minimize the delay between an action and its reinforcement was most surprising to conventional learning theorists. Contiguity – the close association in time between stimulus or response and reinforcement – was considered essential. Figure 5.6 shows the familiar picture from Pavlov's work: CS and UCS must be close or overlap if learning is to occur. If the CS ends too early, so that there is a delay before the UCS signals the reinforcement, then no association is formed. Skinner found, in general, that delays of more than about 8 seconds between a response like bar-pressing and its reinforcement greatly slowed learning. In practical terms the deleterious effects of delayed reward can often be overcome by introducing a **secondary reinforcement**. Suppose a rat learns that a reward is delivered when a light comes on in the Skinner box (light and reward must overlap as discussed above), then it will learn to press the bar to switch the light on. Light becomes a secondary reinforcement or a 'bridging stimulus' between the response and the primary reinforcement – food. Bridging stimuli are useful for training animals, as in a circus when it is often difficult to reward immediately after the response is made.

However, just as there were exceptions to the rule about it being possible to link any response to any reward, so we find that immediate contiguity is certainly not an invariable requirement for learning. There are certain types of associative learning which consistently occur when reinforcement is delayed for a matter of hours. Barnett (1963) describes how wild rats only nibble at small amounts of any novel foods that appear in their territory. If it proves edible, they will gradually take more on successive nights until they are eating normally. If it is poisonous, yet they survive, they avoid it completely on subsequent occasions. This type of behaviour is highly adaptive and makes poi-

Figure 5.6 The effect of the sequence of stimuli upon the formation of a conditioned reflex. In each case the upper, thin line denotes the duration of the conditioned stimulus (CS) and the lower, thick line that of the unconditioned, reinforcing stimulus (UCS). The results are given on the right; note that the CS must not end with or persist beyond the UCS if a positive conditioned reflex is to be established. If it does so, then the CS will remain neutral and may even tend to inhibit the response to the UCS. (From Konorski 1948.)

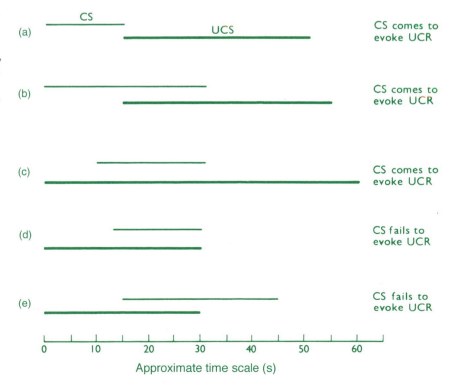

soning rats no straightforward task. The interesting feature for our discussion is the delay that must ensue between a rat tasting poison bait (always made superficially palatable with sweet substances) and any subsequent ill effects. Few rat poisons take less than an hour to produce effects. Laboratory findings have confirmed this ability. Not only will rats learn to avoid tastes associated with sickness that sets in at least an hour later; if deprived of the vitamin thiamine they will learn to choose a diet containing it, although many hours must elapse before they can feel its benefits. Rozin and Kalat (1971) provide an interesting review in which they link the specializations of learning which are involved in such 'specific hungers' with the rat's ability to avoid poison baits.

It is only when a new taste acts as the CS that reinforcement can be so delayed. In one ingenious experiment, Garcia and Koelling (1966) supplied rats with a drinking tube containing saccharin-flavoured water so arranged that when they licked the tube, bright lights flashed on. After these sessions the animals were irradiated with X-rays or, in other experiments, given an injection of lithium chloride; both treatments made them sick about an hour later. Subsequently the rats avoided saccharin taste but did not avoid flashing lights. Conversely, if they were given flashing lights plus immediate electric shock to the feet whenever they licked the tube, they subsequently avoided the light but still licked at saccharin. Rats are in some way 'prepared' to associate taste with sickness after a single trial and a long delay, but visual stimuli and sickness are not so connected. Conversely, light and electric shock are easily associated if they occur close together, but taste and shock are not.

Visual stimuli and sickness can, however, be associated in birds, and again this exemplifies how natural selection has been able to shape the way learning operates to suit the requirements of very different animals. Birds are visual hunters, in contrast to rats and many other mammals, and if delayed sickness is produced with lithium, they associate the *appearance* of novel food with this effect and avoid it later (Martin & Lett 1985). In the wild, they learn rapidly in this way to avoid poisonous caterpillars, for example.

Hitherto, although its timing clearly can vary with circumstances, reinforcement itself seems central to the learning process. Certainly, in the situations covered by standard Pavlovian conditioning or trial and error, there can be little doubt that it is a crucial factor. If Pavlov ceased the supply of meat powder, his dogs rapidly ceased salivating to the stimuli he was presenting, a process he called **experimental extinction**. Extinction also follows in the Skinner Box situation but, for reasons which are not clear, it usually takes much longer than with classical conditioning. Once a rat has learnt to press the bar for food, the proportion of reward to presses can be reduced to as low as 1 in 100 in some cases, and the rat will go on pressing. If rewards are stopped altogether it is a long time before the response finally extinguishes.

Pavlov realized that an extinguished CR did not just disappear and leave the animal as it was before conditioning started. In the first place, if we simply leave the animal alone for a few hours and then give it the CS again, the CR returns, i.e. it shows **spontaneous recovery**. This recovery is not back to the original level and the response extinguishes more rapidly, but this process of a pause followed by spontaneous recovery can be repeated several times.

A second way of reviving an extinguished response is to give a novel stimulus along with the CS. A dog which has had its conditioned salivary response to a bell extinguished, salivates again if a light flashes as the bell is sounded. Similar results have been obtained with rats in Skinner boxes. Pavlov called this process **disinhibition** because he regarded extinction as another new learning process which inhibited the original CR. Neutral stimuli presented with the CS early in the original acquisition of the CR often 'inhibit' it temporarily and reduce its strength. Similarly, perhaps the neutral stimulus disinhibits an extinguished CR by inhibiting the new learning that takes place during extinction.

In the heyday of rat and pigeon experimental psychology, reinforcement was a central concept and was often equated with 'drive-reduction'. Animals were said to have needs for food, water, etc. and would learn tasks which reduced these needs or drives. Without drive reduction learning would not take place. This immediately raised the question of how many drives there are and how distinct they are from each other – questions which have already been touched on in Chapter 4. Reduction of 'anxiety' or some similar concept was needed to explain avoidance conditioning (see p. 269), because once a rat has learnt the shuttle-box problem it gets no more conventional reinforcement (i.e. foot shock).

The drive-reduction hypothesis also runs into trouble with what we might call exploratory learning; in Thorpe (1963) and in Munn (1950) it is called **latent learning**. Rats given an opportunity to explore a maze at will, without any inducement from hunger or any other drive, turn out to learn it quicker than other naïve rats when both groups are later running the maze for food. No biologist is surprised by this result – animals learn the vital features of their home ground in this way. Initially, however, it caused great controversy amongst the adherents of drive-reduction, for what drive is reduced during exploration? We could equate it with anxiety if we wished, but there seems little point to the exercise. Monkeys will learn to press a bar in order to move a shutter aside, giving them a few seconds' view into another room with a toy train set in action. Reduction of boredom seems the best hypothesis! It is better to accept that conventional reinforcement is not the only route to learning and recognize the adaptiveness of the result and the biases – sometimes highly specialized ones – which may have become built into the learning mechanisms, whatever they may be, as a result of natural selection.

Specialized types of learning ability

If we look beyond the familiar rat and pigeon, prime subjects for laboratory studies of animal learning, we must expect to find species whose learning systems are specially adapted to match their very different life histories. We will look at two very different examples.

Honey-bees

Learning is particularly crucial to worker bees during the second half of their brief 6 weeks of life. This is the time when they act as foragers, making regular excursions outside the hive to collect nectar and pollen from flowers. Von Frisch (1967) and Lindauer (1976) give good accounts of the role learning plays in the life of bees and the types of features in their environment which they learn. Menzel and Erber (1978) and Menzel et al. (1993) review experiments on the learning process and its underlying mechanisms. Honey-bees learn rapidly and retain remarkably well the effects of even a single association between a colour and food reward. They make excellent subjects for experiments on their learning ability and will readily learn to visit a dish of sugar solution on a table.

In one set of experiments, Menzel and Erber (1978) arranged to have a food dish lit from below with light of variable wavelength, forming in effect an artificial flower whose colour could be switched at will (Fig. 5.7). Since it takes

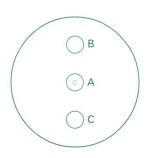

Figure 5.7 Plan view of a table with artificial 'flowers' provided in Menzel and Erber's experiments on learning in honey-bees. Marked bees feed at the table whose surface carries three ground glass discs which can be illuminated from beneath with either pure blue or yellow light. In the training situation only disc A is illuminated and a bee feeds at sugar solution from a tube which opens at its centre. The colour of A, blue or yellow, can be changed instantly at will by shifting interference filters beneath the table. When the bee next returns from the hive disc A is covered and discs B and C are now illuminated, one blue one yellow; again, colours can be changed at will. It is the bee's choice between these two which indicates what colour association it has made whilst feeding on A. Further details in the text.

a bee a minute or two to fill its crop with sugar solution, they could easily arrange to change the colour of a 'flower' during a visit. Using naïve bees as subjects, they changed flower colour between blue and yellow at various stages and then gave the bees a choice of these two colours to see which colour they now associated with the sugar reward. The results were quite dramatic and are illustrated in Fig. 5.8.

The association period is very brief, between 4 and 5 *seconds* only, and colour changes made outside this period are ignored. The timing of the period is also very precise and we can best measure it from the exact moment at which the visiting bee's proboscis touches the sugar solution, i.e. the moment when reinforcement begins. Then the colour stimulus must be present within 3 s before and 0.5 s after the start of sucking if an association is to be made. Sucking has to continue for 2 s for the bee to retain the memory, but the colour signal can be changed 0.5 s after sucking begins – it will still be associated with the reward and the bee will fly to this colour on the next visit.

So if the flower is, say, yellow as the bee approaches, switches to blue 3 s before it alights and begins to suck, switches back to yellow 0.5 s after sucking starts and remains yellow for all the time it is filling its crop – perhaps 1 minute in all – and is still yellow as it leaves, the bee subsequently associates blue with reward. This is extraordinary if we think in terms of conventional

Figure 5.8 Strong association between colour and reward is dependent on exact timing. As the bee approaches the disc its colour has no significance until about 4 s before the bee begins to suck sugar. Then there is a sharp peak of strong association, which has disappeared within 2 s following this onset of sucking. Apart from this brief period the colour the bee sees during the approach or for the whole time it takes to fill its crop, is of no importance. (Further explanation in text; from Menzel and Erber 1978.)

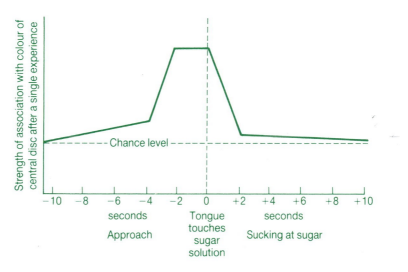

275

vertebrate learning. There we might well assume that the enduring association would be with the colour most recently and most persistently associated with the reward (note, for example, the response of rats given vinegar after saccharin flavour preceding induced sickness, which is described below on p. 280). However, for the specialized honey-bee learning system the brief flash of blue at the beginning is the significant feature, this degree of contiguity is enough. By confining a worker bee to the hive after such a trial, so that it was prevented from learning any other associations, Menzel and his co-workers could demonstrate that a single experience of this type was retained for several days. Three such trials on the one colour was retained for the whole of the rest of the bee's life! (Fig. 5.9). The same kind of time scales for both learning and retention hold true for conditioning to scents as well as to colours (Menzel *et al.* 1993).

In nature, honey-bees use both colour and scent in guiding their responses to flowers and learn both characteristics. Colour is effective during the early approach, but once they are close, the flower's scent when associated with reward is a powerful stimulus for bees to alight and probe. We must remember that honey-bee foragers are often directed to a flower crop by the dances of their hive mates (see Chapter 3, p. 179) and they also learn the position of each crop in relation to the hive. Most flowers do not secrete nectar

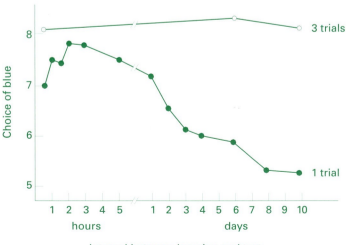

Figure 5.9 The time course of a bee's memory after training to associate colour as described. The results are expressed as choice of blue over yellow after 1 or 3 trials with disc A coloured blue. (Prior to any training blue and yellow are chosen equally.) One trial alone strongly biases choice for a day or so, three trials fixes it indefinitely. (From Menzel *et al.* 1993.)

uniformly throughout the day and bees are able to match their foraging activities to match the secretory cycle of flower crops and so increase their efficiency. Like many animals which use the sun or stars for navigation, bees have an inherent sense of time against which they regulate their behaviour through the daily cycle. Koltermann (1971; see also Bogdany 1978) showed that it is possible to train bees to associate a particular scent *at a particular time of day* with food reward. More, they could learn up to eight distinct pairings of a specific scent with a specific time of day and switch their choices appropriately as the day progressed. Thus, having been trained to associate orange blossom but not lavender with food between 1000 and 1100 h, and then the converse between 1100 and 1200 h, bees change their choice at the appropriate time. Such beautifully adapted behavioural responsiveness enables bees to be extremely efficient in deploying the huge team of foragers to the best effect and thus extracting the maximum amount of food from their environment.

Food-storing birds

Among the most striking learning specializations of vertebrates is that associated with food storage. When the chance arises, a number of bird and mammal species collect more food than they can eat immediately and hide it. Thus they can take advantage of a rich food source while it is there and reduce the degree to which they have to share it with others. Some animals make a single larder to which they return regularly, but most interesting from our point of view are those species which hide food items singly or in caches dispersed around their territory (see Sherry 1985, for more details).

For example, in Europe a small woodland bird, the marsh tit (*Parus palustris*) will store several hundred seeds a day in winter, secreting them in tiny crevices under tree bark or stones, etc. Careful observation in the field and experiments in aviaries have shown that the birds, though not perfectly recovering each seed, can recall dozens of dispersed hiding places for 1 or 2 days with extraordinary precision (Cowie *et al.* 1981; Shettleworth 1983). An American relative, the black-capped chickadee (*P. atricapillus*) can remember hiding places for at least 4 weeks (Hitchcock & Sherry 1990). However, perhaps the most extreme specialists in food-hoarding and retrieval are those members of the crow family called nutcrackers (genus *Nucifraga*).

Clark's nutcracker (*N. columbiana*) lives in the high coniferous forests of the western North American mountains. For winter food it depends almost entirely on caches of pine seeds. During late summer and autumn there is a huge glut of food and the nutcracker hides tens of thousands of seeds in several thousand separate caches in the soil, mostly on south-facing slopes. Each cache is made by forming a small hole with its bill and depositing a few seeds which it collects in quantity and can store in a large, distensible pouch under its tongue. The cache is then covered with soil and sometimes a stone is moved over to cover it. Figure 5.10 illustrates this behaviour in the European nutcracker (*N. caryocatactes*).

Working with Clark's nutcracker, Balda and Kamil (1992) had few problems in getting some captive birds to form caches in a large room with suitable holes drilled in the floor. The exact sites of each bird's caches were recorded and, once made, they were excluded from the room for various periods. Even after a lapse of *40 weeks* or more, the birds sought out the correct locations far better than by chance.

This is a very remarkable development of spatial learning and memory, the more so because close relatives of these food-storing species – other tits

Figure 5.10 A European nutcracker caches a pine seed at the foot of a tree. It has dug a small hole to receive a few food items (2–12 usually) which it will then cover up. It carries a number of seeds in an enlarged pouch (the subgular pouch) opening into its mouth cavity. (Drawing by Nigel Mann.)

such as the great tit (*Parus major*) or other crows such as the North American crow (*Corvus brachyrhynchos*) or the European jackdaw (*C. monedula*) – store little or not at all and, in general, perform less well in spatial learning tasks. Evidence is now accumulating that the brains of food-storers differ significantly from those of their non-storing relatives. The former have relatively more neurons in the hippocampus, an area which is known to be involved with memory formation and most particularly with spatial memory (see p. 312). Comparisons of this type which try to relate brain proportions to behaviour in different species are not problem-free, because it is difficult to be sure that we are comparing like with like. The behaviour of even close relatives can differ in many ways, but the evidence so far is quite convincing, that there has been rapid co-evolution of brain and behaviour in these specialized birds and some hints of the same development in mammals. This is an active field of research and a good example of the way in which comparative behavioural studies can contribute to an understanding of the brain and of memory mechanisms (see Shettleworth 1995 and Krebs *et al.* 1996, for useful reviews).

Knowledge is very incomplete, but if we survey the animal kingdom we can expect to find more specialized learning systems. Natural selection will shape associative learning into diverse forms to match life styles, but it is none the less valuable to look for common features. There is, at a behavioural level, a real equivalence between classical conditioning in a honey-bee, an octopus and a dog – the same logical processes are required.

What do animals actually learn?

It is here that some of the most valuable interactions between experimental psychologists and ethologists have come about. Approaching from rather different viewpoints, both have begun to investigate what it is exactly that animals are learning. As we have seen we can reject any idea that what they form is a simple, almost idealized association between *an* event and *a* reinforcement which occur in some regular sequence. It is more profitable to start from the notion, put forward early in this chapter (p. 265), that what animals must do is learn which events provide the best prediction of important outcomes. We must also remember that animals will come to all new situations with some inherent biases to pick on certain events as signals and to respond in particular ways.

Mackintosh (1983), Rescorla (1988) and Pearce (1996) cite a number of interesting examples of how this operates in practice. Thus in the taste aversion situation, described on p. 272, rats learn to avoid saccharin whose flavour has been associated with sickness some time later. If, between tasting saccharin and being given the injection of lithium which leads to sickness, the rats are given vinegar-flavoured water to drink, they subsequently avoid vinegar but not saccharin. The physical and temporal association between saccharin and sickness remains the same but the rats – sensibly enough, we may feel – choose to pick on the taste of vinegar as the key event. Because it occurred more recently, nearer to the reinforcement, it is taken as the better predictor.

A rather more complicated example shows another side of the same process. Two groups of rats were given experience of electric shocks in a Skinner Box where they were obtaining food by bar-pressing; not suprisingly, when the shock arrived, bar-pressing was depressed. In one group – we may call it group A – the shocks were paired with a distinctive light stimulus; in the other group, B, there was no light. Now both groups were given the light plus a sound tone stimulus together with the shock and this combination of light plus tone was certainly effective. The rats signalled that they had learnt, so to speak, because when light plus tone was given, now without any shock, bar-pressing showed a sharp decline – they anticipated a shock. But had the two groups learnt the same thing? The test came when they were given just the tone by itself. Now group B showed just as strong an effect on bar-pressing as with light plus tone. Group A did not: these rats paid little attention to the tone without the accompanying light. A's previous experience had told them that light was the best predictor of shock, so although the contiguity of tone and reinforcement were the same for both groups, they interpreted the situation differently. Experimental psychologists say that the rats' reaction to the light 'blocks' their reaction to the tone, and there have been a number of studies investigating how blocking works (see Pearce 1996).

Now it may be said that such an effect is just what we would expect. We probably mean by this that the rats are behaving as we would. But this is really the key point; the rats are not behaving like associating automata, they are behaving in a more complicated way: we might say intelligently. This leads us directly to the following question:

Are there higher forms of learning in animals?

Thorpe's (1963) book includes an examination of what he, following others, calls insight learning. He defines it as 'the sudden production of a new adaptive response not arrived at by trial behaviour or the solution of a problem by the sudden adaptive reorganisation of experience'. We can readily see the drawbacks of such a definition (e.g. how do we know a response wasn't arrived at by trial behaviour? etc.) which is not to say that, as humans, we do not recognize the phenomenon. Everyone can recall occasions when the solution to a problem has 'come in a flash', perhaps as the climax to several minutes of concentrated thinking. It is obviously going to be very difficult to demonstrate conclusively that there are similar processes going on in animals.

All we have to go on is what animals actually do in learning situations, but this is no reason to despair. As Dickinson (1980) discusses very well in his book, mental processes *can* be inferred from behaviour, even if it is difficult to do so. In practice, animal workers have used the term 'insight' when they observe animals solving problems very rapidly, too rapidly for normal trial and error – at least, too rapidly for the animal to carry out actual trials, but there is the possibility that it is 'thinking' about them and trying them out in its brain. This would imply that the animal can form ideas and reason; and studies on animal reasoning seem doubtfully distinct from those on insight. Accounts of insight or reasoning in animals are highly heterogeneous and there have not been many clear experimental investigations. It is the nature of the phenomenon which tends to defeat us, but this does not diminish its importance.

Some of the first attempts by experimental psychologists of the 1930s, to get some hold on insight and reasoning, involved 'detour' experiments in maze learning. One point at issue here relates to two contrasting views of how animals learn. The first, particularly associated with the learning theory of Clarke Hull, envisaged a rat learning a maze as building up a chain of stimulus–response (S–R) associations starting from the goal box and reward. The rat associates a corner or a left turn with the arrival at the goal – the first S-R association. As it explores further it learns that a previous corner or turn leads it to the first S–R point, and so it 'attaches' the second S–R to the first. Gradually, by a process often called chaining the rat puts together a chain of S–R associations which can then be used, in reverse order of their acquisition, to guide it from the start box through the maze. Hull therefore envisaged maze learning as essentially a set of simple associations with one leading, rather automatically, to another.

An alternative view of the process is linked with the work of E.C. Tolman. It can be stated much more simply, if less precisely than Hull's, because Tolman (1932) considered that rats build up a mental picture or cognitive map of the whole maze during the exploration and learning phase. They could then use this to find their way through avoiding turnings that they recall lead to blind alleys, choosing the path that leads most quickly to the goal box.

Figures 5.11 and 5.12 with their captions provide two examples of the detour experiments mentioned above, whose aim was to test reasoning in the rat. If suddenly presented with a new situation in a maze, or given the opportunity to take a short cut, how will it respond? Although it does not, perhaps, constitute a rigid test of the two theories given above, most would agree that Hull's 'S–R' view of the learning process would not leave much scope for the rat to take short cuts especially of the type offered in Shepard's complex maze (Fig. 5.12). The fact that some rats do change their path, and very rapidly, strongly suggests that Tolman's view is more sensible and that rats can show signs of reasoning and insight.

Detour studies of this type – more examples are given by Munn (1950) and Thorpe (1963) – often have three features in common. Firstly, they all include a longish period of exploration prior to any testing, so that the animals will already have learnt a good deal about the general features of the situation.

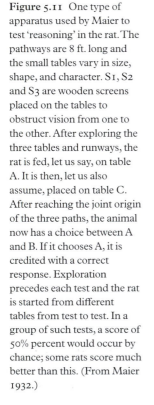

Figure 5.11 One type of apparatus used by Maier to test 'reasoning' in the rat. The pathways are 8 ft. long and the small tables vary in size, shape, and character. S1, S2 and S3 are wooden screens placed on the tables to obstruct vision from one to the other. After exploring the three tables and runways, the rat is fed, let us say, on table A. It is then, let us also assume, placed on table C. After reaching the joint origin of the three paths, the animal now has a choice between A and B. If it chooses A, it is credited with a correct response. Exploration precedes each test and the rat is started from different tables from test to test. In a group of such tests, a score of 50% percent would occur by chance; some rats score much better than this. (From Maier 1932.)

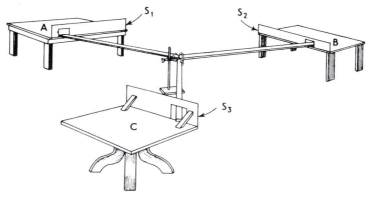

Secondly, performance is judged by speed of solution. If the rat makes mistakes or has to explore further it is assumed that a simpler, trial-and-error form of learning is being used, and not insight. Thirdly, it is an interesting fact that rats and other animals which have been similarly tested are highly variable in their apparent insightfulness; it is often only 20–30% that respond rapidly and we do not know what factors of prior experience or genetic constitution contribute to these differences.

As mentioned earlier, a diverse series of observations have often been lumped together as evidence of insightfulness or reasoning in animals, but sometimes we may confuse capability *per se* with intellectual ability. For example, some remarkable findings on the capacity of pigeons to 'form concepts', as we might say, is often discussed in this context. Using a quite simple Skinner box situation which projects photographs of scenes above the key at which the pigeon must peck, Herrnstein *et al.* (1976) have shown that pigeons can learn to discriminate one set of pictures which share a 'feature' in common from other sets which lack this feature. The features were such objects as human beings, trees or water. What strikes us about this result is that the 'human being' could be a picture of a crowd, a lone figure on a lawn or a close-up of somebody's head, the 'tree' a distant view of a wood, a single tree

Figure 5.12 A complex maze used by Shepard to test 'reasoning'. After rats have learned the maze, the wall indicated by X is removed, thereby causing a previous blind alley to become a short cut. Having discovered the change whilst running along 11c, and exploring from there a little into 4c, some rats entered 4 (and thence 4a, 4b, 4c) instead of 5 on the next trial. (From Maier & Schneirla 1935.)

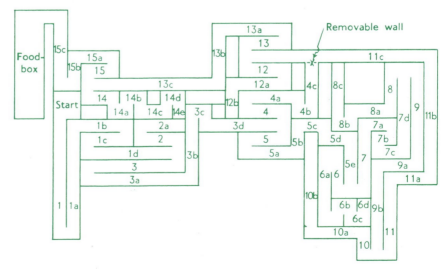

or a pattern of branches. Our first inclination may be to suppose that the pigeon is categorizing objects in the world rather as we do, a rather amazing conclusion because we probably feel it must represent advanced intellectual capacities with which we are reluctant to endow the humble pigeon!

In fact, we are probably misled in this view by our anthropocentrism, which tends to equate human capacities with intelligence. Pigeons, evolved for flight and intensely visual animals, must have a sensory system adapted to rapid recognition of features in a rapidly changing environment. As they swoop over the landscape they must respond to trees in the distance as a massed feature, or close to and singly as they dive into the branches to alight. Viewed in this way, the fact that the pigeon can classify diverse images on its retina into certain common categories is more readily seen as a matter of sensory processing rather than intellect *per se*. We may note that Hollard and Delius (1982) have shown how quick pigeons are to recognize those rotated shapes which form a component of human intelligence tests (see Fig. 5.13). But, again, pigeons are likely to have a specialized nervous system and, in any case, far more practice than we do at seeing shapes apparently rotate below them as they fly over their home areas. This type of explanation might still assume that pigeons are categorizing their visual world in the same way that we do. The categories 'human beings', 'trees', 'water', etc. are human ones and, although it is not easy for us to detect them, pigeons may be classifying pictures very differently. However, Delius (1992) has demonstrated a striking capacity in pigeons to learn and remember a whole series of 'spherical' versus 'non-spherical' shapes and suggests that they categorize these two sets in rather similar ways to humans.

We have no conclusive answers as yet and in the search for intellectual capacities it is probably better to get away from tests which tend to accentuate sensory processing. If we are to acknowledge true insight in animals it must come from the identification of truly novel behavioural associations, such as we referred to in our original definition. It is not surprising that many of the best examples come from primates. One of the most famous is the work of Wolfgang Köhler (1927) on chimpanzees, described in his book, *The Mentality of Apes*. Fond of bananas, they would learn to overcome hurdles of increasing difficulty to get at them. If presented with a bunch too high to reach, they would pile up boxes to make a stand for themselves or fit two sticks together in order to pull down the bananas. Often they arrived at this solution quite suddenly, although they benefited by previous experience of playing with boxes and sticks and showed considerable and obvious trial and error when actually building a stable pile of boxes. Köhler's chimpanzees were using knowledge obtained in one context (something of the properties of sticks and boxes) and applying it in another. There surely is no question but that apes and other primates can show true reasoning on occasions. We may recall the examples of deliberate deception within wild primate groups, mentioned in Chapter 3 (p. 179). Many dog and cat owners will cite examples of reasoning in their pets; this is possible – after all, we have just described experiments which suggest that rats have some reasoning ability – but we must take care to exclude other explanations.

Before we try to extend our discussion of insight learning to consider the nature and the implications of reasoning ability in animals, it is necessary to reflect on why we should assume that primates will be more able to reason than dogs, and dogs more than rats. We are tacitly assuming that learning

Figure 5.13 Test to measure the ability of pigeons to recognize complex shapes even when rotated from the familiar. (a) The experimental procedure, a Skinner box with three keys. Trained to respond to a shape presented on the central key, the pigeon is now given three shapes on the three keys – see the rows in (b) – with a reward for pecks to the side key whose shape is identical to the sample. When the wrong shape is a straight mirror image of the sample, as in the top two rows of (b), the task is quite easy. It becomes (for us) much more difficult when the comparison has to be made with shapes rotated from the sample orientation. (c) Some of the shapes with which pigeons were tested in this way. They are remarkably quick and efficient at such discriminations. (From Hollard & Delius 1982.)

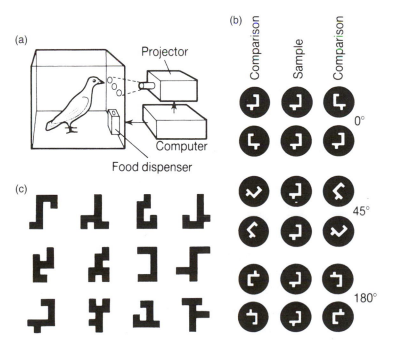

ability and 'mentality' have, like brain size, shown trends during the evolutionary history of the vertebrates.

The comparative study of learning

Comparative psychology has a long history (Dewsbury 1984) and, at least in the early stages, researchers were working with a wide range of animals even if later the rat and the pigeon came to dominate. Psychologists have always tended to concentrate upon studies of learning and recently those with an ethological background have also come to study the diversity of learning abilities, so we have a reasonable body of data to search for evidence of evolutionary changes.

There are all kinds of inherent difficulties in making such a survey. The first and most obvious is our essential vanity concerning human intellectual powers, which often leads us to look for progression upwards amongst animals, leading towards human beings on a very detached, high pinnacle. Any biologist must expect there to be other pinnacles on alternative pathways. Honey-bees are not to be judged by the same criteria as monkeys. Nor is it sufficient to rely on laboratory studies of learning alone; we must try to get some estimate of the role that learning plays in the natural life of different animals and also attempt to distinguish between simple and more advanced types of learning which, as we have seen, is not always easy.

One approach has been to examine the correlation between brain development and learning ability. Certainly the prevalence of learning, the capacity to process information from diverse sources and the general complexity of behavioural responses are greater in mammals and birds than they are in fish and reptiles and we can reasonably associate this difference with the evolution of a large brain – encephalization, as it is termed (Jerison 1985). It is not, of course, just brain size that is important; whales and elephants have larger brains than ourselves but smaller, though considerable, learning powers. It is less easy to construct a series among the invertebrates, which are much more heterogeneous, but advanced insects and cephalopods have the largest brains of their respective phyla, Arthropoda and Mollusca, and also the greatest capacity for learning. However, as we mentioned in Chapter 2, the two most advanced groups of insects, the Diptera and the

Hymenoptera, both have brains enlarged and much modified from the ancestral type, but they differ markedly in learning ability.

Within the vertebrates the most dramatic aspect of brain evolution is the enlargement of the cerebral hemispheres, especially the cerebral cortex, which reaches its greatest extent in the primates. However, we can reject any simple equation of cerebral cortex size with learning ability. The birds have in the past often been underestimated because, although their brains are relatively large, those parts homologous with the mammalian cerebral cortex are small. But the birds have evolved along a line separate from the mammals for some 250 million years. They have evolved a different type of brain structure, yet learning ability in some bird families, notably the titmice (or chickadees) the crows and the parrots, is in some respects second only to that of the primates.

One of the clearest demonstrations of this ability comes from Pepperberg's (1990, 1991) accounts of the remarkable capacities of an African grey parrot (*Psittacus erithacus*). She has developed a very effective way to train this bird – named Alex – to associate words with objects. He acquired some nine words for objects such as wood, cork and paper, which he liked handling with his bill and feet, and a number of other words. Of course, one of the most striking features of the situation is that Alex has the vocal ability to imitate human speech. The fact that he seems to 'speak' to us seems so extraordinary that we may be in danger of overestimating his intellectual powers, an issue we have already encountered when discussing the abilities of pigeons to recognize rotated shapes. Nevertheless Pepperberg has proved that, for example, Alex can count up to six and identify the quality, i.e. shape or colour, which a group of diverse objects have in common. Such abilities are probably beyond most monkeys.

It is clear that brain structure alone is inadequate as a guide to learning abilities, and to study the evolution of learning we need to compare how different animals perform on particular behavioural tests. In selecting representative animal types for our evolutionary analysis, it is all too easy to refer to 'higher' and 'lower' animals. Within the vertebrates, for example, we often find the sequence fish–amphibian–reptile–bird–mammal quoted as an evolutionary scale of increasing complexity of behaviour and increasing learning capacity. The construction of such a scale from living representatives of each class ignores the actual course of evolution. We have just mentioned the completely separate histories of the birds and the mammals. All the living vertebrate species are equally distant in time from their common ancestor and all are specialized for their particular modes of life. It would be naïve to expect the learning capacities of a modern teleost fish (e.g. the goldfish, commonly used in learning studies) to reflect accurately those of the ancestral fish from which the teleosts and other vertebrates diverged some 400 million years ago. Hence the construction of a valid phylogeny of behaviour is fraught with difficulties because behaviour does not fossilize. Hodos and Campbell (1969) provide a vigorous critique of the phylogenies that have sometimes been constructed by comparative psychologists.

A further problem for comparative learning studies is that of devising truly comparable situations for testing different animals which vary so widely in their sensory capacities and manipulative ability. The procedures needed to measure discriminative conditioning in an octopus, a honey-bee and a rat have to be very different and we can no longer be sure that problems are of equal difficulty or that the animals 'see' them in the same way. Motivation and reinforcement present further problems: the level of motivation often affects the rate of learning and may, indeed, determine whether the animal learns at all. How can we equate levels of hunger motivation in a rat and a fish? The latter may live for weeks without food, the former only days. It is just as difficult to equate reinforcement between different animals. A small piece of food may be an excellent reinforcement for a hungry mammal, but mean much less to a fish and less still to an earthworm. It is perhaps easier to equate punishment, since all animals 'dislike' electric shocks, though even here there are difficulties because shock or fear affects the behaviour of animals in such diverse ways. As we discussed earlier (p. 269), animals come to learning situations with a good deal of built-in bias. To equate the effects of punishment we require some knowledge of each animal's natural responses in fearful situations. Whenever possible we must try to use a reinforcement which is directly relevant to the animal being tested. Escape into a darkened area may be best for many small invertebrates. Schneirla (see Maier & Schneirla 1935) and Vowles (1965) found that the best reinforcement for maze learning by ants was to get back to their nest.

In the face of all these difficulties there are those who question the validity of any comparisons of intellectual ability between different animals.

McPhail (1985, 1987), for example, takes up a position which we might call constructive provocation and argues that, 'there are, in fact, neither quantitative nor qualitative differences among the intellects of non-human vertebrates'. Few would agree with this, but recourse to common sense should not be our only reply: we must try to devise valid comparative tests.

Amongst the vertebrates, a good series of learning tests ought to be able to record stages in the evolution of 'intelligence'; speed of learning – a quality much admired in human schoolchildren – might seem to be one useful measure. However, a brief survey of the literature shows that speed alone does not tell us much. Ants and rats, for example, show very comparable speeds when first learning a fairly complex maze. Gellerman (1933) describes in detail experiments in which two chimpanzees and two 2–year-old children were learning that a food reward was associated with a white triangle on a black square and not with a plain black square. One child learnt in a single trial, but the other took 200 trials and both chimpanzees took over 800 trials to reach the criterion of 19 correct trials out of 20. On a comparable test most rats would learn in 20 to 60 trials, though admittedly they would usually be mildly punished for wrong responses as well as rewarded for correct ones. Discrepancies of this kind abound both within and between species and we have no reliable evidence that speed of initial learning for simple associative problems varies within the vertebrates, or even between them and the advanced invertebrates.

However, speed is only one aspect of learning; we might also ask what is learnt. For instance, Fig. 5.14 shows that, although the chimpanzee may take longer to learn the triangle discrimination outlined above, it learns more about 'triangularity' than does the rat. One aspect of 'intelligence' is the ability to strike a reasonable balance between generalization and discrimination in tests of this type. Similarly, if we consider more complex forms of learning then we can at least note some gradation of ability within the vertebrates.

Testing for what have been called learning sets has been useful in this respect. If an animal can form a learning set it means that it can learn not just a problem, but something about the principle behind it, and then can steadily increase its learning speed when given a series of similar problems. Harlow (1949) has described the basic technique with primates. A monkey is presented with a pair of dissimilar objects – a matchbox and an egg-cup, for example. The matchbox, no matter where it is placed, always covers a small food reward, the egg-cup never has a reward. After a number of trials, the monkey picks up the matchbox straight away. Now the objects are changed: a child's building block is rewarded, a half tennis-ball unrewarded. The monkey takes about the same time to learn this, again the objects are changed, and so on. After some dozens of such discrimination tests the monkey learns each discrimination much more rapidly, although viewed as an individual problem it is just as difficult as the first one. Eventually after 100 or so tests the monkey presented with any pair of objects lifts one and, if it yields a reward, chooses it for all subsequent trials. It has learnt the principle of the problem or, in Harlow's terminology, it has formed a learning set.

This is one type of learning set based on successive trials of discrimination. Perhaps a simpler version is the 'repeated reversal' problem. Here we train the animal to select object A in preference to object B. Once learnt, object B is now rewarded and A unrewarded; when this first reversal is learnt, the reward is again given with A, and so on. If the animal gets progressively quicker at learning each reversal, this again implies that it has learnt a principle.

The ability to form learning sets was once regarded as a capability of the more advanced mammals only, but we now know this is certainly not the case.

Figure 5.14 The concept of 'triangularity'. Trained to respond to the top figure, a rat makes random responses to any of the lower figures. A chimpanzee responds to (a) and (b), but makes random responses to (c). A 2-year-old human child recognizes 'a triangle' in (a), (b) and (c). (From Hebb 1958.)

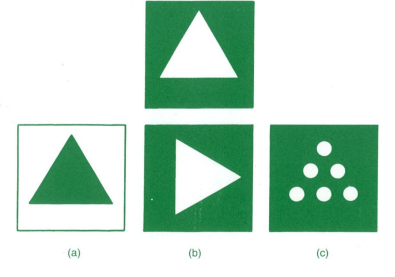

(a)   (b)   (c)

Warren (1965) summarized the comparative data to that date which suggested that all the vertebrates with the exception of fish shared this ability. In fact, later work (Mackintosh *et al.* 1985) has shown that goldfish can form repeated reversal sets and so can one of the most advanced invertebrates, the octopus (Mackintosh 1965).

Although, as we have already seen, the speed of learning a simple discrimination does not vary systematically between different animal groups, the speed with which learning sets are acquired does change dramatically. Figure 5.15 clearly shows this phenomenon for different animals acquiring discrimination learning sets. The subjects vary greatly in the time they take to improve. As Warren (1973) points out, there are all the usual problems in taking this result at its face value; the types of manipulation involved in testing a squirrel are very different from those for a cat or a monkey. Most people would not wish to argue too much about comparative intelligence from this sort of data. Sometimes, though, we can arrive at a safer kind of comparison, even if the subjects are very distant in evolutionary terms. Figure 5.16 compares rats and goldfish solving two of the easier reversal learning set situations, one a spatial task, left versus right, the other a visual brightness discrimination. Both learn the original discrimination and its first reversal – always the most difficult – with about the same number of errors, which suggests that

Figure 5.15 The rate at which various mammals can form the more demanding discrimination learning sets in which each problem presents them with completely novel objects. Thus, at each change the animal's choice on the first trial has to be random, but if it has learnt the principle behind the problems, trial 2 should be correct. Note how long it is before the scores of rats or squirrels on trial 2 become better than chance or 50%. Many monkeys reach almost 100 % within 400 problems. (From Warren 1965.)

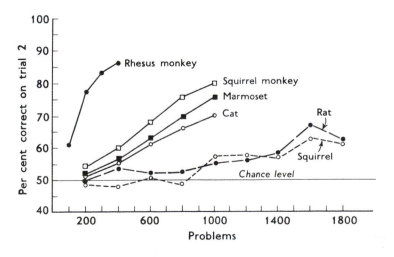

both animals find the problem roughly 'equal' in difficulty. Thereafter, however, rats improve much more rapidly than goldfish. It may be significant that nobody has yet succeeded in demonstrating that fish can ever improve their rate of learning to discriminate between novel pairs of objects as were used in the tests shown in Fig. 5.16.

Many of the uncertainties with which such comparisons are afflicted still remain; nevertheless, with caution we may conclude that cats and dogs do reveal more signs of intelligence than rodents and that the superiority of primates is clear. Perhaps not suprisingly in view of their large brains, dolphins (the only cetaceans small enough for intensive study) have also proved quick to acquire learning sets (Herman & Arbeit 1973). However, it seems safest to regard the superiority of cetaceans and primates as a quantitative one; we have no evidence that they possess any abilities which are not foreshadowed elsewhere among their mammalian relatives.

One cannot help feeling vaguely dissatisfied by this conclusion. Especially for the primates, the rather circumscribed artificiality of many learning experiments does not seem to do justice to the great flexibility and ingenuity of animals such as chimpanzees. Here we are undoubtedly influenced by their similarity to ourselves, particularly because they can manipulate objects much as we do. We may underestimate the intelligence of other animals because their structure and environment is so different from

Figure 5.16 Repeated reversals for learning visual and a spatial discrimination by goldfish and rats. The graph plots the number of errors to achieve the criterion of all responses to the rewarded object. The point o represents the original task; note how scores both for it and for the first reversal (always the most difficult) are very comparable for both animals. Thereafter, with repeated reversals, the rat improves much more rapidly than the goldfish. (From Mackintosh *et al.* 1985.)

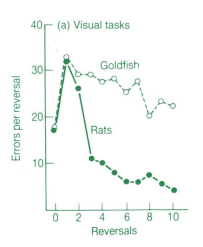

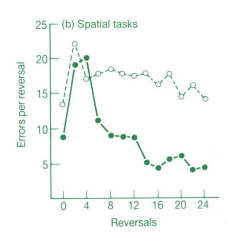

our own. The cetaceans are an obvious case in point; we have only just begun to take some measure of their behavioural complexities.

Can animals think and reflect on their actions?

This issue brings us back to questions which we raised when introducing the topic of insight learning, and addressing them forces us to attempt going beyond mere description of how animals behave in learning situations. It is surely impossible to keep a pet or to observe any animal carefully without wondering whether it 'thinks' or has a 'mind' and 'consciousness'. Those words are in quotation marks because, whilst none of us doubts that we possess them – *Cogito, ergo sum* – they are not easy to define exactly and it is even harder to suggest how we might set about proving them to exist in other animals. After all, each of us has no proof positive that they exist in any human being other than ourself; it is just reasonable to assume it because we are built the same way and behave the same way. As we mentioned at the outset, the information we have which bears on these issues is not very systematic; often it is frankly anecdotal. We may consider three such examples as illustrations.

The first comes from casual observations made over a century ago by the great biologist, Alfred Russell Wallace, who shared with Darwin the original formulation of a theory of evolution by natural selection. He spent much of his life on collecting trips in the Far East and whilst staying on the island of Borneo he had opportunity to observe a captive orang utan. Around its cage there foraged some of the local village's domestic chickens. The orang showed great interest in them and once or twice tried to capture one, but they evaded the ape's long arms. One day Wallace saw the orang scatter grain from its food dish outside its cage, dropping some grain just outside the bars. It then sat quietly as the chickens foraging nearby found the grain and followed it right up to the cage, whereupon the orang flashed out an arm, captured a bird and killed it!

Our second example comes from an interesting article on the behavioural capabilities of sheep-dogs and the way they are trained:

> Dogs experience particular difficulty when faced with recalcitrant ewes with lambs, one such ewe which had split off from the main flock refused to be moved and faced the dog square on, stamping its hooves. The dog returned to the main flock, cut off several sheep, and brought them over to the stubborn ewe. The ewe promptly joined this group and the dog was able to move them all back to the main flock. (VINES, 1981)

The last example records an aspect of honey-bee behaviour which a number of people have noted rather than studied. In the kind of experiments we have already described in connection with honey-bee communication (p. 184) it is often necessary to train them to forage at a dish some hundreds of metres from the hive. It is possible just to put out a dish of sugar solution and wait, but more often one begins by placing a colour-marked dish on the landing board of the hive. Once some bees have begun to feed, the dish is moved a few centimetres back; usually some bees will still find it and continue to feed. The dish is then moved further out, and further. At first, moves of more than a metre or two disrupt the foragers and they have considerable difficulty, but as the day wears on it becomes possible to move the dish in much bigger steps – 10 m, 20 m or more. At this stage, people have reported a remarkable phenomenon. Sometimes as they move the dish out to its next position, they find bees flying around, already there searching. Markings show that these are not new recruits alerted by dancing and searching vaguely in the same direction: they are the regular visitors, 'anticipating' the next position of the dish.

All these examples raise a host of intriguing questions. As is so often the case, the descriptions of the orang utan and the sheep-dog are one-off cases. We might see similar behaviour again, but it is the essence of the situation that the observations are not repeatable. This does not make them any less interesting or important. Some may argue that to have observed such behaviour at all is of profound significance for our view of animals. Others will immediately suggest caution. How are we ever to know their significance? Perhaps the orang spilt its food completely by accident: apes are not tidy feeders. Wallace's – and our – assumption of insight and imagination may be completely misplaced. So we have to face difficulties both with the accuracy of such observations and with their interpretation. Similarly for the sheep-dog: what prior experience had it? Vines seems to suggest that dogs often have trouble with individual sheep. It might have had an opportunity to copy this strategy for dealing with them from older dogs during its training. If so, would this make any difference to the way we interpreted the behaviour? The honey-bee example raises some different questions. The behaviour in itself is not very complex, but it is the fact that it comes from an insect which makes it so extraordinary. We are usually unwilling to entertain the possibility that insects think about the future, or could deduce a rule that tells them the food dish will next appear in that direction.

When discussing the source of the adaptiveness of behaviour in Chapter 2, we noted that through history animals have sometimes been regarded as little better than automata, sometimes as fully sentient beings like ourselves; Boakes (1984) and Walker (1983) both provide interesting reviews of these changes. For most of this century, with the rapid rise of science, the pre-dominant view has not regarded it as profitable to speculate on the nature of animal thinking. Rather, as we have seen, attention has been focused on analysing what animals actually do, in both natural and laboratory situations, and deducing the causal factors, the function and so on. Most serious workers have always had at the back of their minds a rule proposed around the turn of the century by Lloyd Morgan, one of the founding fathers of comparative psychology, now referred to as 'Morgan's canon':

> In no case may we interpret an action as the outcome of the exercise of a higher psychical faculty, if it can be interpreted as the outcome of one which stands lower in the psychological scale. (MORGAN, 1894)

It must be said that Morgan's canon has served us well. It has enabled us to approach the analysis of behaviour sensibly and to avoid the superficial anthropomorphism which led to many absurdities in the past. In Germany around the turn of the century, the eccentric Baron von Osten seriously believed that many animals, and certainly horses, could perform mental arith-metic – all they needed was the right training. His best pupil, 'Clever Hans' became world famous, through being able to give the answer to sums by beating the ground with his hoof the correct number of times (see Fig. 5.17). In fact it turned out that he was responding to subtle unconscious cues from his owner. If the Baron did not know the answer himself, the horse got the wrong answer too. Not long before, another Morgan – Lewis – studied beavers in the northern USA and he believed that they revealed an extensive understanding of hydraulics by the construction of their dams and channels. Both of these classic examples of anthropomorphism are described more fully by Sparks (1982).

However far we can progress without invoking concepts of mind, thought or foresight, there is now a considerable body of examples accumulated – such as those just given – where it is much more difficult to avoid them. We may struggle to explain the sheep-dog's response, in terms of a chain of S–R units with appropriate reinforcements, but such a procedure totally lacks conviction. As Walker (1983) puts it, 'sometimes explanations can be too simple to be sensible'.

We are indebted to Griffin (1984) for re-opening the whole issue in a compelling way (see also the collection of essays in his honour edited by Ristau 1991). A substantial number of animal behaviour workers would now accept that, no matter how difficult they may be to study, it is no longer sensible to deny the possibility of true thought processes and even consciousness of some sort in some mammals and birds. One of the most powerful arguments for this conclusion is an evolutionary one. It is hard to accept that mind and consciousness in human beings have just arisen *de novo* without *any* precursors in animals which were ancestral to us and probably very similar to the non-human primates which we observe now. It is obvious that there would be

Figure 5.17 Baron von Osten with 'Clever Hans' and minder; a photograph taken about 1904. The day's lessons for Hans are set out on the blackboard and an abacus is at hand for aid with arithmetic. (Photograph from Mary Evans Picture Library.)

considerable advantages to possessing even some slight ability to anticipate the consequences of one's actions and predict the responses of others. Such abilities would confer a selective advantage, and probably a powerful one, to animals that possessed them and perhaps particularly to social animals. Humphrey (1976) has argued that the complexities of social life, especially in the primates, provided one of the most important selective forces behind the emergence of thinking ability. Increasing brain size, developing new populations of neurons not designated for sensory or motor processes but free for associating diverse information and for parallel processing, would be the physical basis for such abilities.

Even if we accept such arguments, we must not abandon Morgan's canon. For example, we should not accept the idea that honey-bees have such capacities before eliminating every other possibility. In explanation of the example we cited above, Griffin suggests that in their normal lives honey-bees sometimes encounter a situation when a food supply does extend out in one direction as, for example, when the shadow of a hill or tall trees moves off a flower crop with the rising sun. As the flowers are warmed by the sun, they begin to open and secrete nectar and bees can now extend their feeding range, step by step as the shadow retreats. It could then be argued that the ability to move out along the line of food dishes is a manifestation of an inborn ability. This most certainly does not explain how the bee's nervous system can develop or operate so as to organize such behaviour, but it need not require that the bees consciously reflect on what they are doing. Some kind of unconscious, more automatic rule-of-thumb may be working, as we suppose it does in the bees' remarkable dance communication discussed in Chapter 3.

Similar rule-of-thumb explanations may suffice for many animals, but surely not all. For example, Povinelli *et al.* (1992) carried out a very neat experiment which strongly suggests that chimpanzees have the ability to understand the outcome of their own actions, the way that these affect another individual, and also to put themselves effectively into that individual's place. The apparatus they used is illustrated in Fig. 5.18 and explained in the caption. It requires two chimpanzees to cooperate to get a reward, with each having a distinct and different role. Povinelli *et al.* found that each role-player needed only a brief demonstration of how to operate, working with a human collaborator. After a few operations of the apparatus the experimenters swapped the two animals round so that they had to take on the other's role; three out of four chimpanzees did so without hesitation. We can scarcely explain this away as mere copying of a motor pattern from one to another because the viewpoint is so different. For instance, the operator chimp has to appreciate the significance of the informant's gestures and, after the swap, take on this gestural role. It does so spontaneously and using its own gestures which are not necessarily the same as those it had seen used before by its partner.

Many people have regarded this ability of an individual to recognize its own identity, and that of others outside itself, as a first requisite of consciousness. Accordingly there has been a great deal of attention directed to this issue in primates. Self-recognition, as represented by identifying one's mirror image, or steering actions using a television image of oneself to reach objects which are out of direct sight, has been claimed many times for chimpanzees and orangutans. Even so the claim is contested and the controversy makes for profitable reading, e.g. Kennedy (1992), Gallup *et al.* (1995) and Heyes (1995).

Captive chimpanzees often adjust very well to human surroundings and to human companions. There is now a burgeoning literature on the complexities that their behaviour can reach in such situations. One of the earlier series

Figure 5.18 Apparatus developed by Povinelli and his collaborators to test whether chimpanzees could interpret the role played by another individual and adopt this role when it was their turn. Some attractive food is put, into one pair of trays and is visible to a chimp on the 'information' side (upper tray). In the lower tray, the operator has pulled the handle on the side indicated by the informant and both chimps get a reward. After working in one role, they could immediately adopt the other role when their places were exchanged. (After Povinelli *et al.* 1992.)

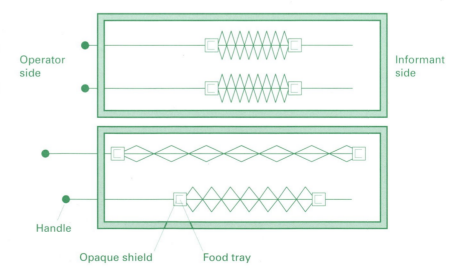

of studies reported by Premack & Woodruff (1978) and Savage-Rumbaugh *et al.* (1978) attempts to take further the demonstration that chimpanzees can comprehend the position of other individuals, in this case humans. Thus, a chimpanzee was shown video tapes of a human being interacting with certain objects, all known well by the chimp. Various situations were portrayed which implied a need. The person was seen shivering and huddled with cold in a room with an unlit stove, or struggling to get out through a locked door. The chimp was then offered a series of photographs of objects, one of which offered a solution, for example the burning wick of the stove, a key for the door. Could the chimp anticipate the need of the person separate from itself and choose for them as it might choose for itself? The evidence suggests it can. One chimp chose the 'correct' picture seven times out of eight. (Fascinatingly, it chose 'neutral' photographs much more commonly with videos of a person it disliked!)

Such studies are difficult to arrange and carry out. In the discussion following the papers referred to, various commentators point out difficulties of interpretation and it remains essential to be rigorous in eliminating explanations which require only simpler associative abilities. We need only refer to the 'chimpanzee language controversy' for a prime example of the difficulties that arise and the heat that they can generate amongst the scientists involved! Kennedy (1992) gives a thorough if totally sceptical review of attempts to train chimps to use symbolic language; Dawkins (1993) is more moderate but shares many of the same doubts. One pioneer study, that of the Gardners with Washoe, used the gestures of American sign language for the deaf; others have used plastic shapes to represent words or computerized versions of them. There is no disagreement regarding chimps' ability to learn large numbers of associations between signs or symbols representing objects or conditions (such as 'more' or 'open'). The dispute concerns their ability to combine them in 'grammatical' ways or in novel combinations. Critics have suggested that unintentional cueing from the experimenters or the experimental situation – as with Clever Hans – produces spurious effects.

Nevertheless, more recent studies with the pygmy chimpanzee, or bonobo (*Pan paniscus*), both in captivity and in the wild – the forests of the Congo basin – reveal complexities and subtleties of behaviour which most people would feel defy such simplifications. For example, one 'home-reared' bonobo could follow verbal instructions given through earphones or via a loudspeaker so that all visual clues were absent (Fig. 5.19). 'Kanzi, put the hat in the refrigerator' is a sufficiently bizarre instruction when given for the first time, that it is hard to believe but that the ape recognized the significance of the whole sentence. He didn't, for example, put the hat *on* the refrigerator. Savage-Rumbaugh and her collaborators (1996) make claims for the language comprehension of bonobos which go well beyond any others so far. For example, they say,

> Understanding of symbolic communication arises spontaneously in the bonobo, . . .
> Human speech can be readily understood by apes when exposure occurs at an early age.
> The correct meanings are ascribed to spoken words spontaneously, without shaping or
> planned reinforcement by human beings. (SAVAGE-RUMBAUGH *et al.* (1996))

This attention to language reflects, of course, a very human interest, but as biologists we must try to get beyond it to consider the abilities of animals in relation to their own life requirements. The ape workers are well aware of this and Savage-Rumbaugh and her group have been following bonobos in the

Figure 5.19 Kanzi, a young male pygmy chimpanzee or bonobo, wearing earphones through which he responds to a whole series of verbal commands when there is no possibility of his getting other clues from the experimenters. (Photograph by Sue Savage-Rumbaugh.)

wild trying to understand how they communicate and whether any hints of their remarkable achievements in a laboratory setting can be related to their natural life. Bonobo groups exhibit a range of social behaviour patterns which help to maintain their cohesion as they move through dense vegetation of their forest habitat. As they forage during the day groups often break up into separate smaller parties which gather together again before settling for the night. Collaborating with skilled African trackers, the workers describe indications that foraging parties deliberately mark vegetation along trails which can indicate their path to others. Sticks were sometimes found stuck into the ground, leaves were pressed down far beyond the extent needed for passage for, in other circumstances, bonobos can move without leaving a trace. Could this be an indication of the symbolic communication which they take to so readily in a human setting?

We urgently need more careful observations of this kind and, inevitably, much of the work will have to be with primates who are most likely to share sophisticated mental processes with us. Nevertheless, it is also important to keep up such investigations with other animals which have their own specialized cognitive abilities. Amongst the birds one thinks of the crow and parrot families, and for other mammalian groups, the carnivores, cetaceans and certainly elephants. Common observation would tell us that these include some extremely intelligent animals. Elephants, in particular, have large brains and complex social life coupled with longevity comparable to our own. Quite often the starting point for a fresh recognition of the complexity of a species' behaviour has come from the new approaches to the study of communication between members of a social group, some of which we outlined in Chapter 3. Thus we are now beginning to learn about the context and the significance of the 'infrasonic' (8–10 Hz) communication between elephants (Poole *et al.* 1988).

However admirable has been the deliberate policy of behavioural scientists to reject subjective explanations of our observations, it may have led us to underestimate some of our animal relatives in the past. It is most important that we no longer continue to do so because everywhere, in the Western world at least, there is increasing concern for the conservation and welfare of animals, both domestic and wild. It is the responsibility of behavioural scientists to provide a sensible and practical basis for our efforts (see Dawkins 1980, 1993; Fraser and Broom 1990).

The nature of memory

Learning is nothing without memory. We must be able to store the results of experience and recall them to our benefit later. It is surely one of the most remarkable properties of the nervous system that it can retain some representation of past events for almost a lifetime – tens of years in some cases.

The nervous system operates by transmitting electrical impulses along defined pathways and certainly the process of learning involves heightened activity in those channels that record sensory impressions and their outcomes. It seems unlikely, however, that heightened activity *per se* could constitute memory, that a memory could be stored in the form of a continuous train of nerve impulses running for years around the same pathways. Such ideas of 'self-reverberating circuits' were entertained at times in the history of psychology, but can now be abandoned. Perhaps the neatest disproof was provided by Andjus and his collaborators (1955). They managed to cool rats down to 0 °C for periods of up to an hour, at which temperature all electrical activity in the nervous system ceases. When warmed up again, these rats retained their memory of events prior to cooling as well as normal animals.

If, on the other hand, the establishment of memory involves *structural* changes of some kind in the nervous system so that some channels are facilitated, then we can readily understand why function is restored when electrical activity begins again. It is now generally accepted that memory storage must be represented in a chemico/physical form and we have enough evidence to understand, in some cases, how such changes are brought about and what is the nature of the store.

Obviously the study of memory mechanisms will involve neurophysiological and biochemical methods, for we must investigate the fine-scale operation of neurons and the synapses between them. Yet it remains the case that much of the most crucial evidence still has to come from behavioural observations. It is what the animal actually *does* which provides our knowledge of what it has learnt, stored and now recalls, and from this knowledge we go on to deduce mechanisms and then check them physiologically and biochemically. Dudai (1989) provides a very clear introduction to the whole field, taking account of all these different levels from which we must approach.

Just as with learning itself, so the study of memory mechanisms has benefited from the comparison of different types of animal. Valuable evidence has come from certain molluscs, the honey-bee, a few bird species and a few mammals. Human memory itself has been a rich source. Although invasive experiments are usually impossible, we have inevitably a stream of cases where the effects of brain damage arising from accidents can be studied, and of course we can get detailed evidence from humans who, unlike animals, can tell you verbally what they can or cannot recall.

Different types of memory

Introspection tells us that not all events are stored in the same way in our memory. We look up a telephone number to make a call, but may not be able to recall it a couple of minutes later. On the other hand, we can recall the numbers of close friends after months or years. Repetition and the greater significance attached to certain events must make a difference to the way they are stored. Perhaps all events go into a **short-term** store but only events of some consequence are held in a **long-term** store.

The idea that short-term memory and long-term memory are somehow distinct is supported by some remarkable psychological and physiological evidence. People suffering from concussion or other severe shock are often unable to recall the events that led up to the accident, but their memory of events in the more distant past is unaffected. Further, if their recent memories recover – and they often do – they do so roughly in order, the most distant first, until gradually almost all memory is recovered. Commonly, though, the few moments before the accident can never be recalled. This phenomenon of **retrograde amnesia** can also be reproduced in animals.

It suggests that short-term memory and the process by which events are moved into a long-term memory store is more labile and sensitive to disturbance than the store itself once in place. More evidence to support this conclusion comes from studies in rats and chicks using a range of drugs which affect the way the nervous system functions (see Andrew 1985). For example, administering drugs which inhibit protein synthesis just after learning a new task does not affect old memories, but the memory of the new task rapidly fades and never recovers.

It has been particularly useful for such studies to develop tasks that can be learnt very rapidly in an approximately all-or-none fashion. Learning a maze, for instance, is a complex task which takes time. How can we ever be sure, at a given moment, how much a rat has learnt? We can get much more precise timings for memory processing with rats stepping across a barrier to avoid electric shock, or chicks refusing to peck at a coloured bead after a single experience when the bead was coated with the intensely bitter substance, methyl anthranilate. As mentioned earlier (p. 276) honey-bees show very persistent recall after a single experience of a colour combined with a food reward. It is interesting to discover that their memory, just like that of vertebrates, also goes through a labile phase – lasting about 3 minutes – when it is susceptible to destruction by cooling or anaesthesia, before becoming incorporated into a much more robust long-term store.

Studies using this kind of task suggest that, at least in the chick and the rat, it is possible to identify *three* types of memory with different time courses which we may call short-term, intermediate-term and long-term; see Fig. 5.20 and Andrew (1985). The evidence for this comes from two sources. Firstly, there are drugs which affect memory specifically at each of three time phases, and are ineffective earlier or later. Secondly, corresponding to these times of drug sensitivity, there are remarkably precise timed fluctuations in the ability to retrieve the memory of an event after it has occurred. Figure 5.20 indicates

Figure 5.20 A scheme for the three stages of memory storage constructed from experimental evidence in chicks. As time since the learning trial increases the chick retrieves memory of the event first from short term memory (STM) then intermediate-term memory (ITM) and finally long-term memory (LTM). The dips in retrieval at 15 and 55 min represent periods when the chick is switching from one store to the next. The three chemical agents listed each affect memory of one type only and lead to subsequent amnesia when administered during the period indicated by the extent of the bars at the left. This is supportive evidence for a three-stage system. (Modified from Andrew 1985.)

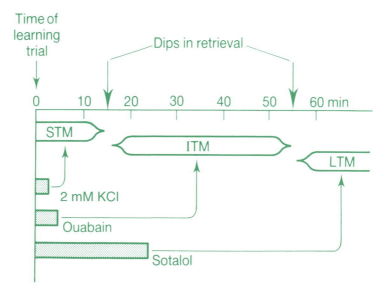

these fluctuations: chickens have pecked at a coloured bead coated with methyl anthranilate. Now different batches of them are tested at increasing delays after this unpleasant experience. Retrieval, revealed by their avoidance of beads of the same colour is, at first, good. Then at 14–15 minutes after the event there is a fading of retrieval. This is not conventional 'forgetting' because a minute or two later retrieval returns, only to take a second dip 55 minutes after the original event from which fall it also recovers and more permanently. Similar evidence comes from rats, which also show two dips in retrieval before long-term memory is established, although the exact timing is different. Honey-bees show a single significant dip in their retrieval ability after one-trial colour learning, which reaches its maximum depth at 3 minutes, recovering thereafter (Fig. 5.21; Menzel *et al.* 1993).

Such precision in timing of the ability to recall events suggests that animals retrieve their memories from different stores in turn. It is the process of shifting from one store which is now fading, to the next, now forming, which manifests itself as a dip in retrieval. We can only speculate about the nature of such processes for we do not know for certain how the different phases of memory interact. Is each a necessary precursor of the next, playing a part in its formation, or does each form and fade in parallel, perhaps interacting and passing information across as it fades and the next comes into

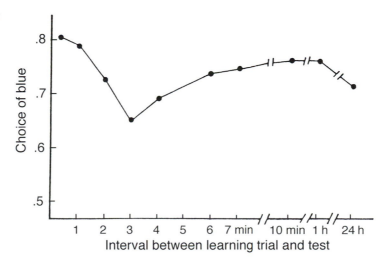

Figure 5.21 The retention of a honey-bee's association with blue after a single trial made as described on p. ooo and in Figs. 5.7–5.9. Retention fades steeply over the first 3 min from the trial but then recovers as the memory becomes established in a longer-term store. (After Menzel *et al.* 1993.)

play? Dudai (1989), reviewing evidence from various sources, suggests that we have to accept the likelihood of a 'multi-step, multi-channel nature of memory in different species'.

However they interact, each phase of memory must be carrying a representation of the events which have been learnt. Long-term memory can be shown to be in place after about an hour in most animals studied. As mentioned above, there is good reason to believe that its establishment involves structural changes in the brain which subsequently facilitate neural transmission along new pathways (we will discuss the nature of these changes below). The processes leading to the setting up of short- and intermediate-term memories probably involves persistent activity in some neural circuits. This is why they eventually decay and why they are so vulnerable to physical shock and to drugs.

The anatomy of memory

We can now identify regions of the mammalian brain involved in these stages of memory formation (see Morris 1983; Mishkin & Appenzeller, 1987). The original evidence came from human subjects with particular types of brain damage. Over a century ago the Russian neurologist Korsakov described patients who were able to recall distant events normally but had lost permanently the ability to form new memories following a head injury, a stroke or severe alcoholism, so-called anterograde amnesia. A common feature of such patients was damage to neural structures at the base of the forebrain and adjacent to the hypothalamus, especially the mamillary bodies and the thalamus (see Fig. 5.22). Since Korsakov's original descriptions, damage to other closely associated brain regions, notably the hippocampus, has been shown to have the same effect on the inner side of the base of the cerebral hemispheres, and the amygdala which lies anterior to it (Fig. 5.22). Bilateral damage to any of these structures can lead to a complete inability to store new memories, and recall beyond a few seconds may be impossible. There is no better description of the behavioural consequences of such damage than the essay, *The lost mariner*, by Sacks (1986). He provides a vivid account of the extraordinary and tragic existence of a person who has lost all ability to memorize events and has only a distant past to which he can relate.

One person who has become very well known to psychologists and neurobiologists is an American known, to preserve his anonymity, as 'HM'.

Figure 5.22 The human brain viewed (top) in straight sagittal section and (bottom) from the same aspect but with brain stem and cerebellum removed so as to reveal more of the structures involved in memory formation. Further explanation is given in the text.

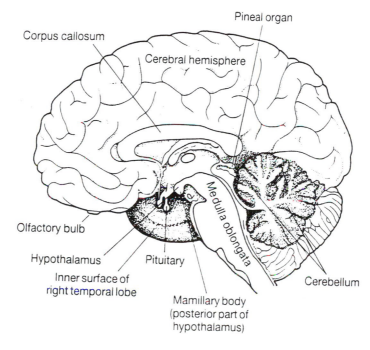

Pineal organ

Corpus callosum

Cerebral hemisphere

Medulla oblongata

Olfactory bulb

Hypothalamus

Pituitary

Inner surface of right temporal lobe

Cerebellum

Mamillary body (posterior part of hypothalamus)

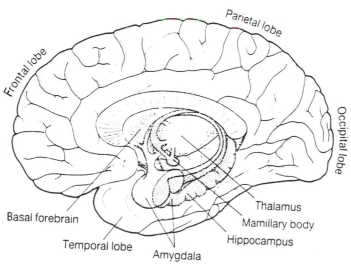

Parietal lobe

Frontal lobe

Occipital lobe

Basal forebrain

Thalamus

Mamillary body

Hippocampus

Temporal lobe

Amygdala

HM suffered complete loss of short-term memory in 1953 at the age of 27 following the unforeseen side effects of a brain operation to prevent epileptic seizures. Batteries of researchers have studied HM, testing the details of his disability in a multitude of ways over many years. This might sound intolerable, but so complete is his inability to aquire new memories that, after a few minutes, a researcher or a repeat of a test situation is greeted as totally novel: HM has no conscious recollection of them.

For normal people, memory does seem to form part of our consciousness; it shapes our self-identification and we know that we know things. This is true whether we or not can recall them at one particular moment – we feel that we will remember later, and in the majority of cases we do. HM very largely lacks this kind of memory and his recall after learning lists of words, or events or faces is almost non-existent.

These are the kinds of memory which are often called 'declarative' or 'explicit' – we can think about them and recall them actively. But there is another rather distinct type, so-called 'procedural' or 'implicit' memory, in which we have stored knowledge to which we do not have the same conscious access. This often involves acquired motor skills – riding a bicycle would be one such example. (Certainly we *know* we can ride, but the processes of balance and steering are unconscious, once acquired.) It is of great significance that HM and other such amnesiac patients *can* aquire new motor skills and retain them more or less normally. Amnesiacs have been asked to learn a 'finger maze' (tracing a path through channels with one's finger, without the help of visual clues) or read novel words in a mirror, so that they appear reversed. They improve their performance well on such tasks and on re-testing 3 months later, demonstrate that they have retained these new skills. Yet they have no conscious recollection of ever being tested in this way. HM improved at mirror reading and retained this skill, but could not retain for even a few minutes the words he had read nor could he remember that he had been tested in this way (see Dudai 1989, chapter 15).

Such evidence shows clearly that not all learning is processed along the same channels before it enters the memory stores. The human evidence has been amply supported by work with rats and primates. Numerous studies using specific brain lesions and drugs are building up a picture of several circuits for different types of memory. Mishkin and Appenzeller (1987) give more detail and also discuss evidence of diversity in the memory system of humans and other primates. Ungerleider (1995) describes a new type of approach using remarkable new imaging techniques – positron emission tomography (PET) – which enable us to 'see' patterns of blood flow in different parts of the brain while a person is actually learning or recalling some event or procedure (see Fig. 5.23). Active, i.e. rapidly firing, nerve cells

Figure 5.23 Position emission tomography (PET) reveals areas in the human brain that become active, and hence suffused with blood, when a subject learns a spatial task (navigating along an urban route, displayed on film). The subject's blood is made very slightly radioactive by an injection of labelled water and the whole brain is scanned to reveal which parts have enhanced blood flow. (a) Sagittal view (comparable to that in Fig. 5.22); (b) transverse, effectively horizontal view. The two light (i.e. active) areas are in the region of the hippocampus and its associated structures. Subjects who viewed similar film but did not learn a route showed no such activity. (From Maguire *et al.* 1996, who give more details of the technique and procedures.)

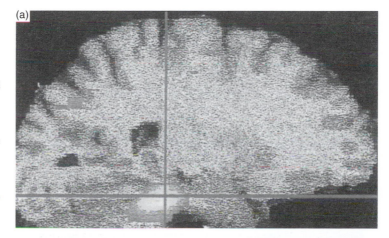
(a)

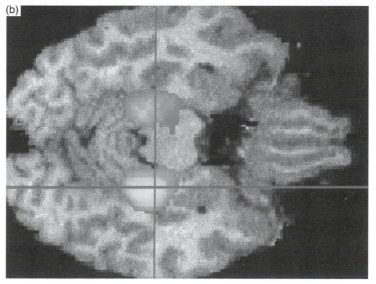
(b)

require more oxygen and energy than in their resting state, which means that blood circulation is switched towards such areas. It is not possible to get very fine definition, we can identify only regions of the brain involved, but nevertheless evidence gained this way confirms and extends other studies.

The brain damage suffered by HM and other amnesiacs involved those brain areas at the base of the forebrain which Korsakoff first associated with memory deficits. We have seen that procedural memory was little affected and, although they cannot establish new memories of an explicit type, they cannot totally lack a short-term memory either. They can hold apparently normal conversations which would be impossible if they could not recall words and sentences which had gone just before. It is only after a few minutes that deficits begin to appear. Imaging and studies with monkeys suggest that the pre-frontal areas of the cerebral hemispheres (see Fig. 5.22) are one of the sites which 'hold' working memory – probably in the form of continuous neural activity in pathways once activated. Lesions to these pre-frontal areas make it difficult for monkeys to perform tasks which require them to hold an event in mind before they act. Figure 5.24 illustrates a simple form of such testing. A reward is put under one of two covers in full view of the watching monkey, but it is not allowed to reach them at once. Delays of more than a few seconds are affected by pre-frontal lesions and the effect increases with the amount of the damage.

If working memory is to be converted into a more persistent state, other areas of the brain become active. Mishkin and Apenzeller (1987) review evidence that for procedural type learning it is the relevant sensory and motor areas of the cerebral cortex which are involved, both for processing and for eventual long-term storage.

Because of their striking memory deficits, it is clear that the forebrain structures around the hippocampus which are damaged in amnesiacs must be involved in other types of memory. The hippocampus itself has been the focus of a great deal of memory research, particularly concerning spatial memory. Some elegant experiments by Morris and his collaborators (Morris *et al.* 1982; Morris 1989) reveal something of its function in relation to other memory systems. They have rats training under two conditions swimming in a water tank to reach an escape platform. In the first situation the rats learn to recognize which of two platforms visible above the water surface will give them a

Figure 5.24 The 'Wisconsin General Testing Apparatus' is now a familiar feature of laboratories where primate learning is being investigated. Below, the diagrams illustrate the views through the one-way screen when delayed-response tasks are studied. The monkey is presented with a tray on which there are three wells. The central one is covered by a conspicuous object – here a disc – which when moved reveals an attractive food item. The opaque screen is now lowered cutting off the monkey's view of the tray. After a delay, which can range from a few seconds upwards, the screen is lifted to reveal a tray whose central well is uncovered and empty, but the outer two are covered, one with the familiar object – the disc, the other with a new object. In 'matching to sample' tests the monkey is rewarded for moving the matching disc; in non-matching tests, it must move the other object. Lesions in the pre-frontal cortex greatly impair the ability to cope with any delay to the choice. (Further details in Dudai 1989, from which these drawings are modified.)

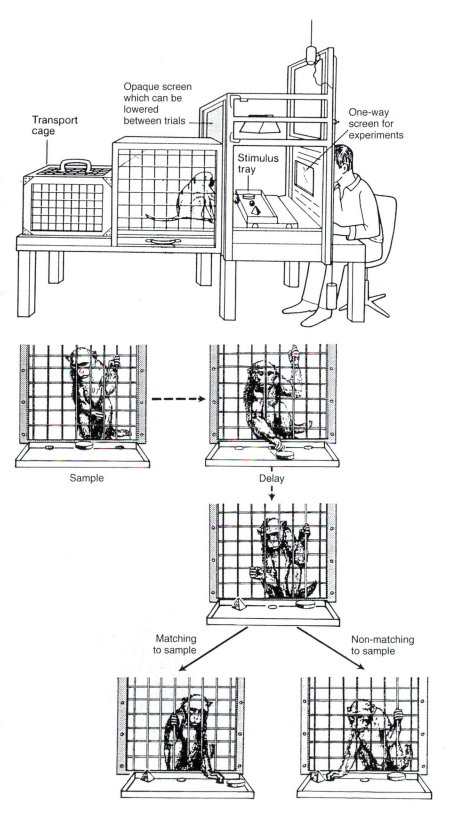

safe place to stand (the other platform sinks beneath their weight); this is a visual discrimination task and they swim straight to the platform which they can see directly. In the second situation rats learn to swim towards a platform which is constant in position but which they cannot see because it is just below the surface and the water has been made opaque by adding milk. This is a spatial learning task, but note that the rats still use their eyes to some extent because they have to take account of landmarks around the tank, using this information to navigate in space as they swim to the unseen platform. Lesions in the hippocampus prevent spatial learning but not visual discrimination learning, and from this and other evidence we can deduce that the hippocampus is involved not just in short-term memory formation but in the organization of spatial memory generally.

There is further evidence supporting this conclusion, from birds. They have had a long evolutionary history separate from mammals and it is necessary to be cautious in homologizing brain regions between the two groups. The avian hippocampus is located far more dorsally in the forebrain than that of mammals; nevertheless, there is good evidence from development, neuron type and its connections with other parts of the brain, that it is homologous with that of mammals. Also it does function in memory and particularly spatial memory, for Bingman *et al.* (1988) have shown that homing pigeons with hippocampal lesions have great difficulty in homing to their loft. It is not their overall homing orientation which is affected, for they set off in the right direction and fly normally. It appears that as they get closer to home, at the point where pigeons begin to use local landmarks around the loft which they have learnt on previous flights, the lesioned birds become disorientated, so their memory for landmarks appears to have been lost.

Earlier (p. 277), we discussed the remarkably specialized spatial memory of food-storing birds. It is therefore very intriguing to discover that there are differences in certain of their brain proportions which seem to confirm a crucial role for the hippocampus in spatial learning and memory. Krebs *et al.* (1989) have examined a wide range of birds and find that food-storing species have a hippocampus which is significantly larger in proportion than that of close relatives which do not store food. Their spatial learning specializations have apparently required neural specializations to match. Similar differences in the size of the hippocampus have been found between mammals which hold territories, and hence require a good spatial sense, and those which do not, even between the sexes of one species where males hold territories and females do not (Sherry *et al.* 1992).

Long-term memory storage

We may turn finally to consider the nature of long-term memory itself. The diversity of memory mechanisms in mammals mentioned above might prepare us for accepting that there is no one special area of the brain where long-term memories reside. The long-term representation of an event stored in the brain is sometimes called an engram. In 1950 the psychologist Karl Lashley published a paper entitled *In search of the engram*, in which reviewed years of work studying memory storage by rats; he found the engram elusive. In one set of experiments rats learnt a fairly complex maze and then had parts of their cerebral hemispheres removed. By the conclusion there was no portion of the hemispheres (and many other brain structures) which had not been removed from one or another group of subjects. The overall result suggested that the degree to which the rats lost their memory depended not on which parts of the brain were removed, but only on how much. But could the engram really be this diffuse?

An obvious problem is that learning a maze is such a complex task and will involve visual, spatial and tactile cues at the very least. Their processing and subsequent storage probably involves several separate pathways and interactions between them. Consequently we might expect memory of a maze to be disrupted by lesions at several different sites. Remembering may be a process which reflects the term's origin – the opposite of *dis*membering – a putting together again of information gathered from several sources. Simpler and more precisely defined learning tasks may give a clearer picture of storage.

One of the best examples of this approach comes from the work of Bateson, Horn, Rose and their collaborators, who have used visual imprinting by newly hatched domestic chicks as a learning and memory system for behavioural, anatomical and biochemical studies; Horn (1985, 1990) and Rose (1992) provide full reviews. The great advantages of imprinting are (i) the rapidity with which young chicks learn, (ii) the fact that they are innately biased to approach conspicuous objects and require no period of deprivation followed by reward (as with rats being trained to run a maze), and lastly (iii) that imprinting occurs very early in life. Thus it may be possible to study localized effects of learning on a brain which has been little 'marked' by previous experience. It is, ultimately, structural effects that we are looking for. We have already emphasized that long-term memory storage must involve permanent, or at least very persistent, changes to the structure of synapses. These are growth processes and, as we have seen, they are blocked by protein synthesis inhibitors such as puromycin (see p. 303 and Fig. 5.20).

For many of the imprinting experiments, chicks were hatched in the dark and then, when about 24 hours old, put into a small running wheel (see Fig. 5.25a) close to an attractive visual stimulus, often a flashing red light. It was possible to vary exposure time and to measure a chick's efforts to approach the light by recording the number of wheel turns. This enabled the experimenters to quantify the amount of experience their subjects had and subsequently to relate this to the extent of imprinting which had occurred. This could be estimated by giving it a choice of two objects – a discrimination test between familiar and unfamiliar – and measuring the degree of effort they would make to approach the preferred object (see Fig. 5.25b).

Using a variety of techniques, a region of the chick's forebrain, the intermediate hyperstriatum ventrale (IMHV) has been identified, which is crucial for storing a representation of the imprinting experience and mediating discrimination. Other parts of the forebrain are not so involved. It is of special interest that differences have been found in the way the left and right hemispheres and their component IMHVs operate. Andrew (1985) describes other evidence that the two sides of the chick brain act complementarily during memory formation and the same is likely to be true for mammals.

Using biochemical and electron microscope techniques, it has been possible to go further and reveal some of the growth and microstructural changes involved in forming the store. The association of events which sets up a novel pattern of neural activity does definitely bring about synaptic changes at specific sites which facilitates activity at these sites again (see Horn 1985).

These findings from the imprinting system of chicks are now paralleled by work on mammalian memory formation. Because, as we have seen, the hippocampus is closely involved in memory processes, it has received a great deal of attention. Much of this has related to a striking property of hippocampal neurons referred to as 'long-term potentiation' (LTP). If hippocampal neurons are stimulated with rapid trains of impulses, i.e. artificially induced to become active, their synapses with other neurons become strengthened and facilitated (potentiated) and thus more ready to transmit excitation the next time a similar pattern of stimulation occurs. This changed state persists for some hours, even if it was initiated by only a single burst of stimuli. If a second or multiple bursts of stimulation are given then LTP can persist, at least in

Figure 5.25 Apparatus used to measure the extent of filial imprinting in young chicks as a preliminary to studying its neural basis. In (a) chicks are placed in running wheels close to attractive objects, in this case flashing lights of different colours. Length of exposure can be varied and the number of turns of the wheel gives a measure of how much time the chicks spend 'following' the object. Then individual chicks are placed in a wheel running on a rail track between two objects (one familiar, the other unfamiliar) and asked to choose between them. This apparatus (b) measures how much effort they will put into approaching the model of their choice. By a clever arrangement of gearing, as the chick in its wheel strives to approach a model, it is carried away from it along the rails. Eventually it gives up and the distance at which it comes to rest is proportional to the degree of its effort to approach. (From (a) Horn 1985; (b) Bateson & Wainwright 1972.)

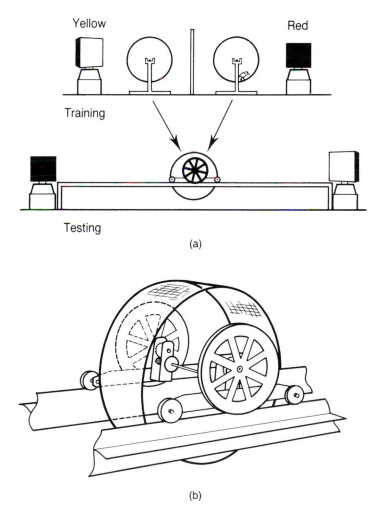

Yellow Red

Training

Testing

(a)

(b)

intact animals, for days or weeks. We emphasize *intact* animals, because a lot of research into the neurophysiology of LTP is carried out on isolated slices of hippocampal tissue, which can be kept alive and active for many hours in a suitable nutrient bath. Working with these slices it has been much easier to check on the particular circuits which are potentiated. It is quite clear that we are not dealing with some general arousal of whole populations of neurons; only the circuits which were facilitated by the original burst of stimulation subsequently respond. So, in this crucial respect, the conditions necessary for the storage of a memory trace – a newly modified neural pathway whose synapses have been strengthened – are fulfilled. This is an active field of research and it is perhaps too early to judge its overall significance as a general model for long-term memory. Dudai (1989, p. 6), Gallistel (1995) and Bear (1997) all discuss this issue.

LTP can be induced in other parts of the brain and, although we know how crucial is the role of the hippocampus for some memory functions, something similar must be taking place in other areas as well. The neural changes which take place during chick imprinting are a case in point. We also know that related types of long-term neural modifications occur during learning in one well-studied invertebrate, the sea-hare *Aplysia*. Kandel and his co-workers have traced in great detail the cellular events concerned with habituation and simple associative conditioning in this mollusc. Changes to the synthesis and release of neural transmitters at identified synapses means that their activity is enhanced and these changes persist (Kandel & Schwarz 1982).

In conclusion, then, even though many questions remain unanswered – and this is a difficult area of research – we can now identify with some certainty crucial events for memory formation and storage. It is perhaps our own human memory system which leaves us with the most wonderful but tantalizing questions. Even allowing for the garnishings of our imagination, the richness of association in human memory is amazing – Hamlet soliloquies, a long dead grandparent's face, the creak made by a particular door in the home of our childhood, the smell of the upholstery in one's first old car. How can it all be stored? Is there any limit to the amount stored? Luria (1975) in his study of a Russian mnemonist describes a person who could memorize staggering lists of nonsense syllables, meaningless mathematical formulae, strings of numbers and, without prior warning, repeat them back again *many years* later. Luria could find no limits to his storage capacity and no evidence that he ever forgot anything. Such an astonishing memory was as much a burden as a blessing. Selective forgetting will always help us, and our animal relatives, with the management of our lives.

6   Evolution

Evolution by natural selection is the great unifying concept of biology, so much so that most biologists now feel that, without it, none of the phenomena they study really make sense. Richard Dawkins (1986) provides an excellent modern survey of the great explanatory power of what has come to be called 'neo-Darwinism'. Natural selection, operating on random, inherited variation has, over the generations, shaped animals to match the environments in which they live. Sometimes adaptation has been achieved through genes, accumulated over many generations, biasing the development of behaviour directly into appropriate responses. In other cases animals inherit only biases to respond adaptively to their immediate environment so as to acquire, individually, appropriate responses by learning. Much behaviour develops by a mingling of such processes and the end result is often very much the same. All through this book we have emphasized the adaptive role of behaviour in an animal's life and so the concepts of 'evolution' and 'adaptation' have been implicit in much of what has already been said. Now we look at them in more detail. We will look at the genetics of behaviour, as it affects its evolution, and at the ways in which natural selection has led animals to cooperate and care for each other as well as to fight and to kill. We will also look at the nature of behavioural changes over evolutionary time and the way they, in turn, influence the evolutionary process itself. We begin with the most basic, and yet in some ways most elusive evolutionary concept, that of adaptation.

The adaptiveness of behaviour

It is easy enough to recognize in a general sense that animals are adapted to their way of life. A camouflaged insect sitting quite immobile on a background that it matches perfectly, for instance, is clearly adapted to conceal itself from predators. However, to appreciate the power of evolutionary processes to bring about adaptation, we may need more detailed studies which show just how perfect this can be.

Mole crickets are large insects whose males dig burrows underground in which they sit and sing to attract females flying overhead (Fig. 6.1). As with many other species of crickets and grasshoppers, the song of each species is very distinctive and plays a role in the sexual isolation between species (see p. 363). The sound of the song is determined in part by the structure and size of the forewings, which are rubbed rapidly together to produce sound. A

318

scraper or plectrum on the hind margin of the left wing rubs along the toothed undersurface of a vein on the right wing (the 'file'). In two species studied by Bennet-Clark (1970, 1987), the fundamental sound frequency within the pulses of the song (see Fig. 6.1) is about 1600 Hz in *Gryllotalpa gryllotalpa*, a species with small wings, shallow teeth on the file and a quiet song, but 3500 Hz in *G. vinae* which has large wings, deep teeth on its file and a much louder song.

Here, then, we can see a direct correlation between behaviour (the singing) and morphology, but the mole crickets' adaptiveness goes much further than this. The male of each species excavates its burrow in a different way (Fig. 6.1b). Bennet-Clark showed that the burrow forms an exponential horn with a bulb at its base so that when the male sings with his head just at the origin of the horn, his song is amplified, as by the horn of an old-fashioned loudspeaker. However, the size and shape of the burrow are quite different in the two species. *G. vinae* builds a double-mouthed, horn-shaped burrow to

Figure 6.1 (a) A mole cricket (*Gryllotalpa vineae*); note the massive forelegs, modified for digging burrows. (Photograph by H.C. Bennet-Clarke.) (b) Side views of males of two species of *Gryllotalpa* sitting in their burrows, head downwards in the singing position. The shape of their burrows approximates to an exponential horn. *G. vineae*, on the right, has a smoother tunnel and a much louder song, but in both cases the shape of the burrow is adapted to their song frequency (oscillograms of which are illustrated below) so as to emit with maximum efficiency. (From Bennet-Clark 1970.)

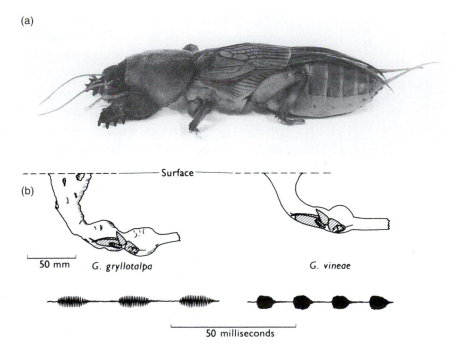

(a)

(b)

Surface

50 mm   *G. gryllotalpa*

G. vineae

50 milliseconds

sing in. The walls are smooth and there is a bulb that acts as a resistive load for the vibrating wings and concentrates the sound in a disc-shaped patch of sound energy above the burrow. *G. gryllotalpa* also builds a horn but it is much larger and is very effective at amplifying the much lower pitched sound that this species makes. Thus the wing structure, singing and burrow-digging behaviour of the mole crickets are all precisely co-adapted with each other so as to provide the most efficient sound production, 'aimed' at the females, which are attracted to the males of their own species.

Similar co-adaptation between morphology and behaviour is exemplified by the 'eye-spot' patterns on the wings of a number of butterflies and moths and the ways in which the insects display them. Many predatory birds, especially hawks and owls, have large, prominent eyes. They need them to detect their prey, often under low light conditions. It can be no coincidence that small birds respond with intense alarm to large eye patterns and, for example, are far more disturbed by painted eyes on a plain background than by a real owl whose eyes are covered. The development of colour patterns on the wings of Lepidoptera often takes place from foci, which means that colour tends to be laid down in whorls or circular shapes. These intrinsic forms have offered butterflies and moths an 'opportunity' to discourage small birds which try to prey on them, by presenting a sign stimulus which resembles that presented to birds by their predators – owls.

There is experimental evidence (e.g. Blest 1957a) that quite crude circular markings are already better at alarming small birds than other patterns, but most Lepidoptera go much further than this. Figure 6.2 illustrates how quite elaborate eye-spots on the hind wings of a hawk moth, hidden beneath the forewings when at rest, are suddenly flashed into view if the moth is touched. The extent to which natural selection has been able to perfect this particular adaptation is shown by the eye-spot pattern of the Brazilian butterfly, *Caligo* (Fig. 6.3). *Caligo*'s markings mimic large bulging vertebrate eyes set in deep sockets. There are even some flecks of white set in an arc, representing highlights reflected off the glistening eye surface. Because changing the pigment of those scales on its wings that go to make up the eye-spots has presumably only trivial costs for *Caligo*, there has been no barrier to selection operating over hundreds of thousands of generations to bring the eye to ever more perfect levels of adaptiveness. Each tiny advance will have brought only a tiny advantage, but there has been time for even the tiniest advantage to pay off in terms of slightly more effective bird scaring and hence result in the survival of a few more of those butterflies which bear it.

Such attention to detail, as it were, has many more strictly behavioural counterparts. For example, a pattern of behaviour exhibited once per year and then only for 5 minutes might seem to be of trivial significance. This is the habit of black-headed gulls to remove empty eggshells from their nests soon

Figure 6.2 The moth *Automeris coresus* has cryptically coloured forewings, which provide camouflage when it is settled. However, if touched, as by a predator, it flashes open the forewings to reveal bright eye-spots on the hind wings, draws in its legs and rocks rythmically. The effect is quite startling and induces fear in small birds. (Modified from Blest 1957b.)

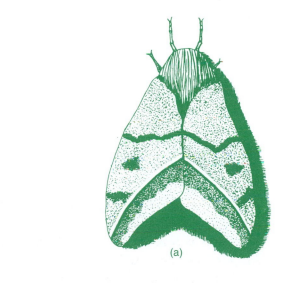

(a)

(b)

after their eggs hatch. Even gulls breeding for the first time carry away eggshells promptly, although they never pick up eggshells in any other situation. Tinbergen and his colleagues (1962), in a classic study, clearly showed that eggshell removal is vital for keeping the nest camouflaged. By placing eggs out on the sand-dunes with or without eggshells near them, they showed that chicks or unhatched eggs with eggshells near them were much more vulnerable to predation by crows and herring gulls. Both aerial and ground predators discover nests and chicks more easily if empty shells, with their smell and white shell insides, are nearby. Natural selection has operated to ensure that parent gulls always perform this brief but important behaviour. Clearly, we must never write off as functionless – however trivial it may seem – any piece of behaviour which we regularly observe in a natural situation.

It is worth pausing for a moment to consider how we can feel completely confident that natural selection has been operating on a trait. We have no

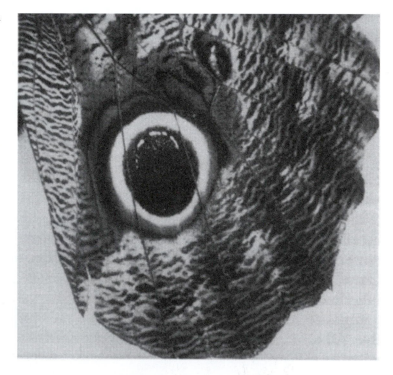

Figure 6.3 The-eye spots on the underside of the hind wings of the butterfly *Caligo* show how far natural selection has been able to shift colour patterns which presumably once resembled those shown in Fig. 5.3. The resemblance to large vertebrate eyes, such as those of owls, is remarkable.

direct evidence of what happened in the past; all we can do is look at a situation which exists today and then try to argue how it came about, making the assumption that the result is adaptive and aids survival. Often, as with the gulls removing eggshells, we do have evidence for this, but we should try to approach each case objectively and accept the possibility that not everything we observe is the direct result of selection for the best adaptation. Gould and Lewontin (1979) have accused ethologists of ignoring this possibility on occasions. In their article they refer to Dr Pangloss, a character in Voltaire's satirical novel, *Candide*, who believed that 'everything is for the best in the best of all possible worlds'. 'Our noses were made to carry spectacles, so we have spectacles.' From such a viewpoint we should have no difficulty in seeing adaptiveness all around us! The moral for ethologists is that some aspects of an animal's form or behaviour may, like wearing spectacles on our noses, be a simple consequence of processes – the development of a projecting nose – which had no functional connection with the current situation and thus are not a result of selection for adaptiveness.

However, if we return to the case of the mole crickets, we have two pieces of evidence of absolutely genuine adaptation to function – the shape of their burrows and the difference in shape between the burrows of the two species. A human engineer, asked to design a simple device for amplifying sound, would probably come up with an exponential horn (as in ear-trumpets or the early gramophones). The fact that the mole crickets dig burrows of precisely this shape strongly suggests that sound amplification was the feature favoured by natural selection. Crickets that built burrows of a different shape would not be able to emit such a loud sound (an engineer could tell us that) and, by implication, would be less successful at attracting females. The burrow appears to have been 'designed' to amplify the male's song.

This conclusion is confirmed by the second piece of evidence – the comparison between the two species. The sound engineers know that the exact 'best' shape for an exponential horn depends on the sound frequency it has to transmit. If a mole cricket's burrow is 'designed' to amplify with maximum efficiency, then we should expect species with different song frequencies to have burrows of different shapes with each 'tuned' to match the song frequency. This is exactly what we find. In fact, comparisons of this type between species, each of which is successful but achieves success in a slightly different way, has become one of the most important methods by which ethologists can document the adaptiveness of behaviour.

As we have compared the two mole crickets, so we can compare ground-nesting gulls that remove eggshells and the cliff-nesting kittiwake gulls that do not. On looking more closely, we find that kittiwakes nest on tiny cliff ledges where no predator can land and, as E. Cullen (1957) pointed out, this probably explains why their nests can be conspicuous and messy without any danger and why they do not waste time removing eggshells. But we can gain even more information if, instead of just comparing two species, we examine trends in the habits, morphology and environments across a range of many different species (e.g. Clutton-Brock & Harvey 1984).

For instance, a pair of goshawks may copulate up to 100 times before egg-laying and, in trying to understand the adaptive significance of this time-consuming and potentially dangerous behaviour, Birkhead *et al.* (1987) looked at the mating habits not just of goshawks, nor even of hawks in general, but of 131 species of birds from many different families. They found that copulation rates were highest in those genera where the female was most likely to be mated by more than one male, for example, in colonial species where there are many opportunities for extra-pair mating and/or in species where the male leaves his mate alone for long periods of time. In solitary nesters or where the male guards his mate for most of her fertile time, copulation frequencies are lower. This suggests strongly that the adaptive significance of a very high copulation rate is to prevent a female's eggs being fertilized by the sperm of another male, as the highest rates are found where there is the greatest risk of such sperm competition. Sperm competition explains the data much better than an alternative hypothesis, namely, that more copulations are needed to fertilize larger clutches of eggs since there is no correlation between clutch size and copulation frequency. In fact, the phenomenon of sperm competition is clearly widespread and there may be something of a war between the sexes involved. Males will try to exclude the sperm of other males, but it will also be advantageous for a female to be able to select who fathers her offspring. She may do this in a variety of ways (some of which we discuss later in this chapter), but one mechanism is to select which sperm in her reproductive tract are favoured during fertilization (Birkhead & Møller 1992, 1993; Eberhard 1996).

To return once more to the black-headed gulls' eggshell removal, notice that in this case Tinbergen and his group did not rely primarily on comparisons between species to reveal adaptive significance. They did an experiment with eggs and eggshells, providing a situation in which they could see what would happen if black-headed gulls did not remove eggshells (the problem is, of course, that all existing gulls do so – so well has their behaviour become adapted!). When such experimental manipulations are possible, they provide another way to remove the study of adaptiveness from the realm of 'just-so' stories (another of Gould and Lewontin's phrases, harking back to Rudyard Kipling's classic children's stories telling how animals became the way they are now!) and putting it firmly to empirical test: artificially creating a situation that does not exist naturally.

But however effective natural selection can be in the shaping of behaviour, it often results not in the absolute best for a particular function, but in the best compromise; the point being that almost always there are other functions involved because one single aspect of behaviour can rarely evolve in isolation. The eye-spots of the butterfly *Caligo*, are about as nearly perfect as we can envisage precisely because, for once, no compromise was necessary: they could evolve in isolation. Few aspects of behaviour can have this luxury. The elaborate displays or loud songs which many male birds use to attract their mates, for example, are also likely to attract predators. Hence most males reduce or even stop singing as soon as their mate begins to incubate and some lose their breeding plumage at the autumn moult. There is obviously a compromise between reproductive success and danger from predation.

The neotropical Tungara frog demonstrates this compromise in a particularly dramatic way (Ryan *et al.* 1982). Females are attracted to the sounds of males calling at the edges of ponds. But frog-eating bats use the males' calls to home in on the males and eat them (Fig. 6.4). Unfortunately for the frog, those aspects of male vocal behaviour that are most attractive to females (calling more intensely and producing calls with 'chuck' sounds) are also the ones that most increase the predation risk.

Genes and behavioural evolution

The evolution of any adaptive trait by natural selection implies that there exists – or at least there has once existed in the past – genetic variation between individuals. Evolution is about changes in gene frequency; one genotype becomes more frequent in the population because it makes the body it is in taller, fatter, able to run faster than another, or succeed in some other way. In Chapter 2, we have already considered some of the special problems of bringing a genetic analysis to bear on behaviour. We had to recognize that, whilst it is possible to study some examples of how genes control the way a nervous system develops, behaviour itself remains much more elusive. We have every reason to believe that behaviour evolved alongside morphology in the evolution of adaptive responses, but it may be difficult to study how the genes are operating.

In part, it depends on the nature of our inquiry. Suppose we ask the question, is aggressive behaviour inherited? In the first place, this could mean, are genetic factors involved in building a nervous system with the potential, given the right stimulus situation, to perform certain types of behaviour, attacking, biting, etc.? The answer is certainly, yes: there is often plenty of quantitative variation within a population of animals, but all individuals will usually fight at a push and, when they do so, they fight in the same way. As a consequence we rarely have any inherited variations which we can work with to investigate how a nervous system with this aggressive potential is put

Figure 6.4 The frog-eating bat locates male Tungara frogs by picking up their courtship calls. (From Butlin & Ritchie, 1989.)

together. Note that exactly the same considerations would apply if our question asked if learning has a genetic basis.

However, there is a second and more straightforward interpretation of the original question, which is, are there inherited differences between individuals in their aggressiveness (or learning ability)? Again the answer is yes, but this time we can get considerably further.

There is a good deal of evidence, both experimental and from comparative studies, on how genetic changes affect the modification and expression of aggressive behaviour patterns or of learning once they are in place. Some genetically based variations in learning ability were discussed in the last chapter; here we shall concentrate on aggression. For example, although the males of the different inbred strains of mice will all perform the typical mouse fighting patterns – tail-rattling, sideways threat posture, biting and chasing – the extent to which they will do so and the frequency of such attacks differs greatly between strains. We can easily show by crossing high and low aggressive mice that these differences are largely due to the genes they carry. The hybrids show intermediate levels of aggressiveness and it is clear that the high and low lines differ by a number of genes which affect the expression of the behaviour. (To be sure that the differences we find really do have a genetic basis, we have to arrange by cross-fostering that all males from the parental strains and the hybrids are reared by the same sort of mothers. Such a procedure may not eliminate environmental influences affecting the way differences in aggression develop, but it will greatly reduce them.)

Quantitative variations of this sort could certainly become raw material for the evolution of behaviour. Related species often share a common repertoire of instinctive behaviour patterns, but differ in the frequency with which they are performed. For example, all the gulls of the sub-family Larinae in the family Laridae perform a display called 'choking' at the nest site during pair formation. It is not seen very frequently in most species, but in two of them, the cliff-nesting kittiwakes of northwestern America and northern Europe, it is the dominant display (see Fig. 2.7, p. 48) used in every context of pair formation, courtship and territorial affirmation. Suitable cliff ledges are difficult to obtain and are fiercely fought over. It makes sense that the nest site display of the gull family should be the one which has become dominant in the kittiwakes. It is another feature of their behaviour which has been affected by the shift to cliff-nesting – like the absence of eggshell removal – a shift which was perhaps made originally as an anti-predator adaptation (Cullen, 1957). The assumption is that the ancestral gulls all shared a set of genes in common which influenced the development of a nervous system which organized the potential to perform the gull repertoire of displays, including choking. However, different populations of gulls varied in the genes they carried, which affected their thresholds for performing the different displays. If it conferred an advantage of some kind, natural selection would favour birds carrying particular sets of these genes and thus the frequency with which the displays were performed would change. Gradually, then, the different gull species diverged and their common repertoire of displays came to be used in different ways.

This may all sound very speculative and admittedly we have no knowledge of genetic variation affecting behaviour in gulls, nor can we know how this may have changed in their past history. Nevertheless, it is the case that our assumptions fit very well with the kind of experimental evidence we have on genetic variation in some species and how this affects the expression of behaviour. We can mimic, and greatly speed up the effects of natural selection by artificial selection, by selectively breeding from extremes in a population which shows variation in the expression of some behaviour.

Because they are convenient and fast-breeding laboratory animals, fruit flies (*Drosophila*) and mice have been common subjects for selection experiments. Most stocks of *Drosophila* show considerable variation in the speed with which pairs of flies mate when introduced to each other. Using this as raw material, Manning (1961) selectively bred for fast and slow speed of mating over a number of generations (Fig. 6.5). Mating speed is a complex character which involves the interaction of male and female, and more than just their sexual behaviour. However, amongst the other changes, the behaviour of both sexes had been altered in a quantitative way. The males from fast-mating lines performed high intensity courtship movements more frequently than those from slow lines. Conversely, females from fast lines were more easily stimulated by their courtship to accept males, from their own or other lines, than were slow-mating females. The artificially induced genetic changes here had produced results very comparable to those found between closely related species in nature.

In a 'wild' insect species, a cricket of the genus *Gryllus*, Cade (1981) used selective breeding to demonstrate genetic variation which may well be the basis for two different behavioural strategies which we can observe naturally. Males are either 'callers' that sing a great deal thereby attracting females, or 'satellites' that remain relatively silent but may intercept females on their way to the calling males. Calling males attract more females than satellites but suffer more from parasitic flies, which are also attracted by the males' songs. Cade (1981) reared males in isolation in the laboratory and recorded how much each male sang. By measuring the mean calling rate per night from the 7th to 16th day of life, he showed that there was a bimodal distribution in the amount of singing. He then selected two to four males from each end of the distribution and mated them with non-sister females whose brothers were of the same type, and continued this process for four generations. The result was a divergence in the amount of singing shown by the two groups (Fig. 6.6), showing that there must have been a considerable genetic component to the

Figure 6.5 The mating speed of groups of 50 pairs of *Drosophila melanogaster* from two lines selected for fast mating, FA and FB, and two selected for slow mating, SA and SB, compared with unselected controls. These samples are from the 18th selected generation. Some 80% of the fast lines have mated before the first of the slow lines begin. (From Manning 1961.)

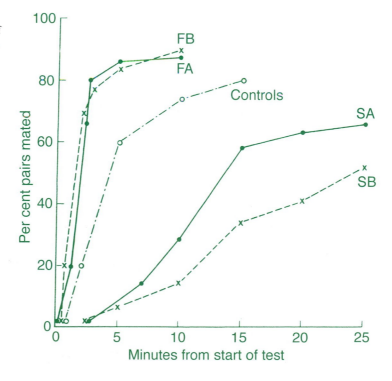

variation in the original population. Genetic differences are not the sole determinants of calling rate, which is certainly not an all-or-nothing trait. The
amount a male calls can be influenced by his environment. Callers that are
surrounded by other callers tend to sing less than when they are on their own,
and 'satellites' may be quite vocal if no other cricket is singing nearby. Yet
Cade's results show how, if favoured by selection, genetic changes can change
the bias in the way calling behaviour develops.

We mentioned above how inbred strains of mice differ widely in their
aggressiveness, and again in this mammal artificial selection has revealed how
common it is to carry genes which affect the level at which such behaviour is
expressed. Lagerspetz and Lagerspetz (1974) were able to produce a big
divergence from a parent population of mice selectively bred for high and low
aggression. The genes that had accumulated in each line strongly influenced
the directions along which behaviour developed. Yet the inevitable interaction
with environment was there and it was found that naturally aggressive mice
would cease fighting if they had been repeatedly defeated in staged fights early
in life. Conversely, non-aggressive mice became much more aggressive if they
were consistent winners in early fights.

Again, comparable types of genetically based quantitative variation have

Figure 6.6 The length of
time per day spent in bouts
of calling by male crickets
selectively bred for high
and low calling over four
generations. (From Cade
1981.)

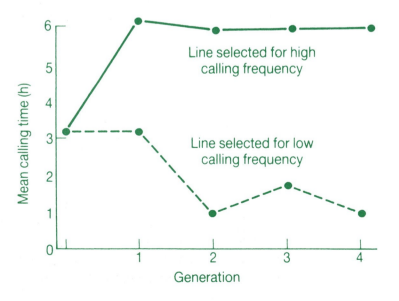

been observed in natural populations. For example, Berthold and Querner (1981) have investigated the phenomenon of migratory restlessness in different populations of the blackcap. Blackcaps fly to Africa for the northern winter, but breed in summer from the Canary Islands northwards over much of Europe. This wide distribution obviously means that the length of migratory flights varies greatly between different populations of blackcaps. Berthold and Querner found that birds from different populations differ genetically in how far they are prepared to fly. They took nestling blackcaps from four different breeding areas, (southern Finland, southern Germany, southern France and the Canary Islands) and hand-reared them. The young, inexperienced birds were kept in aviaries away from adults. When, in their first autumn, the time came when they would naturally be undertaking their migration to Africa, the captive birds showed migratory restlessness, fluttering and jumping in the aviaries. The interesting thing was that there was a correlation between how far the birds would have had to travel had they been left in their natural populations and the degree of migratory restlessness (its intensity and how many days it persisted). Birds from the Canary Islands, which are close to the wintering areas in Africa, showed much less migratory restlessness than birds from the Finnish and German populations (see Fig. 6.7). The genetic basis of these differences was clinched by cross-breeding birds from different populations. Hybrids showed intermediate levels of migratory restlessness.

In extensions to this work (see Berthold 1993; Berthold & Helbig 1992), artificial selection experiments have revealed just how rapidly natural selection could change migratory behaviour. In populations of blackcaps breeding in the south of France, most birds migrate south in winter but about 25% stay put. Berthold and his collaborators selectively bred from aviary-reared birds which either showed migratory restlessness or alternatively did not show it. Within three or four generations they had populations which were totally migratory or totally non-migratory (Fig. 6.8). Obviously blackcaps in nature are genetically 'poised' to respond rapidly to climate change!

In summary, whilst we have rather little information on the origins of behavioural systems like aggression or learning ability, once they are in place we can quite easily explain the behavioural differences between individuals or species in genetic terms. We can also understand how natural selection brings about behavioural change as it 'searches' to produce the best fit to an animal's environment.

Optimality and behaviour

But in practice, does natural selection produce the very best fit possible? Is the adaptiveness of behaviour so good that it can be described as 'optimal'? There is certainly a tendency among biologists to attempt to do this. A good example is the optimal foraging model we discussed in Chapter 4. Here, the behaviour of a great tit looking for food was described by a simple optimality model which was then used to predict the behaviour of the bird – how much it should

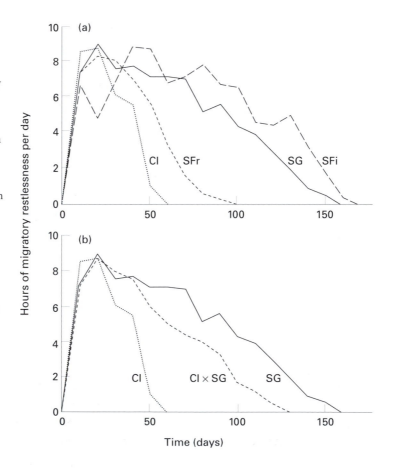

Figure 6.7 Duration of migratory restlessness in populations of blackcaps from different parts of Europe and from the Canary Islands off the west coast of North Africa. The lines represent mean values for groups of birds ranging from 6 to 26 in number, adjusted so that the first nights of restlessness for each bird coincide. Note that the length of time birds spend 'migrating' reflects the distance to their winter quarters in sub- Saharan Africa. (b) Shows that hybrids between Canary Islands and southern Germany blackcaps show an intermediate level of migratory restlessness. (Modified from Berthold & Querner 1981.)

eat in one place, how soon it should fly on to the next tree, and so on. These predictions were made by assuming that the animal was optimizing some function – a simple one would be net rate of energy gain. This would mean that its decision rules should be geared to always ensuring that there is maximum 'energy in' for minimum 'energy out' – in other words, that the bird should optimize the energy value of the food it is eating relative to the energy it has to expend in flying toward or searching for that food.

Real animals rarely conform perfectly to the predictions of such simple models, but to the extent that they do, we can conclude that optimizing net rate of energy intake is an important part of the animal's decision-making process. For example, suppose a bird forages on a particular tree and moves on to the next one when it has depleted the food in the first one by a certain amount but moves on slightly sooner than predicted by the optimality model, we might conclude that the amount of food present relative to what the animal could find elsewhere was indeed an important factor controlling its behaviour. At the same time, to the extent that the real animal's behaviour departed from that of the simple model, we can conclude that other factors such as the need for a balanced diet or to keep a look out for predators are also influencing its

Figure 6.8 The dramatic result of two-way selection for blackcaps which migrate (i.e. show migratory restlessness in the northern autumn and spring) and birds which do not migrate. Starting with a population from southern France which showed about 75% of migrants and selectively breeding, within 3 generations the migrant line was fixed with all birds migrating. Within 6 generations the low line showed virtually no birds migrating. Clearly the French population was highly variable for genes affecting the migratory tendency. (Modified from Berthold & Helbig 1992.)

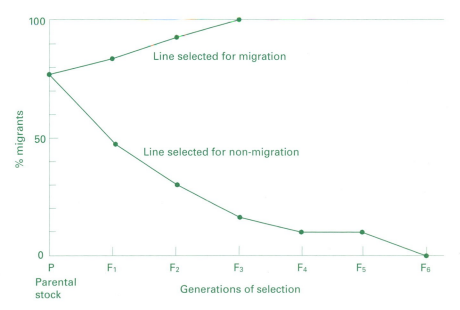

behaviour. This should come as no surprise. We saw in Chapter 4 how complex animal decision-making is and how many factors animals take into account when they 'decide' what to do. More complex optimality models – for example, ones that incorporate the state of the animal's food reserves, time of day, etc. – may be needed to account for real behaviour.

What is very confusing at this point is what we can or cannot legitimately conclude from optimality studies. Building optimality models and then testing them is an instructive way of finding out what an animal's decision-making rules might be. Such models therefore have a use in answering *causation* questions about animal behaviour. But they also have a use in answering *adaptive* (function) questions since the decision-making rules will themselves reflect the action of natural selection. An animal that foraged optimally according to the predictions of a simple optimal foraging model would do so only if optimizing energy intake was very important to its survival and reproductive success. In fact, it would really forage optimally only if optimizing energy intake was the *only* factor influencing survival and reproductive success, since if other factors were important, they would inevitably compromise the optimization of food intake.

As we have seen repeatedly through this book, animals are very rarely subject to just one selection pressure but instead are caught up in a tangle of compromises between different ones. We therefore expect the behaviour of real animals to be an optimal compromise between different selection pressures. In other words, we should not expect an animal to forage optimally at all times in its life because there will be some periods (for example, when it is not short of food and is being chased by a predator) when starvation is not the greatest threat to its survival and reproduction. On the other hand, there will be times when getting enough to eat quickly and efficiently will be quite critical: small birds in winter may have to give the very highest priority to finding food or they will quickly die of starvation, for example. But usually, we should expect that all the various components of an animal's success (feeding, avoiding predation, avoiding dehydration, finding a mate, etc.) will all individually be somewhat less than optimal in the interests of the overall optimality of the whole animal. Animals can therefore be optimal in the sense of achieving optimal compromises for maximum lifetime reproductive success but less than optimal when individual parts of their behaviour (e.g. feeding) are compared to an optimality model designed only to represent one component. A great deal of confusion and controversy can be avoided by keeping this distinction firmly in mind (for a fuller discussion see Dawkins 1995).

Evolutionarily stable strategies

A major weakness of many optimality models is that they focus on the behaviour of individual animals without taking into account the effect of what other animals are doing. But this is hardly realistic. Even our optimally foraging great tit will have its food supply affected by the foraging strategies of other birds. If all of them are attempting to optimize energy intake, they will interfere with each other in such a way that it may become more profitable to feed in a poor quality tree alone than in a food-rich patch in competition with other birds.

In other areas of animal behaviour, the effect of what other animals are doing becomes even more apparent. In aggression, for example, the advantage of attacking an opponent will depend critically on whether the other animal is more likely to fight back or run away. Our study of adaptation therefore cannot be complete until we have found some way of taking into account what other animals are doing. The way we do this is to think in terms of the Evolutionarily Stable Strategy (ESS), a term we owe to Maynard Smith (1982). A 'strategy' is simply a specification of what an animal does, such as 'remove eggshell from nest' or 'always attack if challenged'. An evolutionarily stable strategy is one that, if adopted by most members of the population, cannot be bettered by any other strategy. We can see the significance of this definition by considering the example used by Maynard Smith and Price (1973) when they introduced the idea of ESSs into animal behaviour. Their example was that of aggressive behaviour in which there were two strategies: 'Hawk', which always attacked and 'Dove', which always retreated if challenged. In a population consisting of Hawks and Doves, there would be four kinds of encounters (Fig. 6.9). When a Hawk met another Hawk, there would be a fight. Both Hawks would have a chance of winning but also a high chance of being injured. When a Hawk met a Dove, it would always win and the Dove would get nothing. This would make it seem that it was always better to be a Hawk, but this is not so as although Doves often retreat with nothing, they also don't get injured. And when a Dove meets another Dove, it will win on half the occasions. It will be clear that the advantage of following either a

Hawk or a Dove strategy will depend on how many Hawks and Doves there are around. For example, in a population consisting almost entirely of Doves, Hawks will have a huge advantage. They will be able to drive the Doves away from any resources they may want and they will not get injured because the Doves do not fight back. However, because they are so successful, they will over evolutionary time become more numerous in the population. And the more Hawks there are, the less successful the Hawk strategy will become. The greater will be the likelihood that a Hawk encounters another Hawk and the more likely it is to be injured. As the Hawks begin to suffer more and more injuries as the result of their encounters with other Hawks, they become less successful and the Dove strategy becomes the better option since Doves obtain resources on many occasions and do not suffer injuries. As a consequence, Doves begin to increase in frequency until they too become common, at which point Hawks begin to have an advantage again. It is thus clear that the success of both the Hawk and Dove strategies is negatively frequency dependent – that is, the more of the same kind of strategist there is around, the lower the success of that strategy. Both Hawks and Doves are most successful when they are rare and least successful when they are common. This keeps Hawks and Doves in equilibrium in the population and means that neither Hawk nor Dove is an ESS, since when either becomes common in the population, it is always bettered by the other. The only thing which is evolutionarily stable is a mixture of Hawks and Doves, giving an Evolutionarily Stable State of a mixture of the two types, known as a Mixed Evolutionarily Stable State.

	HAWK	DOVE
HAWK	$1/2V-C$	V
DOVE	0	$V/2$

Figure 6.9 Payoffs (gains or losses in fitness) resulting from the four possible encounters betwen Hawks and Doves; V=rewards of victory, C=costs of losing. When two hawks meet, they always fight and have an equal chance of winning as well as an equal chance of incurring costs (e.g. through being injured). When a Hawk meets a Dove, it takes everything that there is to be gained and the Dove gets nothing. When two Doves meet, they have an equal chance of winning but incur no loss. In this version of the Hawk/Dove game, Hawks always fight and Doves always retreat and the participants do not adjust their behaviour depending on the size or other attributes of their opponent. Such games are called 'symmetrical'.

Sometimes, however, a single strategy can be an ESS. The Hawk and Dove strategies we have just described are obviously over-simplifications as real animals do not go around fighting or retreating regardless of the size or strength of their opponents. They assess each other's potential fighting ability (Chapter 3) and adjust their behaviour accordingly. Maynard Smith and Parker (1976) described such an adjustable strategy as 'conditional', meaning that what the animal did was conditional upon external circumstances such as whether the opponent was bigger or smaller. A single conditional strategy such as 'attack if opponent is smaller, retreat if opponent is bigger', which could be described as 'Conditional Hawk', is an ESS: even if most of the population are Conditional Hawks, then the Conditional Hawk strategy continues to be the most effective. It does not have the built-in negative frequency dependence that would reduce its success when it becomes common.

It is important to stress that ESSs are not a single theory but a framework for looking at a variety of different situations. There are many different sorts of ESS. Some are conditional and allow the animal to change its behaviour depending on circumstances; some have no conditional element built in and simply assume that the animal will always behave in the same way. What they all have in common is an attempt to see what happens when a mutation changing behaviour is let loose in a world where it has to compete with other individuals following either the same or different strategies. A strategy considered in isolation might seem highly advantageous. It might even appear to be optimal. But unless it can hold its own in interactions with other strategies, it will not persist in evolutionary time.

ESS ideas have now been applied to a wide variety of situations including animal signalling (e.g. Grafen 1990), stealing food from other animals, and the decision of digger wasps to dig their own burrows or use those already dug by other wasps (Brockmann *et al.* 1979). ESS models are as useful for situations of conflict such as fighting as they are for situations of cooperation (Axelrod & Hamilton 1981). In each case, having to specify exactly what a strategy does and the precise circumstances in which it gains or loses helps us to see whether it is likely to spread through the population. The quantitative discipline that the ESS framework imposes on our thinking has made it an indispensible tool for the study of adaptation.

Kin selection and inclusive fitness

'Success' in evolutionary terms means leaving offspring that themselves reproduce, but the success of an individual is short-lived and ephemeral. In sexually reproducing species, an individual does not survive for more than one generation. It is genes that are passed from one generation to the next and our adult bodies could simply be regarded as the elaborate packaging that protects them. As far as a gene is concerned, the body it happens to be in at a given moment is a useful, if temporary, vehicle for getting itself passed on into the next generation, as R. Dawkins (1989) puts it. Samuel Butler's aphorism that 'A chicken is an egg's way of producing another egg' can be rewritten as 'An animal is a gene's way of producing more copies of that gene'.

Many people object to this way of looking at animals and feel uncomfortable with the idea of dispelling the sovereignty of the individual in favour of a gene-centred view of evolution. But consider something as basic and fundamental as parental care. We take it for granted that parent animals should feed and protect their young, but parental care itself is the result of strategies genes have for perpetuating themselves. Genes that help to make the bodies they are in more effective at defending their young will perpetuate themselves in the bodies of the protected young. Variation in genes affecting a tendency to defend young – perhaps mediated through variations in the level of a hormone – will result in variation in the numbers of offspring that survive to pass on the favoured genes, and so on down the generations.

But there is a twist to this tale. The direct line of parents to offspring is the only way that genes are passed on into the future, but direct parental care is not necessarily the only genetic strategy that will be successful. Helping a brother, sister, or other relative to reproduce may also enable genes to perpetuate themselves. Full brothers and sisters share, on average, half their genes (although the vagaries of Mendelian segregation mean that particular pairs of siblings may have much more or much less than this). So a genetic tendency to help a sister to reproduce could be favoured by natural selection because the sister, being so closely related, has a high chance of having the same genetic tendency. 'Genes for'* helping sisters thus help copies of themselves in the sisters of the body they are in and perpetuate themselves through the children of those sisters. W.D. Hamilton (1964) showed how genes for 'care of relatives' (not necessarily direct offspring) could spread and Maynard

This phrase does not imply any simplistic link between a gene and a pattern of behaviour. We use it as a shorthand way of indicating a gene or genes which, given a suitable environment in which to operate, can bias behavioural development along particular lines.

Smith (1964) suggested the term 'kin selection' to describe selection which takes account of other relatives as well as immediate descendants.

There is a very important point which has to be made here about kin selection. Helping a given relative to reproduce will only be favoured (genes for it will only be spread) if the benefit – that is, the increase in reproductive chances of that relative as a result of the help – more than makes up for the cost – that is, the decrease in reproduction the helper incurs as the result of its action. For example, there is no point in helping a brother if the help does not enable the brother to have any more children and at the same time prevents the helper having several children of his own. Genes for brother-helping clearly cannot spread under these circumstances: parental care genes would do much better. Hamilton generalized the circumstances in which relative-helping of various sorts would evolve into the equation: $rb - c > 0$.

r is the coefficient of relatedness and expresses how closely two individuals are related to each other, b is the benefit and c the cost of the relative-helping genotype. The net benefit minus the cost must be positive and greater than zero for the behaviour to be favoured. In order to work out whether or not the equation does work out greater than zero, we have somewhere to calculate values for the three terms, r, b and c.

'r' does not usually cause too many problems. From basic genetics we can work out that full siblings, and parents and offspring have a 50% chance of sharing a given rare gene ($r=0.5$); nieces and nephews have 25% chance of sharing with an uncle or an aunt ($r=0.25$), and so on.

In the past, it was very difficult to discover how closely related animals are to each other in nature. They had to be studied over a long period of time to find out which were the offspring of which adult and even then there was often considerable doubt, particularly about paternity. Now, however, DNA fingerprinting has revolutionized the determination of r. All that is needed is a small sample of blood, or even a hair or a feather and in many cases it is possible to determine the relationships among animals with considerable accuracy. This is made possible by the fact that throughout the genomes of many species there are regions of DNA in which the patterns of base pairs are repeated over and over again but the repeat patterns are slightly different in different individuals (Queller *et al.* 1993). The results have often been quite surprising. For example, in a recent study of starlings, which have always been considered like most birds to be monogamous, DNA fingerprinting showed that one brood of chicks was fathered by no less than three different males (Pinxthen *et al.* 1993). Behavioural observations had not revealed any extra-pair copulations and it was only the genetic evidence which showed that these must have been occurring.

It is important to remember, however, that it is not just the degree of relatedness that matters but the number of relatives that are helped. Helping a large number of distant relatives could, if there were enough of them, be just as beneficial as helping a direct offspring – hence J.B.S. Haldane's famous after dinner remark, 'I am prepared to lay down my life on behalf of four grand-children or eight first cousins'!

Although the r component of Hamilton's equation is now readily accessible, b and c are somewhat more problematic. If we observe one animal helping another to rear its young, how do we know that the parent wouldn't have been just as successful without the help? And how do we know what the cost to the helper was in terms of the offspring it would have had if it hadn't been helping someone else? We seem to be dealing with mythical potential offspring that don't exist (the cost of helping) and extra offspring that do but are indistinguishable from the others (the benefits of helping). We can look at some examples to see how estimates of b and c are arrived at in practice.

In his two original and very important papers, Hamilton (1964) addressed himself to a problem which had long puzzled zoologists. The social insects, the Isoptera (termites) and many of the Hymenoptera (ants, bees and wasps) show extreme altruistic or helping behaviour. There is usually just one reproductive female (the queen) and large numbers of sterile workers. The workers – both males and females in termites, but solely females in the ants, bees and wasps – perform all the tasks of the society such as foraging, rearing young, nest construction and defence, and do not reproduce at all themselves. The remarkable fact is that in the Hymenoptera this ultimate form of self-sacrifice – for the sterile workers do not, of course, reproduce themselves – appears to have arisen independently at least 11 times and perhaps more. Clearly it has not been easy to evolve social life or more different types would be expected to have done so, yet somehow this one order of Hymenoptera seems predisposed to achieve it.

Hamilton drew attention to the significance for r – the coefficient of relatedness – of the Hymenopteran unique form of sex determination. Ants, bees and wasps exhibit 'haplo-diploidy' which means that the males are haploid and develop from unfertilized eggs whereas the females are diploid and develop in the normal fashion from fertilized eggs. All the sperm from one male are therefore identical – a simple copy of the male's own haploid chromosome set. When the queen bee fertilizes eggs with this sperm, all the resulting daughters receive the same paternal chromosomes and so have half their genes in common (the half donated by their common father). In addition, they share, on average, half the genes inherited from their common mother and so their degree of relatedness, $r = 0.75$ (0.5 from father plus $0.5 \times 0.5 = 0.25$ from mother).

This means that although the workers are sterile themselves, many of the genes that they share with the young queens (which are also their sisters) will be passed on to the next generation. The sterile workers benefit, not because they help other workers which are closely related to them but are a reproductive dead-end, but because they care for the small number of their sisters which will develop into young queens. As Hamilton puts it, in gene terms, it is more advantageous for a female Hymenopteran to stay and help to rear her closely related reproductive sisters than to leave and attempt to rear less closely related daughters of her own.

While haplo-diploidy appears to predispose the Hymenoptera to the high degree of sociality they show, it is clearly not the only factor because, as we have mentioned, a similar degree of sociality is also shown by the termites. They have an ordinary diploid mating system and sterile workers of both sexes have a degree relatedness to the young reproductives of only 0.5. King and queen termites are long-lived and monogamous. A queen may lay up to 36 000 eggs a day and in some species live for 60–70 years. The king and queen together may have literally millions of offspring in their lifetime, the vast majority of which will never reproduce at all. In order to understand why so many termite workers should be sterile, we have to remember that there are three variables in Hamilton's equation. A high r predisposes towards helping because it increases the effective benefit, but even a relatively low value of r would lead to helping if the benefits were high enough and the costs low enough.

The cost to each worker of being sterile is the loss of those offspring it would have had if it had not been helping the colony. Given that a pair of termites on their own would probably not survive, let alone reproduce even modestly, the costs of helping must also be small (no hope of reproducing means no cost to losing it). The monogamous nature of the termite breeding system means that the workers are guaranteed a long series of full brothers and sisters to take care of. Since their own parents offer no care to them whatsoever, without the workers the young termites would die. The workers, therefore, make a substantial difference to the survival chances of close relatives, even though the benefit, b, accruing to each worker is only a fraction (because it is shared with the other workers which have helped) of the output of each reproductive.

Nevertheless, it appears that the benefits of helping are significant, the costs low and r at least as high as between diploid parents and offspring. Termites often live in deserts or in very dry regions and can only do so because the mounds (Fig. 6.10), built by the collective labours of millions of workers, enable them to create their own microenvironment. A single pair of termites, removed from this specialized microenvironment, would stand no chance, and this fact effectively reduces the costs and increases the benefits of helping the royal pair in the home colony.

Until quite recently, worker sterility and a caste system of sociality were thought to be features exclusive to insects. Now we know that at least one mammal, the naked mole rat (*Heterocephalus glaber*), also has a very termite-like social system (Jarvis 1981). These extraordinary looking rodents (Fig. 6.11) live in underground colonies in dry desert regions of southern Africa. They are almost hairless and, as a result, poikilothermic, but their burrows provide them with a safe environment of almost constant temperature so that their inability to regulate their own body temperature is not a problem. Their burrow systems can be extremely extensive – up to 3.5 km – and shared by 70 to more than 200 individuals. There is usually just one breeding female and up to three reproductive males. The rest of the colony consists of workers, most of which never reproduce throughout their lives. Instead, they defend the colony and, working like a chain-gang, dig tunnels through the hard earth to find roots and tubers on which to feed the rest of the colony. The reproductive female does not take part in the food-gathering activities and has food

Figure 6.10 The ventilation chimneys built by the colonies of some termite species can reach an astonishing size. (Photograph by Martin Speight.)

Figure 6.11 Naked mole rats live underground in colonies of closely related individuals. The workers cooperate to tunnel through the hard earth in search of food. (From a photograph by David Curl; drawing by Priscilla Barrett.)

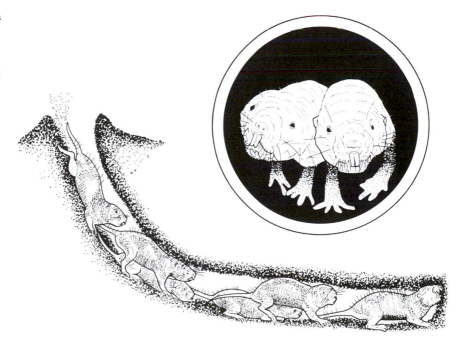

brought to her by the workers. Food is so hard to come by that a pair of mole rats on their own would probably not survive at all. By cooperating with the colony and contributing to the reproduction of others, at least the workers get a 'part-share' of the offspring produced. Monogamy plus a considerable degree of inbreeding results in an average relatedness of 0.81. The reproductive female produces huge litters of up to 27 pups, giving the sterile workers lots of closely related young to look after. The harsh environment makes reproduction hazardous without the help workers can give, and these factors seem, as with the termites, to have tipped the cost–benefit equation in favour of the ultimate in self-sacrifice, life-long worker sterility.

A second species, the Damaraland mole rat (*Cryptomys damarensis*), offers an interesting comparison here (Jarvis *et al.* 1994). This species is larger, covered with hair and lives in subtropical areas of Africa where the soils are softer and the burrow environment more variable. Although it is, like the naked mole rat, eusocial and has non-breeding workers, its colonies are much smaller, usually numbering about 16 individuals. This seems to be related to the fact that the soils are softer and do not take so many individuals to dig them as in the rock-hard desert soils of the naked species. The benefit that Damaraland mole-rat workers gain by staying to help their parents rear young seems to be less and the workers have a much greater chance of leaving the colony and founding one of their own where at least some of them become reproductive. Their cost–benefit equation is still tipped in favour of some worker sterility, but less strongly.

Comparisons with various species of birds show how fine the balance can be. None, as far as we know, goes to the lengths of being totally sterile, but in over 200 species, the parents are helped in some way by other individuals, often their own young from previous years. Later, these helpers frequently become parents in their own right but the selective advantage to their juvenile helping bears a strong resemblance to the helping seen in termites and mole rats. Firstly, most birds not only nest in the same territory year after year but are also monogamous (although, as we have seen, not always perfectly so) which means that the coefficient of relatedness between siblings of different years is high. Secondly, helpers can contribute significantly to the viability of their siblings. They may help in nest-building, territorial defence and feeding, and they are often very important in keeping predators away. Emlen & Wrege

(1989) showed that for white-fronted bee-eaters (*Merops bullockoides*), each helper enables the parents to raise, on average, half an extra chick. Woolfenden (see Woolfendon & Fitzpatrick 1984) has shown that in Florida scrub jays, which also have sibling helpers, the majority of helpers are males and are the sons of the birds they help (Fig. 6.12). They appear to increase the number of chicks the parents can raise to fledging largely through their active defence of the nest. So these helpers gain through having an increased number of young siblings and they seem to have a further benefit in that they tend to take over part or all of their parents' territory in the future, when they themselves start to breed. The cost of helping, that of postponing reproduction for a year or so, may not be all that great, since there appears to be a shortage of nest sites, making it very difficult for young birds to breed at all. Until a young male can secure a suitable breeding territory, he appears to be better off helping his parents to rear extra siblings than trying, and then failing, to reproduce himself (cf. situation of the honey-bee worker).

So we can see that kin selection, of which 'helping at the nest' is a clear example, is not a separate, special sort of selection, in some way different from

Figure 6.12 Three adult Florida scrub jays at the nest. All are colour-ringed and can be identified as the two parents and a yearling bird from a previous brood. This latter is now helping to rear its young siblings in the nest. (Drawing by Priscilla Barrett, from Wilson 1975.)

natural selection, but a logical extension of it. But whereas for 'core' natural selection we can measure success or fitness as the number of offspring reared to reproductive age, with the extension of kin selection the measure of success is slightly more complex. We have to know not just the number of offspring produced by an individual, but the effects of its behaviour on how many offspring its relatives have and how many it does not have itself as a result of its helping. Hamilton (1964) used the idea of 'inclusive fitness' as a way of calculating the conditions under which a gene might spread, taking into account the effects that bearers of that gene might have on different sorts of relative. The concept of inclusive fitness has been quite widely misunderstood (Dawkins 1979; Grafen 1984). In particular, it is often wongly thought that it is a simple weighted sum of all the offspring and other relatives an individual has, taking into account their degrees of relatedness. But this is incorrect. Just because an animal has a sister does not mean that the sister should automatically be included in its inclusive fitness. Only if the animal has in some way helped its sister to survive or reproduce should the sister be included in this way. The reason is that, as we have seen before, natural selection operates at the level of genes, not individuals. What is clear is that sometimes not reproducing oneself but helping close relatives to reproduce instead can, under some circumstances, be the best strategy.

Kin recognition

It will be obvious that if kin selection is to operate, animals must have some way of selectively directing aid towards animals they are related to. Unfortunately, this process has come to be known as 'kin recognition', a misleading term because it might seem to imply that some active process of singling out relatives is taking place. In practice, however, relatives may be helped not because they are recognized as such but because they just happen to be physically closest to the helper. In species that do not disperse very far, an animal which helped its near neighbours would be more likely to find itself helping its relatives than non-relatives, without any other special mechanism being involved. A parent bird that used a simple 'rule of thumb' to recognize its young (such as 'feed anything in the right nest') would in most cases end up feeding its own young rather than non-relatives. Once relatives have been responded to in one location, they can then be recognized as familiar and responded to in other places. This method is, however, open to error if the nearest animals turn out not to be relatives: for example, when nest parasites such as cowbirds and cuckoos deposit their eggs in the nests of host species or when 'egg-dumping' occurs between members of the same species.

To combat the possibility of such errors, some animals have more sophisticated ways of recognizing kin, although on close scrutiny many of them rely ultimately on location or familiarity to work. In a process known as **phenotype matching**, for example, an animal uses its experience of its siblings, parents or even itself to build up a template of what familiar relatives look, sound or smell like and then is able to recognize an unfamilar animal as kin because it looks, smells or sounds similar. In other words, the animal learns something about the phenotype of its relatives and can then respond to unfamiliar individuals if they match it . But the original recognition of siblings and parents as kin in turn often depends on location – siblings are 'recognized' as such because they are familiar through being in the same nest. An interesting version of phenotype matching is the possibility that animals might use themselves templates. For example, an animal could learn what its own smell was like and then pick out other animals that smelt like itself, a possibility that Richard Dawkins called 'the armpit effect'.

As we saw in Chapter 3, mice (*Mus musculus*) may 'match' each other on the basis of whether they are genetically similar or different in the Major Histocompatibility Complex (MHC). The MHC is a large genetic region found in vertebrates and is critically concerned with the recognition of 'self' and 'non-self' through genes that are concerned with cell–cell recognition. Indeed, it was orginally identified because it played such a major role in whether a body rejected or accepted tissue and organ transplants. Its key feature is that it contains a large number of genes that are extremely variable, some loci having as many as 50–60 alleles, giving the possibility of very fine discrimination between individuals. The cell surface proteins coded for by MHC genes produce breakdown products that appear in the urine of the mice, and appear to act as chemical cues for social recognition, including recognition of relatives (see Brown & Ekland 1994, for a review).

A combination of different processes seems to lie behind the kin recognition shown by Belding's ground squirrels studied by Holmes and Sherman (1982). They observed the behaviour of litters in which the mothers had been fertilized by several different males and where there were, therefore, full and half-siblings in the same litter. Even though they grew up in the same burrow and had even developed in the same uterus, full siblings were less aggressive to each other than half-siblings, suggesting an ability to discriminate individuals of different degrees of relatedness even when experience of them was apparently similar. On the other hand, if baby ground squirrels are transferred to new litters when very small, they show reduced aggression to their unrelated foster-siblings compared with unrelated but unfamiliar ground squirrels of the same age. They are also less aggressive to their full genetic siblings in the original litters, even though they have had no experience of them since just after they were born. It seems, then, that ground squirrels show a combination of mechanisms for kin recognition. Familiarity obviously plays a part, since individuals are less aggressive to individuals they grew up with than to strangers. But there also seems to be some recognition of full siblings – apparently based on olfaction – even when they are reared completely separately from each other.

Bees, too, can discriminate full and half-sisters. In the sweat bee, *Lasioglossum*, Greenberg (1979) showed that the willingness of the bees guarding the entrance to the nest to let in other bees was directly correlated with the degree of relatedness between guard bee and intruder (Fig. 6.13). In neither of these cases do we have a complete explanation of exactly how the discimination among relatives is accomplished but chemical odour cues, perhaps learnt from self or individuals encountered early in life, seem to give the animals a very remarkable ability to distinguish not just relatives from non-relatives, but individuals of different degrees of relatedness as well.

Conflict and infanticide

Relatives will not always help each other, nor will parents always put themselves at risk for their offspring. Although Hamilton's equation shows us the conditions under which altruism between relatives can occur, there is a darker side to the same equation, even for relatives as close as parents and offspring.

Since parents are not genetically identical to their children, the interests of the two may not always precisely coincide. Sometimes the best interests of the parent may even be best served by killing or at least withholding care from some offspring if, as a result, the parents are enabled to have many more offspring in the future. Owls and other birds of prey often lay more eggs than they can normally rear as an 'insurance policy'. All the chicks are reared if food is plentiful but the smallest one dies in times of scarcity. It is only a small step from withholding food to passive or even active killing. Mock *et al.* (1987) showed that great egrets (*Casmerodius alba*) often practice 'siblicide' – active killing of nestlings by their older brothers and sisters, with the parents looking on and not intervening. A chick may stab a sibling with its beak, peck it and even push it out of the nest. In the closely related great blue heron (*Ardea herodias*), however, siblicide is rare, and while the young may jostle each other for food, they do not harm each other. Both species are monogamous, so their nestlings are closely related to one another. Mock *et al.* argued that the difference between them was in the size of the food items the parents bring back to the nest. Egrets bring small boluses of fish, enough for one chick at a time, so that an aggressive chick can monopolize everything the parent brings. The result is that the aggressive chick that disposes of its siblings gets more of

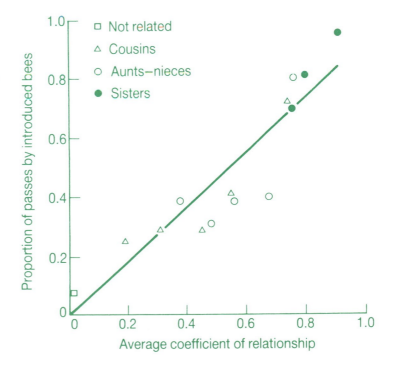

Figure 6.13 Relationship between the proportion of sweat bees (*Lasioglossum*) allowed into a nest by guard bees at the nest entrance and the degree of relatedness of the potential intruder to the guard. (From Greenberg 1979.)

the food that the parent brings back to the nest than it would if it had to share it. In the great blue heron, by contrast, the parents tend to bring whole fish to the nest, far more than one chick could eat alone. Here, killing siblings does not have the same advantage.

It is important to stress that the siblicidal behaviour of the egrets is just as explicable by Hamilton's equation as is the sibling altruism and helping shown by social insects. Although an egret chick is related to the sibling it kills by $r=0.5$, the extra food it gets for itself and the better survival chances it gives itself as a result more than make up for the loss of a full sibling. It is effectively trading more offspring of its own, through increasing its own chances of reaching reproductive age, for the life of its sibling and the nieces and nephews the sibling might have gone on to produce had it survived. Of course, if it could have both (offspring plus nieces and nephews) it would do even better, but if food is scarce and it is a choice of one or the other, the balance between 'b' and 'c' may tip in favour of selfishness. In herons, the selfish benefits of siblicide appear to be much lower and the balance tips back again.

Hamilton's equation also expresses the genetic conflict that exists between parents and offspring. Every time a parent feeds its young, even though $r=0.5$ it is expending effort and energy that could have been spent on rearing other offspring. If by withholding that care, the parent could rear two or more other offspring, then withholding care might be the most beneficial strategy. As Trivers (1974) pointed out, conflict between parents and offspring is most apparent at weaning, when it is in the offspring's best interest to demand a bit more food and in the parent's best interests to conserve resources for future children. For example, Hinde (1977) showed that as a baby rhesus monkey becomes older, the mother rejects her infant's advances and stops it making nipple contact with steadily increasing frequency until, at the end of the period of investment, all the offspring's advances are rejected (see Fig. 2.25).

In burying beetles (*Nicrophorus vespilloides*), this conflict is even more apparent as the parents actually kill some of their offspring, depending on the size of the corpse (Fig. 6.14). These beetles locate a newly dead animal, such as a mouse, and then proceed to bury it as food for their young. The number of larvae that can survive on a the corpse increases with corpse size and if there are too many of them, the parents simply kill off the surplus, thus ensuring the survival of the maximum number of their young (Bartlett 1987).

Figure 6.14 A burying beetle feeding one of its larvae which is reaching up from the cavity of 'the crypt'. This neat spherical structure has been formed from the carcass of a mouse. It is here shown uncovered but would normally be below the surface. The beetles dig a depression beneath the body which collapses into it. They then bend the corpse into a rough circle and remove all the hair. They coat the body with anti- bacterial secretions in their saliva and line the cavity with the hair some of which also sticks onto the surface of the treated body. Feeding on the flesh, a cavity is produced into which eggs are laid. The larvae feed directly on the flesh but are also, as here, fed by one or both of the parents. (Photograph by Linda Partridge.)

Killing of unrelated young is even easier to understand in evolutionary terms. In a number of species, males kill infants that are not their own in order to further their own reproductive interests. For example, the permanent members of lion prides are the females and these are from time to time joined by groups of two or three males, usually following a fight that ousts the previous males (Fig. 6.15). At each takeover, the new males tend to kill all the young cubs in the pride, particularly those that are still taking milk from the females. The new males gain from this infanticidal behaviour because breaking off lactation brings the females more rapidly back into oestrous. Pusey and Packer (1987) found that females whose cubs had been killed at a takeover conceived again about 4½ months later, but those whose cubs had survived did not conceive again until over 20 months later. As these new conceptions would be by the new males it is quite clear that the more cubs a male can kill, the more quickly he will be able to father offspring of his own. Cub-killing by male lions is particularly associated with taking over a pride and ceases by the time the lionesses are bearing the new males' own cubs.

Figure 6.15 Three lionesses from the same pride share a wildebeest kill. A half-grown cub, still showing signs of dappling on the flanks, is on the left. It is the females who form the permanent core of a lion pride. (Photograph by Karen McComb.)

A similar explanation – sexual selection between males to sire as many infants as possible – has been put forward to account for infanticide among primates (Struhsaker & Leland 1987). As with the lion, male langur monkeys have been regularly observed to kill infants when they move into a new troop. Showing a rather curious role-reversal, in the wattled jacana, it is the females that kill the young. In these birds, females are larger than males, defend territories against other females and have several males incubating their eggs. Emlen *et al.* (1989) showed that when a female takes over a new territory, she kills young already there, and so hastens the time when the males will be incubating her own eggs.

Cooperation between non-relatives

'Nature red in tooth and claw' seems to be an apt summary of many of the examples we have discussed so far in this chapter. Apart from altruism between relatives, competition between genotypes has some fairly ruthless outcomes, and even relatives can be sacrificed if the balance between cost and benefit tips in a particular way. But this cannot be the whole story because we also find examples of help and cooperation even between animals that are not related to each other at all. Mutual grooming or preening are quite common, with one animal grooming parts of the body, such as the head, neck and back, that the animal itself cannot reach. This may occur between relatives, but is also found in animals that are unrelated but very familiar associates. It is particularly well developed in the primates where, as we shall discuss in Chapter 7, friendly contact helps to cement bonds which may have all sorts of other payoffs within a social group. Both parties gain from the interaction and both can break it off if the other does not participate.

An even more interesting form of mutual help is called reciprocal altruism, where one animal helps another but may not then receive assistance back itself until some time later. This 'mutualism with a time lag' is therefore potentially open to cheating, since one animal could apparently receive benefit itself and then not reciprocate. For this reason, we do not find reciprocal altruism evolving in situations where there is a 'one-off' advantage with two animals being unlikely to meet again. Rather, it evolves in situations where the same animals associate over a long period of time and where a 'cheat' will be penalized because benefits will be withheld in the future if it does not reciprocate. It is most highly developed in animals which have the capacity to remember which other individuals are reciprocators and which are cheats, such as the chimpanzes described by de Waal (1989) or baboons by Packer (1977). What is of interest is that their repeated, long-term cooperation leaves both parties to the pact better off than either would be alone.

Wilkinson (1984) has described a remarkable example of reciprocal altruism in vampire bats which feed on the blood of mammals, particularly domestic animals such as cattle and horses (Fig. 6.16). A bat will inflict a tiny (3 mm) wound on its host with its sharp teeth. Its saliva contains anticoagulants and the bat flicks its tongue rapidly in and out to take the blood. These bats are active during the darkest hours of the night and return to communal roosts in hollow trees or caves by day. Sometimes a bat will return to the roost without having fed and if it fails to find food for three successive nights, it may starve to death. But a bat that has not found food itself will often be fed by another bat regurgitating a blood meal to it. Starving bats are more likely to be fed than well-fed ones, and although bats feed their kin, they also feed unrelated bats, *particularly those that have fed them in the past*. Regurgitation of blood meals was seen only between bats who regularly (over 60% of sightings)

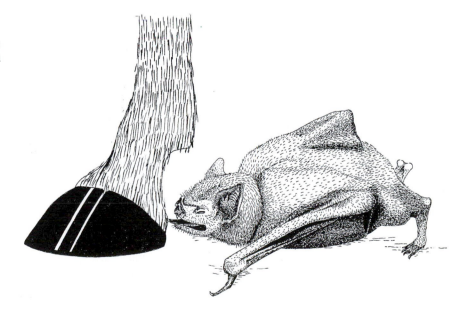

Figure 6.16 A vampire bat laps at a tiny slit above the hoof of a horse. It has cut this with razor-sharp incisor teeth and its saliva contains a powerful anticoagulant. (Drawing by Melissa Bateson.)

associated with each other. The bats fulfil the criteria for successful reciprocal altruism. They return to the same roost day after day and associate with particular individuals over long periods of time, remembering benefits given to them. This means that they can detect cheating individuals that do not reciprocate. The bats are also sensitive to whether another bat is starving or well fed and are able to help a starving bat without too great a cost to themselves. A starving bat loses weight at a high rate, but a well-fed one at a much lower rate, so that the giving of a blood meal from a well-fed to a starving bat gives the recipient more time until starvation than the donor loses. Finally, failing to find food on its own appears to be something that can happen to any bat at some time, so all bats benefit from the 'insurance policy' of being fed by others in time of need. But those which donate food will be a in a more favourable position to receive it when they themselves are in need of it. Cooperation, in the long run, is selfishly the best policy.

Acts of reciprocity have been placed into an ESS framework, which has helped to clarify the conditions under which reciprocal altruism is most likely to evolve (Axelrod & Hamilton 1981). A very important condition is that there must be repeated interactions between the participants so that failing to cooperate on one occasion (cheating) has a penalty in the future through not having the cooperative act reciprocated next time. If vampire bats never re-encountered the bats to which they had given food, we would not expect them to behave so altruistically. For this reason, we expect this sort of reciprocal cooperation to be characteristic of animals that stay together over long periods of time, long enough for the roles of donor and recipient to be exchanged many times. A second condition is that 'cheats' can be recognized and penalized. This does not necessarily mean that the cheat is singled out and victimized. It could just mean that if an animal fails to reciprocate, its partner simply moves away and starts to interact with other individuals.

The particular problem with explaining how reciprocal altruism has evolved is the time lag between one cooperative act being performed and when it is returned by another. Until the altruist has been repaid, it has gained nothing – in fact it is temporarily worse off than if it had never done anything. It is therefore not surprising that documented cases are relatively rare. What is much more common is to find cooperation between animals where the benefit is mutual and immediate: for example, horses simultaneously grooming each others' necks or flocks of birds all benefiting from each other's vigilance in being able to spot the approach of a predator. We will return to this important topic in the last chapter as it forms a way of bringing together many of the different aspects of animal behaviour we have talked about throughout the book. In the meantime, we turn to what might be called the end-point of all adaptation: successful reproduction.

Sex and sexual selection

It will be clear by now that it helps in understanding the adaptiveness of behaviour if we look at genes as well as individual animals and that if we do this, we can even understand the evolution of behaviour that may be positively damaging to the individual doing it. There is one area, however, where even a gene-centred view appears not to give a complete explanation. This is the realm of sexual reproduction, which is so commonplace that we tend to take it entirely for granted and fail to realize why it should be difficult to explain.

The reason that sexual reproduction is such a problem is that, in species where the male makes no investment in the young, a female could apparently pass on twice as many of her genes to the next generation if she reproduced asexually and made identical copies of herself as she could if she reproduced sexually and combined her genes with those from a male (Maynard Smith 1978). Her sexually reproduced children are only half related to her ($r = 0.5$), not fully ($r = 1$) as they would be if she reproduced without sex. From both the gene and the individual point of view, therefore, there would seem to be a loss, a twofold disadvantage of sex. So we would expect there to be some major corresponding advantage to compensate for this and explain why, despite the disadvantages, sexual reproduction is so widespread among both animals and plants. Although the last 20 years have seen a great deal of theorizing by evolutionary biologists, we still do not fully understand what that advantage might be (Hurst & Peck 1996). The most popular explanation is that sexual reproduction is an advantage when the environment is changing very rapidly: for example, when parasites are evolving resistance to their hosts' defences. As we know to our cost, parasites and disease organisms are capable of such rapid evolution that they develop resistance to drugs that initially kill most of them. They also develop resistance to the natural defences of their hosts and when this happens, sexual reproduction may be the best counter-measure the host can take, because it leads to variation and thus to the possibility that at least some of the sexually reproduced variants might just happen to have the right combination of genes for resistance to the new strain of parasite (Hamilton *et al.* 1990).

Whatever the selective advantage of sexual reproduction, however, it has extremely important implications for animal behaviour. Not only does it result in the evolution of two sexes so different in appearance and behaviour that they may be mistaken for two different species. It has also resulted in the evolution of elaborate and conspicuous mate-attracting ornaments and displays that may actually endanger the life of the animal – usually a male – possessing them. For example, male guppies (*Poecilia reticulata*) have bright orange spots that are very attractive to females: males with the most and brightest spots are preferred by females. However, these are just the males that are most likely to be eaten by the two main predators of guppies, killifish (*Rivulus*) and the cichlid *Crenicichla* (Endler 1980, 1983). It is usually males that develop such ornaments because male animals produce large quantities of sperm and often contribute little in the way of resources and parental care to raising their offspring. As a result, a successful male can potentially have far more offspring in his lifetime than any female, and consequently males compete for the available females. Females are expected to be more discriminating in their choice of mate than males because a mating that leads to inviable or unhealthy offspring represents a larger proportion of the female's lifetime reproductive output than it does for a male. A male can more easily mate again than a female.

In 1871, Darwin proposed that elaborate male ornaments such as the classic 'train' of the peacock evolved to attract females under a process he called 'sexual selection', by which he meant competition between members of one sex – usually, for the reasons given above, males – to mate with members of the opposite sex. He believed that this would lead to the evolution of traits that either helped the male to fight off other males, or made him particularly attractive to females, or both. Darwin further believed that sexual selection could be such a powerful force that such traits could evolve even though the male possessing them might not survive all that long. As long as he survived long enough to mate and leave a large number of progeny, his characteristics would be successfully passed on to the next generation. Darwin thus saw sexual selection as a special case of natural selection, with the emphasis on mating success.

For a long time Darwin's theory of sexual selection, particularly its emphsis on female choice, was treated with scepticism. Then, in the 1930s, it was given new impetus by R.A. Fisher, who argued that exaggerated sexual ornaments such as long tails could indeed arise from female choice and that, over evolutionary time, the tails would become longer and longer (Fisher 1958). This was because the more females chose long-tailed males, the more male offspring there would be with long tails like their fathers and the more female offspring there would be with a preference for long-tailed males, like their mothers. This has become known as 'the runaway' theory. Note that on the Darwin/Fisher theory of sexual selection, there is nothing else that is special about long-tailed males. It is not that they are bigger or stronger or more fertile than other males. It is simply that they possess an ornament or are able to perform a display that attracts females because the females have a particular choice convention. We have already seen (Chapter 3) that seemingly arbitrary choice conventions may be quite widespread because of the tendency of animals to respond to sign stimuli and supernormal stimuli. Females that mate with males possessing such ornaments will benefit because, once there is a critical proportion of other females in the population with a similar preference, they will have sons with similarly attractive features and so gain grandchildren through their sons' ability to attract mates.

More recently, a number of alternative theories of sexual selection have been proposed, many of them centred around the idea that elaborate male ornaments and displays allow females to assess a males's 'quality' (health, physical vigour, etc.), particularly genetic quality that he would be able to pass on to her offspring. The female's problem is, of course, how to choose a male possessing such 'good genes' when all she has to go on is what he looks like. Zahavi (1975) put forward the idea that elaborate male ornaments are actually a handicap and that males with such ornaments are demonstrating their physical quality by showing that they can survive despite having such a handicap. He claimed, somewhat paradoxically, that a female choosing a male with a long tail is guaranteed a high quality mate because only high quality males can afford to carry the handicap of a long tail. A low quality male would simply be unable to survive with such an encumbrance. Although Zahavi's ideas were initially greeted with scepticism and declared unworkable they have gradually come to be more widely accepted and shown to be at least theoretically plausible (Grafen 1990).

But there is a second way in which males can signal their 'quality' without going so far as to handicap themselves. Hamilton and Zuk (1984) proposed that male ornaments may enable healthy males to advertise the fact that they are free of diseases and parasites. Since disease is a major source of juvenile mortality and may indeed have been the driving force behind the evolution of sex itself (p. 355), Hamilton and Zuk argued that females should preferentially choose males that have genes for resistance to parasites, which would then be passed on to their offspring. However, they did not claim that an ornament such as a long tail was a handicap in the sense that it actually lowered the male's resistance to disease. They claimed simply that it revealed a male's state of health because only healthy males would have glossy, well-kept tail feathers; or only a really healthy male could keep up a display for a long time, thus 'revealing' that he was in good condition. Unfortunately, there is a great deal of confusion in the literature about exactly how ornaments signal quality and even about whether they do so at all. We can see how difficult it is to distinguish between the various 'quality' theories and Fisher's runaway theory by looking at some examples of female choice in action.

This difficulty can be illustrated by two studies that we have already discussed in Chapter 3 in connection with animal signalling. One of these is Anderson's (1982) study on long-tailed widow-birds (*Euplectes progne*) in which, you will remember, he cut off the long tail feathers of some of the males and then glued onto the feather stumps either extra long tail feathers or much shorter feathers than normal or, as a control, feathers of the same length (see Fig. 3.33, p. 169). He found that males with the artificially elongated tails attracted the most females in to their territories. But while this study shows clearly that females prefer males with long tails, it can be interpreted in several different ways. It could be explained by Fisher's theory – that the females are simply attracted to the tail for its own sake or it could be that the long tail is a handicap, making flight so difficult for the males that only the super-fit ones can cope.

The other study which gave a similar result is Møller's (1988) demonstration that female swallows, like female widow-birds, prefer males with artificially elongated tails. Males with artificially long tail streamers actually had more young fledged than males with normal or shortened tails. By being able to attract females more easily, they started their first broods earlier and consequently had more time to father second broods.

Møller (1990) then went further and provided evidence that the female preference for long tails could be more easily explained by the handicap theory than by Fisher's runaway theory. He showed that many swallows are infested with blood-sucking mites which have the effect of slowing down the growth of the chicks. Interestingly, the mites also affect the length of the tail feathers grown by the adults, the key feature for female mate choice. Møller did a cross-fostering experiment in which he kept half of each brood with its natural parents and put the other haf in the nest of foster-parents. He found that the level of mite infestation was related to the level found in the natural parents rather than in the foster-parents, indicating a heritable component to parasitism level. This suggests that female swallows choosing a male with a long tail thereby ensure their offspring have a high chance of inheriting genes for parasite resistance. However, this may not entirely account for the length of the male swallow's tail. More recently, Norberg (1994) showed that the long tail streamers may help rather than hinder a male's normal flight. He put male swallows in a wind tunnel and was able to demonstrate that the long outermost tail feathers function aerodynamically to enhance the lift force of the tail and so to improve manoeuvrability. This would be the exact opposite of a handicap, as it would suggest that the male derives a direct personal advantage from his long tail, as well as a mating advantage.

Into this confusing and ever-growing body of studies of the evolution of mate choice has recently come yet another sexual selection theory, known as 'sensory bias' which is particularly concerned with the evolutionary origins of female preferences. Sensory bias refers to the curious tendency of females in some species to respond to male features in ways that seem to be inherited from their ancestors and have nothing at all to do with male quality. For example, swordtails (*Xiphophorus*) are so called because they have coloured, sword-like extensions of the caudal fin (Fig. 6.17). Not surprisingly in view of what we now know about sexual selection, female swordtails prefer males with the longest swords (Basolo 1990). But what is surprising is to find that females of the closely related platyfish (*X. maculatus*) also prefer males with long swords even though male platyfish have no swords at all. Platyfish females are therefore more attracted to males of a different species than they are to males of their own! It appears that female platyfish have inherited a 'bias' to respond to sworded males from the common ancestor of swordtails and platyfish, even though the character has been lost in the males of their own species (Basolo 1995). In the wild this is unlikely to matter because the two species rarely meet and the sexual preference seems to be an evolutionary remnant from the past that only shows up in the unnatural conditions of the laboratory.

Sometimes, however, we can see that such sensory biases operate in natural conditions and that it is of advantage to a female to be biased in this way. In our discussion of communication (Chapter 3), we have already seen that the most effective signals are those that grab attention by stimulating the sense organs and brains of receivers particularly strongly and may even be super-normal. The female water mite and the female jumping spider that initially respond to their males because the males attract them with food-like stimuli are responding to a pre-existing bias. Indeed, the success of the males'

Figure 6.17 Male and female swordtails. Given a choice, females prefer to mate with the male whose sword is longest. (Drawing by Nigel Mann.)

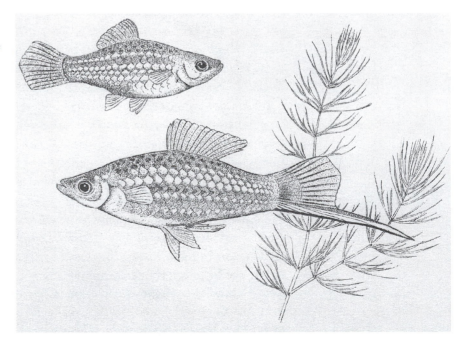

signals lies exactly in the fact that it does exploit the tendency of the female to respond to food. And the female gains too: either she gets real food or, failing that, a willing male of the right species.

Ornaments and resources

We have assumed so far that male displays and ornaments are directed primarily at females but, as Darwin (1871) pointed out, other males may also be the targets. It is possible that many sexual displays are really male–male assessment displays and females simply choose on the basis of their ability to win fights or dominance over other males. This of course would be an adaptive female strategy since males that are able to compete with other males in this way are likely to be physically healthy and strong. In many ways, prowess in battle is a more direct indicator of male 'quality' and likelihood of possessing 'good genes' than growing a long tail. The huge male elephant seals, which are three or four times the size of the females (Fig. 7.10, p. 396) gather harems on the breeding beaches. Only the very strongest can dominate a large area and in some breeding seasons, a mere 4% of the males are responsible for 85% of the matings (Le Boeuf 1974). Since the males are so much larger than the females, the females apparently have little 'choice' if a male decides to mate with them. However, female elephant seals can manipulate the situation to their own advantage. When a male mounts a female, she calls loudly and the call alerts nearby males. If the male that has mounted her is a subordinate one, he is soon chased off by the bigger dominant male. The female's protest behaviour ensures that she mates with the biggest, strongest male around (Cox & Le Boeuf 1977). Many female animals, including domestic hens (Graves *et al.* 1985), prefer dominant males, which also ensures that they mate with males that are fit and healthy enough to win and maintain their position over other males.

While, as we have seen, many females choose mates on the basis of characteristics of the male himself, it is also clear that the resources a male holds may be just as important and, in some cases, even more so. Since the female's chances of successfully rearing offspring may be critically dependent on a reliable food supply or a safe place to rear her young, this is not surprising. A good example of the importance of resources in female mate choice is that of the pied flycatcher (*Ficedula hypoleuca*), which is a summer visitor to northern Europe. The males arrive on the breeding grounds about a week ahead of the females and set up large breeding territories. The males that arrive first tend to be the oldest males with the blackest plumage. They also have the pick of the available territories, so the fact that they are more successful at gaining a mate than late-arriving males could be either because the females like their territories or because they are attracted to some characteristics, such as the blacker plumage, of the males themselves.

Alatalo *et al.* (1986) devised an ingenious way of separating these two factors in the females' choice. They made use of the fact that pied flycatchers readily nest in artificial nest boxes and can be persuaded to nest in different areas of a wood depending on whether there is a nest box there. They forced males to defend particular territories by restricting the number of nest boxes available at any one time, and only putting up more when the first ones had been occupied. When the females arrived, Alatalo and his colleagues noted the order in which the different males attracted a mate. This time – used as a measure of success in obtaining a mate – turned out not to be correlated with how early a male had arrived or with his age or the blackness of his plumage or with any other characteristic of the male himself, but with the quality of his territory. Males that had a low density of birch trees in their territories and safe nest sites high up in trees with thick trunks were the ones that obtained mates first. Normally, older blacker males would choose such territories themselves, but when they were experimentally denied the opportunity to do so, the females could be seen responding primarily to the territory, not the owner.

This idea – that a female may sometimes be more interested in a male's territory and resources than in the male himself – has some very interesting consequences for how we can expect females to behave. If the distribution of resources is very uneven, so that some males have very much better territories than others, it may actually be to a female's advantage to breed in a superior territory where another female has already settled than to choose a vacant, but poor quality territory where she would be the only female. Provided that the disadvantage of having one, two or more other females nesting in the same territory is not too great and does not disadvantage her own young too much, she may be better off mating polygamously in a territory with good nest sites and plenty of food than mating monogamously with a male that holds inferior resources.

Verner and Willson (1966) and Orians (1969) developed a useful way of looking at such a situation called the 'polygyny threshold model', in which they proposed that females would switch between mating monogamously and mating polygamously, depending on the relative advantages of being the sole female in a relatively poor territory or sharing a better territory with several other females. Even though the polygyny threshold model does not always explain the observed patterns of female settlement (Davies 1991), it is an important model because it shows how females as well as males may gain from polygamy, depending on circumstances. The difficulty with testing it is that exactly what those circumstances are may vary from species to species and even within one species. For example, in red-winged blackbirds (*Agelaius phoeniceus*), there are some populations in which the males help with caring for the young and others in which they do not (Orians 1980). Where male care is given, the effect of sharing a male with other females will obviously be greater than where the male gives none. In other cases, severe predation can, quite independently of male care, favour females nesting close together. By nesting in the same territory as other females, the females can gain advantages through cooperative nest defence (Picman *et al.* 1988). In the next chapter we will discuss in more detail how the distribution of resources such as food and territories affects the whole social organization and mating patterns seen in a species.

Species isolation

An important element of mate choice is choosing a mate of the right species. Hybrids between two species, although they may show some signs of 'hybrid vigour' are usually sterile or less fertile than the offspring of within-species matings. Mules (horse × donkey) are a good example of this. Even if they are fertile, they are rarely as successful as either parental type, whose genes have been selected over many generations as the best for their own environment. Dilger's lovebirds (p. 54) show how confused hybrids can be in their behaviour. In addition, their chromosomes may be incompatible, so that there is a mechanical breakage of the process of meiosis.

Consequently, there will be a strong selective advantage, particularly for females, in being able to recognize a mate of the same species. Behaviour, and particularly sexual signals, play a major part in this. For example, there are two closely related species of fruit fly, *Drosophila melanogaster* and *D. simulans*, that look so alike that humans find it difficult to distinguish them. However, the flies themselves clearly do discriminate, apparently by a combination of acoustic signals produced by wing vibrations and pheromones (Bennet-Clark & Ewing 1969).

Searcy and Marler (1981) used synthetic versions of bird songs to find out exactly what it is about the songs of their respective males that the females of two similar species of bird – song sparrows and swamp sparrows – use to distinguish between them. The synthetic songs consisted of either song sparrow or swamp sparrow units or 'syllables', presented in either song sparrow-like or swamp sparrow-like order (Fig. 6.18). For females of both species, the song had to have both the right syllables and be in the right order, whereas the less fussy males seemed to show no particular preference for the temporal pattern of their own or the other species. Again, in red-winged blackbirds, females clearly distinguish between real red-winged blackbird song and the imitation of it produced by mockingbirds, while males seem to be completely taken in by the deception (Searcy & Brenowitz 1988).

As in other examples of animal communication (Chapter 3), we often find mutual adaptation between signal and response. Male fireflies signal to females by producing flashes of light that are different in different species, and females respond only to the pattern of flashes given by males of their own species (Lloyd 1965, 1975). The flashes also vary in colour. Lall *et al.* (1980) showed that there is a good match between the colour of light emitted by the male of a given species and the colour sensitivity of the female's eye in the same species. A comparable coevolution, this time within one species, is described by Ryan & Wilczynski (1991) for the cricket frog. They studied two popu-lations of this species in North America, one in New Jersey and the other 2500 km away in South Dakota. Not only did the calls of the male frogs differ in frequency between the two populations (the dominant frequencies of the calls were 3.56 and 3.77 kHz, respectively), but so did the sound frequency that most excited the auditory systems of the two sets of females. The basillar papilla – the inner ear organ that is used in the reception of calls – responded maximally to 3.52 kHz in New Jersey females and to 3.94 kHz in South Dakota females. Such differences between populations, particularly where there is geographical separation, pave the way for the eventual evolution of new species.

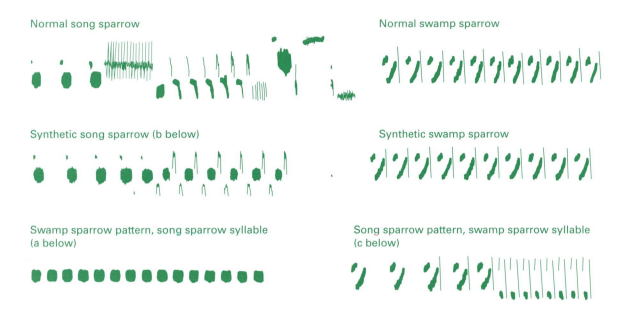

Normal song sparrow

Synthetic song sparrow (b below)

Swamp sparrow pattern, song sparrow syllable (a below)

Normal swamp sparrow

Synthetic swamp sparrow

Song sparrow pattern, swamp sparrow syllable (c below)

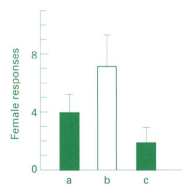

Figure 6.18 The responsiveness of female song sparrows to artificial songs. Females were played these songs via a loudspeaker in their cage and their response measured by the frequency and intensity with which they adopted the copulation invitation posture with raised tail and shivered wings. Above are illustrated some of the song types used. At the top are examples of natural song and swamp sparrow songs. Below are synthetic songs of various types. The song sparrow females respond most strongly (b) to synthesized song sparrow song with both syllables and temporal patterning 'correct'. They are significantly less responsive to (a) a swamp sparrow pattern made up from song sparrow syllables or (c) the converse, song sparrow pattern made up from swamp sparrow syllables. (Modified from Searcy & Marlow 1981, and Searcy *et al.* 1981.)

The phylogeny of behaviour

Over long periods of time, new species evolve and others become extinct. The raw material for such evolution is genetic change in the form of mutation, magnified by sexual reproduction and the crossing-over that occurs at meiosis. Whether these different variants survive in a given environment and which other variants they choose as mates will be among the factors affecting the genetic composition of future populations. Eventually, some of these populations may accumulate large numbers of genetic differences that make them distinct from other populations. It may be that part of a population moves off into a new area, where it gradually becomes so different from the parent population it left behind that two new species are formed, recognized by the fact that they cannot interbreed. (With rare exceptions, it is generally thought that some sort of geographical barrier – such as a mountain range or a stretch of sea – is necessary to keep populations separate enough for long enough to allow two distinct species to evolve.) The new species may look and behave somewhat differently from the parental population it evolved from, although if the separation occured relatively recently, it may still be possible to see similarities and to deduce that one evolved from the other. A courtship behaviour in the descendent species, for example, may be an exaggerated version of courtship in the other, or use an ornament of a different colour or a sound of a different frequency. With many of the species that are alive now, however, the parental species from which they evolved are now extinct and so not available for comparison. But we can still ask how patterns of behaviour arose and what they evolved from. Asking about the evolutionary origins and antecedents of behaviour in animals we see today is in fact to tackle the last of Tinbergen's four questions about behaviour – that of phylogeny. We want to know how, back in the mists of evolutionary time, behaviour patterns evolved. What was the behaviour of the ancestral species like and what were the inter-mediates between them and our present-day species?

In attempting to reconstruct the phylogenetic history of behaviour, we are severely hampered by the fact that (with one or two notable exceptions such as fossilized dinosaur tracks), we have no fossil record of behaviour. But we do have a very important method at our disposal. This is to look at closely related species that are alive today and to try to identify the evolutionary changes that took place in the past from the pattern of diversity in the present.

Such an approach has recently been revolutionized by new techniques for identifying phylogenetic relationships between species and has given a whole new impetus to Tinbergen's 'fourth question'.

Many of the early ethologists studied phylogeny by looking in detail at behaviour patterns to see whether the elements making them up could provide clues as to where the whole pattern came from. Julian Huxley (1914) studied the various courtship displays of the great-crested grebe, one of which involves both partners rearing up out of the water and presenting nest material to each other (Fig. 6.19). Huxley argued that this display originated from elements of the grebe's nest-building behaviour which were then incorporated into the sexual display. Huxley coined the term 'ritualization' to describe this evolutionary change as a non-signal movement (nest-building) became used in a display context. Tinbergen (1959), too, used analyses of what animals do now to indicate evolutionary origins. By showing that the threat postures of the lesser black-backed gull had elements of both escape and attack, he argued for the 'dual motivation' origin of threat displays (Chapter 3).

More extensive studies of closely related species can be used to show how a behaviour pattern can change gradually by a change either of form or of frequency. Hunsaker (1962) studied the striking *Sceloporus* lizards. When they meet other males, the males of these lizards perform rhythmic head-bobbing movements. Each of the species in the group has a characteristic pattern of bobbing (Fig. 6.20) which is produced by rhythmic contractions of the muscles that extend the front legs, thereby raising and lowering the head and shoulders. There is evidence of change in amplitude, speed and length of movement from species to species, brought about by changes in the two groups of muscles that control the bobbing movements.

Obviously, in order to construct a phylogenetic picture of the evolution of behaviour, it is essential to know which is the ancestral form of the behaviour pattern. Did the common ancestor have a simple display that was then separately elaborated into divergent complex displays, for example, or did an elaborate display in one species become simplified? Questions about the order in which evolutionary events took place have recently become much easier to answer because of the molecular phylogenies that are now available for many groups of species. In the past, evolutionary relationships between animals

Figure 6.19 One of the illustrations from Julian Huxley's classic study of the courtship behaviour of great-crested grebes. This mutual display, sometimes called the 'penguin dance', involves each bird collecting some nest material and swimming towards its mate. Both then rear up vertically, vigorously paddling with their feet to maintain their posture and rhythmically moving their heads from side to side, with crest and neck ruff raised. The nest material is held firmly in the bill. (From Huxley 1914.)

Figure 6.20 The specific head-bobbing movements of some *Sceloporus* lizards. The movements of the head are represented as a line with height on the vertical axis and time on the horizontal axis. Changes to amplitude, speed and length of movements are all clearly shown. (From Hunsaker 1962.)

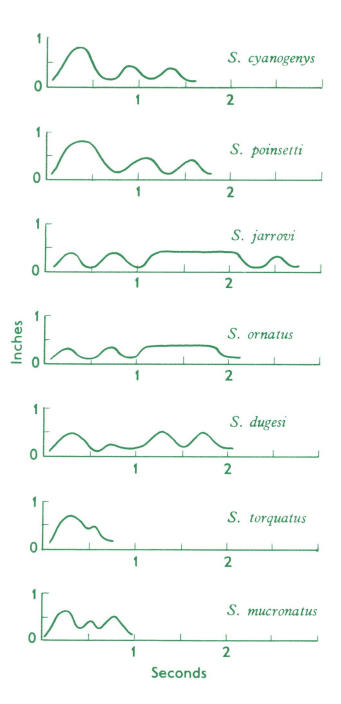

have been based on several different sorts of information such as how many characteristics they have in common, whether these are likely to have been inherited from a common ancestor, whether there is fossil evidence linking them, and so on. For example, despite their differences, rabbits, whales and chimpanzees share common characteristics (such as body hair, mammary glands, and similarities of the skeleton) which mean that they are all classified as 'mammals'. Molecular phylogeny gives yet another clue to ancestry. By comparing the details of composition of certain molecules in different species it is possible to work out how closely related they are. On the assumption that the molecular composition of, say, DNA changes at a constant rate, the differences between DNA in different species can be used as a sort of 'molecular clock'. (The relevant DNA changes are, of course, in non-functional parts of the chromosomes; genetic changes that have a phenotypic effect would not change in this clock-like manner). The bigger the differences in the DNA of two species, the longer ago they had a common ancestor as the more time there has been for the differences to accumulate.

Molecular techniques have been used to shed light on the evolution of a behaviour that has caused some controversy. Earlier in this chapter we discussed the idea of 'sensory bias' in relation to sexual selection. One strong piece of evidence comes from mate choice experiments in swordtails and platyfish where female platyfish were shown to prefer males with swords even though the males of their own species have no swords at all. This raises the question of which came first. Did male swords evolve first and then become lost in the platyfish, or did female preference evolve first and 'drive' the evolution of the male sword? By reconstructing the phylogeny of swordtails and platyfish, Basolo (1990) argued that female preference for male swords evolved before the swords did and that therefore the female bias for swords drove the evolution of the male's ornament. If true, this would be an important idea for the evolution of sexual ornaments but it obviously depends critically on assumptions about the evolutionary relationships between the different species. Using a DNA phylogeny, Meyer et al. (1994) argued that Basolo's phylogeny was wrong and that swords were the ancestral character, casting doubt on the sensory bias explanation. However, this is not the end of the matter, as yet another molecular phylogeny now favours the idea that female preference for swords may have come first (Basolo 1995). Whatever the eventual outcome of this debate, it is clear that the phylogeny of behaviour – that is, the sequence in which behaviour evolved – is an important aspect of our understanding not just of behaviour but of the dynamics of evolution as well. Tinbergen's fourth question, for a long time neglected in favour of function, causation and development, has now been given a new impetus by modern molecular techniques.

7   Social organization

In this final chapter, we draw together the threads of evolution, learning, cooperation, communication and kinship that we have touched on already, and see how they underlie one of the most striking of all features of animal lives – their tendency to be social. In some ways, the chapter can be seen as as an attempt at a synthesis of the different approaches we have adopted in this book. We have emphasized Tinbergen's four questions because they have stood the test of time and are as important today as they ever were. Indeed, the fragmentation into different disciplines brought about by the great growth in our knowledge of animal behaviour makes it even more important to ask different sorts of question about the same behavioural phenomena. Behavioural ecology (an offshoot of ethology) has tended to concentrate on adaptive questions about why animals are social and, at least until recently, has neglected those to do with the causation and development of their social behaviour. Neuroethology has made major advances in the study of causal mechanisms, such as how sense organs decode communication signals, but has often not seen these in the context of recent advances in understanding signal evolution. There is everything to be gained by studying causation along with adaptation and asking adaptive questions about causal mechanisms. Since virtually all animals exist in pairs or groups for at least part of their lives, social behaviour provides an obvious base for a synthesis across a wide range of species.

When we observe huge flocks of birds, swarms of insects or vast herds of antelope, it is easy to lose sight of the individual in the crowd and overlook the fact that the cohesiveness and coordination shown by the group as a whole is in fact the result of natural selection acting upon individual animals and the genes that they carry. A genetic tendency to group with others and to interact with them in certain ways will, in the right circumstances, give the animals bearing it an advantage and the genetic tendency will become more common in the population. What we see as social interactions between animals and describe as 'social organization' at the population or group level is the net result of such selection acting on individuals and genes. But, as we discussed in Chapter 6, the social environment that results can itself impose new selective forces on the individuals involved and affect both the direction and pace of their subsequent evolution.

The term 'social organization' refers to a wide range of phenomena defined as how members of a species interact with each other; for example, whether they group together in herds, space out and defend territories, have a monogamous or a polygamous system of mating, and so on. In some instances – the various social insects, for example – social organization is fairly rigid and species-specific. In many vertebrates, on the other hand, it is a much more dynamic phenomenon and may vary with changing conditions. Use of the term is not even restricted to what we might call highly social animals. Tigers, which usually live and hunt alone in large territories, avoiding contact with others except for breeding, and honey-bees, which spend their entire life in a dense colony, both provide examples of social organization, even though they are very different.

As we will see in the course of this chapter, social organization among animals takes very diverse forms. Female elephants may live in the same family unit for 40 or 50 years. They clearly know and react to each other as individuals and the stability of their relationships suggests that their groups should be called 'societies'. On the other hand, the organization within many flocks of birds or schools of fish is much less complex, although individuals may stay together for months. And when we look at a swarm of water fleas gathered in some rich food area, or a mass of fruit flies collected on some rotten fruit, then clearly the word 'society' would be quite inappropriate. Fruit flies and water fleas form only 'aggregations' because they are attracted to a common food source, not specifically to each other. Even they can be described as showing some social responses, however, because they react to one another's presence by spacing themselves out so that they do not touch.

Advantages of grouping

All animal groups, whether aggregations, flocks, schools or what we may wish to call true societies, result in the individuals that are part of them being better off than they they would be on their own. As we discussed in Chapter 6, however, it is not always easy to discover exactly why (in an adaptive sense) the animals benefit, because being better off implies that we can compare their survival and reproductive success in a group with that in some other situation, such as being on their own. And group living is often so beneficial that animals on their own are usually not around to make the comparison possible.

Discovering the advantage of group living, then, often means putting together evidence from several different sources. Specifically, we can use experimental evidence (artificially creating groups or placing individuals on their own), look at naturally occurring variation (within a species some animals may be more or less likely to group than others or adopt different types of grouping), and make comparisons between species that have adopted solitary or social lifestyles.

Allee and his colleagues (Allee 1938) used simple experiments to show how even loose aggregations can benefit the individuals that comprise them. They showed that water fleas cannot survive in alkaline water, but that the respiratory products (CO_2) of a large group of them are sometimes sufficiently acid to bring the alkalinity down to viable levels. Thus a group can survive where a few individuals could not. Fruit fly cultures do not do well if there are too many eggs because the resulting larvae are undernourished, but they fare equally badly with too few eggs. This is because reasonably large numbers of larvae are needed to break up the food supply, encouraging the growth of yeast and making the food soft enough for all the larvae to feed easily. It is thus advantageous for a female to lay her eggs close to those of others because her own offspring will benefit.

Flocks of birds and schools of fish exemplify groups that are much more than simple aggregations because there is often a high degree of social interaction between individuals. Some, such as geese, may even migrate as families. Even here, physical factors may still count, as in the case of emperor penguins which huddle closely together as they stand incubating their eggs during the Antarctic winter. Heat is conserved and birds on the outside move more than those in the centre, leading to mixing and a reasonable distribution of shelter.

One of the most obvious advantages of a cohesive group whose members respond to each other's behaviour is protection against predators (Fig. 7.1). With a number of animals on the alert, the approach of a predator is less likely to go undetected and one alarm signal will suffice for all. Furthermore, each individual animal will be able to spend more time feeding, protected by the many pairs of eyes around it. Lazarus (1979) showed this effect in action by watching how effectively red-billed weaverbirds spotted a goshawk flying overhead when they were on their own or in groups with other birds. He found that solitary birds often failed to respond at all, but where there were two or more birds, the hawk was much more likely to be seen and responded to. Elgar (1989) reviewed over 50 other studies that show that birds and mammals spend less time in vigilance and more time in feeding, the bigger the groups they are in, but also warns against using such studies uncritically to draw conclusions about the advantages of grouping. He points out that large groups of animals may, for example, congregate where there is the greatest amount of food and so be able to feed faster for reasons unconnected with being in a group. Saino (1994) showed that in flocks of carrion crows (*Corvus corone*), feeding rates were indeed very strongly correlated with food density but that, independently of the amount of food present, there was

Figure 7.1 The defensive formation of a group of musk oxen on the Canadian tundra. When a predator approaches, they bunch with the older animals at the front, facing the threat. (Photograph by D. Wilkinson from Information Canada Photothèque.)

also a strong effect of group size on vigilance rates. Birds in larger groups looked up less and fed at a higher rate than birds in smaller groups or on their own. In purely observational studies on natural groups, however, the separation of different factors has to be done by sorting out statistical correlations and it may not always be possible to disentangle correlation from causation. This is why experiments, although difficult to do, also have a very important role.

Powell (1974) experimentally manipulated group size in starlings in aviaries, making sure that all other factors were the same. As a predator, he used not a hawk but a hawk model to study the anti-predator responses of starlings when they were feeding. In his controlled but somewhat artificial environment, he was able to confirm the results of field studies. He found a more rapid response to danger when there were several birds rather than just one. He also found that in groups of 10, individual starlings spent significantly less time in surveillance than did individuals on their own. In other words, starlings in the larger flocks increased the amount of time they spent feeding (through not having to keep looking around), but the combined efforts of many pairs of eyes nevertheless enabled them to be more effective than single birds in detecting predators when they did appear.

It has been argued that 'cheating' could be a problem in such flocks. If some birds never do their 'fair share' of surveillance and instead spend all their time feeding, it might appear that they would be at an advantage compared to more altruistic birds that sometimes stop feeding and look around. There are, however, advantages to spotting the predator directly rather than relying on the response of other birds. Elgar *et al.* (1986a) filmed flocks of house sparrows as they were feeding and then suddenly introduced a predator stimulus. He showed that the birds which happened to have their heads raised at the moment when danger appeared took off faster than those that happened to be feeding. Certainly for birds at the edge of a flock, which are most vulnerable to the approach of a predator, it would be extremely dangerous to spend all the time feeding and never to look up. Some degree of vigilance is, therefore, of advantage to the individual and a 'cheat' is likely to be worse off than a bird that occasionally looks up for its own safety. Each individual can, selfishly, adjust its vigilance rate downwards in a large group because it does derive some protection from the animals around it. But selfishly each individual

Figure 7.2 An adult meerkat keeps a vigilant eye out for predators. This member of the group remains near the nest hole 'baby sitting' whilst the others are away foraging. All members of the group and not just the parents take turns at protecting the young in this way. (Photograph by Ashleigh Griffin.)

should look up sometimes, and this results in the group as a whole detecting danger for the mutual benefit of all.

Sometimes, however, individual selfishness does not seem to explain what animals do, at least in the short term. In meerkats, which are socially living mongooses, vigilance is undertaken by particular individuals which take turns to go to a high look-out point such as a tree and keep watch for predators while the others feed (Macdonald 1986; Fig. 7.2) Such individuals look as though they are being altruistic since they do not feed while on sentinel duty, but meerkats live in close-knit groups where the animals know each other as individuals and where there are therefore potentially long-term benefits from repeated social interactions. As we saw in Chapter 6, under such circumstances, short-term losses may be incurred if there are repayments in the future, such as surveillance by another member of the group or simply enhanced status.

Even when a predator has been spotted, being in a group offers further advantages to the individual because the group can take concerted evasive action, confuse the predator by having individuals scattering in all directions, or actually physically attack the predators. Figure 7.3 illustrates the evasive action from a flock of starlings and very similar behaviour is shown by some fish which bunch together at the least alarm. Predators rarely attack an individual in a close group and their commonest strategem is to make swoops towards the group which may cause them to scatter, when the predator singles out an isolated animal. Hamilton (1971), in a paper graphically entitled *Geometry for the selfish herd*, showed that if each animal in the group attempted to put at least one animal between itself and the predators then tight formations would be the inevitable result.

Colonial nesting birds – gulls and terns, for example – may provide formidable opposition to an invading predator such as a fox by mobbing it, even hitting the predator with their feet. Even though each bird is responding individually to defend its own nest, the proximity of other birds all doing the same thing means that their combined efforts can be much more effective than that of a single bird on its own. As a result, as Göttmark and Andersson (1984) showed, the nesting success of gulls in a large colony is considerably greater than that of gulls that nest singly or in small groups (Fig. 7.4).

Protection from predation is only one of the advantages gained by living in groups. Another major factor is utilizing food sources found by other animals. We have already mentioned that birds in flocks can spend more time foraging and less time being vigilant because they rely on the vigilance of others. They may also steal food from other individuals or be led to a clump of food located by another animal. Krebs *et al.* (1972) showed that when one member of a tit flock finds a food item, the others rapidly alter their searching strategies and concentrate their attention both on the general area and the type of niche in the trees where the food was found. Sometimes this may be to the short-term disadvantage of the bird that finds the food, but by staying in

Figure 7.3 The response of a flock of starlings to the approach of a bird of prey. (From Tinbergen 1951.)

the flock, it gains a longer term advantage of being able to utilize food found by another bird and the increased vigilance the flock provides.

At other times, when the food source is so rich that there is enough for all, there is no disadvantage at all to the original finder. Brown (1986) showed that cliff swallows follow individuals that have located a rich source of insect food. Birds returning to the colony after a successful foraging trip were likely to be followed on their next trip out by birds that had previously been

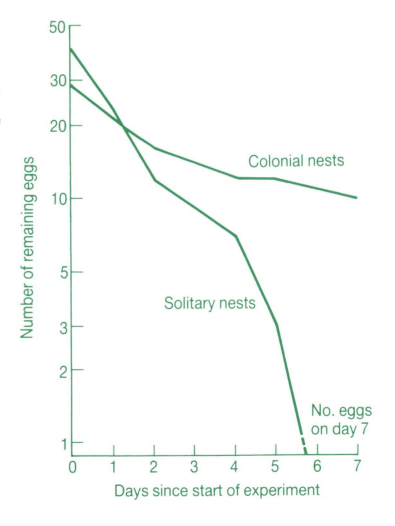

Figure 7.4 The temporal pattern of predation on experimental eggs laid out near (2 or 5 m) nests of colonial and solitary pairs of common gulls. Predation was much higher near the solitary nests and by day 7, there were no experimental eggs at all left near the solitary nests. Eggs placed near the colonial nests were protected by the massed mobbing of the gulls in the colony. (From Göttmark & Andersson 1984.)

unsuccessful. But the food source – a 'plankton' of insects – was so abundant that the finders were at no disadvantage if they were followed. In yet other cases, it may even be of advantage to the finder to have other individuals around. Gannets are often found fishing in groups, and Nelson (1980) argues that a group of birds diving together so confuses the fish that they are more easily caught. Göttmark *et al.* (1986) showed a similar effect in black-headed gulls. Group fishing leads to more successful feeding for each individual than a bird could accomplish on its own because, with more gulls around, the fish are more vulnerable and more likely to be caught.

Lions, hyaenas and Cape hunting dogs are all examples of predators that hunt together. Their strategems may involve some individual driving prey towards others hidden in cover or, as with the hunting dogs, taking it in turns to run down an antelope to the point of exhaustion. Spotted hyaenas (see Fig. 3.30, p. 165) are to be found in different sized groups depending on the size of prey they are hunting. Kruuk (1972) showed that when hunting zebra the mean number of hyaenas in a group is 10.8, when hunting adult wildebeest the mean is 2.5 and when hunting gazelle fawns the mean group size is 1.2 individuals. What is more, the differences in the size of the packs are often apparent long before the hyaenas have sighted their prey. When hyaenas set out in a large pack Kruuk could be reasonably certain that they would end up hunting zebra even if they had to walk for miles through herds of wildebeest to find the zebra. This strongly suggests that hyaenas set out to hunt certain kinds of prey. It is not just humans that can predict hyaena behaviour based on pack size: zebra ignore single hyaenas and only become alarmed by large packs.

We should not assume, however, that just because animals hunt together, they are necessarily cooperating to bring down prey. Packer (1986) argued that although lions pursue prey in a group, they may not actually be helping each other. Rather, the advantage of group hunting for the lions lies in the fact that after the kill, there will be more individuals to keep scavengers and other potential thieves away from the carcass. In other words, the hunting may not be cooperative (more individuals do not necessarily lead to greater success) but the defence of the kill is. Extra individuals are tolerated because they help keep away vultures and hyaenas, even though they eat some of the meat themselves.

It will be clear from the preceding discussion that group living confers some clear advantages but at the same time some undoubted disadvantages. Being near other individuals means increased competition for food, increased risk of disease transmission and greater conspicuousness to predators, as well as greater risks of cuckoldry, mixing and cannibalism of young, and so the balance between the advantages an individual gains as a result of being part of a group and the disadvantages it inevitably suffers from the same source is a very fine one. Hoogland and Sherman (1976) give a long list of disadvantages suffered by bank swallows as a result of their habit of nesting communally in sand banks. These range from increased chances of picking up fleas, to having nest burrows collapse as a result of other birds nesting close by. Nevertheless, the birds benefit from nesting in a colony because they derive safety from the massed attacks of all the other birds on potential predators. This is more than enough to outweigh the evident disadvantages of being close to other birds. The behaviour that will be favoured by natural selection – whether it is territoriality or extreme clumping – will be that which favours the reproductive interests of the individual in the long run. We now discuss some of the types of social groups that have evolved as a result.

Types of social groups

Groups of animals differ widely in the complexity of their organization and the types of interactions individual animals have with one another. This means that it is not enough to describe a group simply by its size. The sex ratio, the degree of differentiation into roles, relationships with other specific individuals in the group, kinship and other factors all affect the costs and benefits of group living. A look at some contrasted types of groups from very different animals will illustrate what this means in practice.

Eusociality: division into castes

The most extreme form of sociality is called 'eusociality' and is characterized by reproductive division of labour, where some members of the group lose their reproductive capacity altogether and become members of a worker 'caste'. Although sterile themselves, their efforts increase the reproductive output of the colony as a whole through helping members of the reproductive caste. In Chapter 6, we discussed the evolution of this reproductive division of labour and saw that it is associated with a high degree of relatedness between members of different castes and overlap of generations. The most highly organized insect societies are found in the termites (Isoptera) and the ants, bees and wasps (Hymenoptera) and the differentiation into castes may become extreme. In some species of ants, for example, some of the workers are called 'repletes' because they store so much liquid food in their crops that they swell up like balloons and are unable to move. They remain motionless in a protected part of the nest, hanging permanently as living honey casks (see Hölldobler & Wilson 1990). Until quite recently, eusociality was associated exclusively with the social insects – ants, bees, wasps and termites. Then Jarvis (1981) described eusociality in a mammal – the naked mole rat (p. 342). Even more recently, eusociality has been described in a snapping shrimp (*Synalphaus regalis*) that lives in sponges on tropical reefs (Duffy 1996). These shrimps live in colonies of up to 300 individuals but have only one reproductive female (the 'queen'). Like termites and mole rats, the shrimps are diploid and the 'workers' are full sibs. As food seems to be abundant, the main task of the workers is to defend the sponge. On the crowded reef, sponge 'homes' are sought by a wide variety of species and it takes the combined efforts of all the shrimp workers to keep larger species at bay.

Even though a relatively small number of species have taken social living to the extreme of eusociality, we can see some elements of this type of social organization in many different animals. As we saw in Chapter 6, for example, it is quite common to find young birds in their first or second year of life staying to help their parents before starting to breed themselves. Such 'helpers at the nest' have been recorded in over 200 species of birds and in mammals such as black-backed jackals (Moehlman 1979) and dwarf mongooses (Creel *et al.* 1993).

In insects, too, we can see two essential elements of eusociality – overlap of generations and cooperative brood care – in species that do not show the caste divisions of fully eusocial species. The aggregations formed by insects such as cockroaches and earwigs are a good example. The life span of a cockroach may be a year or more, three-quarters of which is development through a series of nymphal stages, becoming more like the adult insect. During this period, cockroaches of all ages live together in a loose aggregation near sources of food and shelter. Some species of cockroach incubate their eggs inside the female's body and bear live young which remain in contact with the mother for some hours after birth. This contact may be important for the survival of the young, because not only are they extremely vulnerable to cannibalism when first born, but they may pick nutrients from the mother's body surface.

We know that this nutritional factor is also of major importance in the termites (which are related to the cockroaches) because, feeding largely on wood, they rely on symbiotic protozoa living in the gut to digest cellulose. These protozoa are acquired by the young termites when they feed on fresh faecal matter from the adults, and they can be transmitted only in this way. Perhaps this method of feeding was one of the factors predisposing these insects to the evolution of full eusociality.

Whatever the route, the final result is truly amazing, with the social insects showing a degree of coordination that allows some species to build complex nests, grow their own food, take 'slaves' from other nests and show communal defence of the nest. There is a huge literature on these insects, which have attracted human attention for centuries. For an introduction to the whole literature, Wilson's masterly survey (1971) is still unmatched, Hölldobler & Wilson (1990) give a definitive account of ants, and Gould & Gould (1988) describe the behaviour of honey-bees.

The term 'caste' is well suited to describe the division of labour in eusocial insects. It implies a rigid, limited role in society largely determined by upbringing, which is what seems to be the case here. One of the most important factors determining caste is what the insects are fed when young. In bees, wasps and termites, all eggs laid by the queen are potentially equal, but most larvae are fed a restricted diet and develop into workers. There is evidence that when queen ants are laying rapidly, their eggs are 'worker biased' and develop accordingly no matter how the larvae are fed. But for most of the time their eggs are also equipotential and only richly fed individuals develop into the reproductive castes.

Pheromones are also important determinants of caste. They are secreted by the insects themselves and coordinate development and social behaviour. The development of worker termites is controlled by pheromones produced by the king and queen. Similarly, the queen honey-bee secretes a pheromone ('queen substance') which both suppresses the ovaries of the workers and prevents them from rearing new queens. The queen is always surrounded by attendant workers who lick her body, subsequently offering food to other workers and with it, the pheromones from the queen.

The level of the pheromone must be kept up and once the source is cut off, its concentration rapidly drops, which happens if the queen becomes ill or dies. The effectiveness of the incessant food sharing in circulating queen substance is shown by the fact that some workers in the brood area of the hive exhibit changed behaviour within an hour or two of the colony losing its queen. They begin construction of 'emergency' queen cells in which some of the youngest larvae, destined in the normal course of events to become workers, are fed royal jelly throughout their larval life and become queens which will eventually fight one another to replace their mother. As colonies grow, the dilution of queen substance below a critical level is one of the factors which leads to swarming in honey-bees.

The honey-bee is exceptional because it reproduces its colonies by swarming, but in most other social insects, new colonies are founded by a single queen (or a pair in termites). The queen begins the construction of the nest and rears the first batch of workers herself. These then take over the tasks of extending the nest and bringing food and the queen usually stays in the nest, laying eggs, from this point on. The tasks performed by the worker castes vary greatly in detail, but in most colonies they cover the main categories of foraging, rearing the young, nest construction, attending the queen and guarding the colony. In termites and ants this last task is sometimes the sole responsibility of a special soldier caste which has enlarged jaws and other weapons.

A honey-bee worker lives for about 6 weeks as an adult and her activities are roughly synchronized with her physiology. Thus she spends the first 3 days cleaning out cells, then begins feeding the older larvae on a mixture of pollen and honey. From about the 6th to 14th day of her life, the worker feeds 'royal' jelly (secreted from the pharyngeal or 'nurse' glands on her head) to the younger larvae and any queen larvae in the hive. (Royal jelly is fed to all larvae for a brief period early in their development, but those destined to become young queens are fed royal jelly throughout.) The worker then gradually changes her behaviour from feeding larvae to cell construction as her pharyngeal glands begin to regress and the wax-secreting glands on her abdomen

Figure 7.5 Lindauer's complete record of the tasks performed by one individual worker honey-bee throughout her life. The records are classified according to the type of task. One can recognize the age-determined succession of cell cleaning, brood care, building, guarding and foraging. Note, however, the large amount of time spent in patrolling the interior of the hive and in seeming inactivity. During such periods the worker may be acquiring information on the situation in the colony and adjust her behaviour accordingly. (From Lindauer 1961, with permission Harvard University Press. © President and Fellows of Harvard College.)

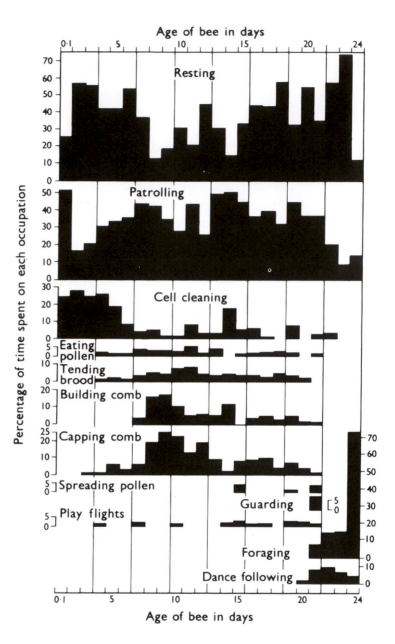

become active. From the 18th day, she may be found guarding the hive entrance or venturing outside for a few brief orientation flights. From 21 days onwards, the worker is primarily a forager, bringing back nectar, pollen and water and will remain so for the rest of her life – about 2–3 weeks. Figure 7.5 illustrates this sequence of behaviour through the brief life of a worker.

This is the general sequence of events but it can be modified to suit the needs of the colony, depending on the flower crop, temperature, age of the colony or other factors. Modification is possible because of the sophisticated communication systems that exist between the members of a honey-bee society (Chapter 3). We have already mentioned the importance of food-sharing as a method of communication, both as a way of circulating pheromones, but also to keep each worker directly informed of the state of the food supplies within the colony. The 'dance-language' of honey-bees (p. 179) is the most sophisticated example of communication because of the detailed information it conveys about the distance and direction of a food source.

This high degree of social organization within social insect colonies and the control it gives them over their environments are based on a relatively simple series of responses to other workers in the colony and to the nest itself. There is no suggestion that they are intelligently working out what to do. To a certain extent, the social organization of the insects is flexible, as when honey-bee workers change their normal sequence of tasks in response to a sudden requirement in the colony. However, the degree of flexibility is limited. Consistency of social organization within species is very characteristic of the insects and we do not find it so well marked in other groups.

Territory in the social organization of vertebrates

In contrast to the social insects, the social organization of vertebrates is not rigidly species-specific. Thus the answer to the question, 'What is the social organization of the house mouse?' is not fixed. It depends upon the nature of the food supply, the density of the population, its age, sex structure and a number of other factors. An important key to understanding the social organization of vertebrates is to remember a point we made earlier: that group living brings costs as well as benefits. Other animals are competitors for food, for mates, for nest sites, and we have already seen, in the section on aggression (p. 208), that animals have well-developed mechanisms for 'seeing off' unwelcome intruders. Not surprisingly, therefore, some animals adopt a solitary way of life or live in monogamous pairs. Many defend territories to keep other animals *away* from them. Some, like the great tit, are both territorial and gregarious at different times of the year. During the breeding season, great tits are strongly territorial and vigorously defend the area around their nest against other great tits. But during the winter they group into large flocks both with members of their own species and of others such as blue tits and nuthatches. This means that when they have hungry young to feed in the spring, they defend a private supply of insect food and in the winter, when food is scarcer, they gain the advantages of locating food and protection against predators that being part of a flock gives them.

Defence of a territory of some sort is in fact a common feature of the social organization of many vertebrates (as well as some invertebrates). A typical case is represented in Fig. 7.6, which shows the spatial distribution of willow warblers. Each male defends a substantial area which will include food for himself and eventually a mate and young, and he will rarely leave the territory during the breeding season. As will be seen, each territory impinges closely on those of other males and, where habitat permits, nearly all the available ground will be occupied. In birds, territorial defence is usually achieved by song and visual displays. In mammals, which may also have densely packed territories, the territorial boundaries are often defined by scent posts marked with urine, special glandular secretions or faeces.

Figure 7.6 A territory map of a population of willow warblers in an area of birch woodland. The map was constructed by observing singing males and the positions at which they gave way during boundary clashes. The size of territories varies from less than 1000 to over 5000 square yards. The warblers did not colonize the shaded areas where there were taller trees and less undergrowth suitable for nesting. (From May 1949.)

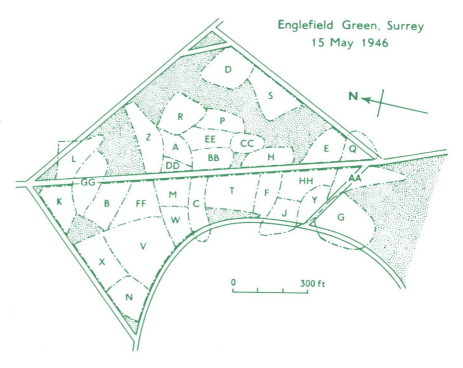

Englefield Green, Surrey
15 May 1946

Defending a territory against constant intrusions is clearly costly because an animal has to expend time and energy seeing off rivals and may make itself vulnerable to predators while doing so. Brown (1969) introduced the idea of 'economic defendability', pointing out that animals should only go to the time and trouble of defending a territory if the resource they were defending (the food in it, for example) was worth defending and, indeed, could physically be defended. A very scattered food source might take so long to defend that the animals would lose more energy than they would gain by doing so. Gill and Wolf (1975) applied the idea of economic defendability to the territories defended by the tiny golden-winged sunbird, a nectar-feeding bird found in East Africa. They calculated the energy used by a sunbird defending its territory against intruders and also the energy available in the *Leonotis* flowers they were defending (Fig. 7.7). If flowers are defended they become a better food source, because the nectar stocks rise if they are not depleted by other birds. Gill and Wolf found that the energy the birds expended in keeping other birds out of their territories was more than compensated for by the increased levels of nectar in their 'private' flowers. Defending a territory was, therefore, energetically worthwhile.

A study of pied wagtails (*Motacilla alba*) by Davies and Houston (1981) shows how finely tuned this energy balance can be. The wagtails were defending territories near a river bank (Fig. 7.8), where they fed on insects washed up on the shoreline. They appeared to exploit this fluctuating food source extremely efficiently. Moving along the water's edge, a territory owner would patrol round the rest of its territory, the round trip taking approximately 40 minutes, by which time more insects would have been washed ashore. Given that the advantage a wagtail gains from defending a territory is the exclusive access to any insects that are washed up, it is somewhat surprising to find that some territory owners tolerated another male in their territory for long periods of time. Such 'satellites' also walked around the territory, following the same path as the owner but half a circuit behind so that each section of river bank was left for only 20 minutes before being revisited by one of the birds. The owners appeared to be collecting only half the food they could collect if they had occupied territories on their own and had not allowed the satellites to feed. The solution seems to be that the satellites 'earned their keep' by helping with defence of the territory.

Figure 7.7 The golden-winged sunbird defends nectar-filled flowers against intruders if it is energetically worth while. (From Alcock 1984.)

Davies and Houston found that when an owner was alone on the territory, 60% of intruders were spotted immediately. The remaining 40% took some food from the territory before being chased. When a satellite was present, however, 85% of intruders were spotted and chased immediately and only 15% took any food. When food was sufficiently abundant, owners could increase their own feeding rate by up to 33% by having a satellite to help them with defence, which more than compensated for the food eaten by the satellite. When food became scarce, however, and there was not enough food for two birds, the owners became aggressive to the satellites and evicted them from their territories.

It is clear, however, that such considerations of energy gained and lost

Figure 7.8 Pied wagtails on riverside territories. The birds patrol around their territories collecting food washed up by the river.

do not apply in every case of territorial behaviour. The territories of ground-nesting sea birds such as gulls, terns and gannets may be as little as 1 m across and contain no food at all (Fig. 7.9). They consist simply of a small defended area around the nest which is always occupied by one member of the mated pair and are important because of the cannibalistic nature of other gulls in the colony, some of which specialize in eating eggs and chicks.

Yet another kind of territory is seen in the 'lek' system of birds such as the sage grouse, ruff and prairie chicken and in mammals such as the Uganda kob and the hammerhead bat. Here the males gather in a tight group or lek and, within the group, each male defends a small territory. There is no food in any of the territories and the function of these leks is not fully understood (Balmford 1991). One possibility is that leks occur when females are very spread out, perhaps because their food is widely dispersed, and the males are unable to defend a large enough feeding territory to attract them or keep them in one place (Emlen & Oring 1977). The lek therefore becomes a sort of 'marriage market' where the normally widely dispersed females come to choose a mate, perhaps because the sight and sound of many males all displaying together attracts females from a large area. Ryan *et al.* (1981) studied the Tungara frog (*Physalaemus pustulosus*), where the males aggregate into leks or choruses and showed that with bigger choruses, more females approach. Fortunately for the males, they are also safer from predation by frog-eating bats (p. 326) in bigger groups.

Another possible advantage of leks is that females can make a direct comparison between different males (p. 13) and choose between males on the basis of their display or the sounds that they make. With the display of the of the sage grouse that we discussed in Chapter 1, for example, Gibson and Bradbury (1985) showed that it is the males that display for the longest and in the most vigorous way that are chosen by the females. In black grouse (*Lyrurus tetrix*), another lekking species, the females choose the males that have the highest chance of still being alive in 6 months' time, suggesting that the females are using the lek to assess the health and viability of the displaying males (Alatalo *et al.* 1991). It is not clear what the females are basing their choice on, but it works!

Figure 7.9 Part of a huge colony of nesting gannets on the Bass Rock, a precipitous rock stack in the Firth of Forth near Edinburgh. (The gannet was named *Sula bassana* by Linnaeus after this colony, which has been famous for many centuries.) The even distribution of the nesting territories is remarkable. The gannets obviously gain protection by nesting together on islands and from mutual defence against aerial predators, but each pair still requires to have a space, albeit little more than a square metre, which it defends against hostile neighbours. The result is an even spread of birds across all the suitable nesting areas. (Photograph by Bryan Nelson.)

Mating systems and social organization

The case of leks shows how many different factors can interrelate to produce what we call 'social organization'. Thus, where food is very spread out over a wide area, females may also be widely dispersed and this will in turn affect a single male's ability to keep them together and defend them as a group (Emlen & Oring 1977). Leks or clusters of males may be one response to dispersed female resources. Another may be for males to space out and vigorously defend the resource the females are interested in, as in the case of the pied flycatcher discussed on p. 362, or the bluehead wrasse (*Thalassoma bifasciatum*), where the males defend individual protruding coral heads that make good spawning sites (Warner 1987).

Another very major factor in determining social organization, however, is the role adopted by the two sexes in reproduction and, specifically, how much investment the two sexes make in the care and nurture of the offspring. Once she has mated, for example, the female sage grouse has no further use for either the male or his territory. She leaves the lek area to rear her young alone. The male's contribution to his offspring is thus purely genetic and he will continue to try to attract other females without giving any of them assistance with rearing the young. Or, as in baboons, the male may not assist with the care of individual offspring but will defend the whole group. This contrasts with many other birds such as gulls, mammals such as marmosets and gibbons, and many species of fish where the male gives a great deal of care – feeding and protecting his offspring.

It used to be thought that there was a very close correlation between the amount of care given by the two parents and the type of mating system that evolved. For example, monogamy (where individuals mate exclusively with one partner) was thought to exist where both parents cooperated to care for the young, whereas polygamy (where individuals mate with two, three or many more individuals) was thought to be associated with one parent (usually the female) giving a great deal of parental care and the other parent (usually the male) contributing little but sperm and mating many times. Polygamy was further subdivided into polygyny and polyandry. Polygyny, where an individual male mates with many females, was the mating system described for most mammals and a few birds such as chickens, peacocks and, as we have seen, sage grouse. Polyandry, where individual females mate with several males, was described as very rare and confined to a few species, such as the jacana or lily-trotter (*Jacana spinosa*) where a large female defends a territory containing nests belonging to several different males (Jenni 1974). The female moves round to the different nests, laying eggs in each one, but the males carry out all the duties of incubation and parental care.

In the last 10 years, however, DNA fingerprinting (p. 339) has shown that the number of individuals a male or female mates with may have very much less to do with the amount of parental care given by the two sexes than was previously thought. By giving us a much clearer picture of which individuals are the fathers or mothers of which offspring, molecular techniques have revealed a quite unexpected amount of multiple mating, even in species that had previously been described as monogamous and in which both sexes care for the young. The starlings we discussed on p. 339 provide a good example. Both parents care for the young, and the pair could be described as 'socially monogamous', but certainly not genetically monogamous since three or four different males may be the fathers of one brood. What is more, females in many species actively solicit copulations from males (Birkhead & Møller 1992), so it is not simply that more powerful males are coercing females into mating with them

While it is clear that males gain an advantage from multiple mating in that they can increase their number of offspring by mating with several females, it is less clear why females should gain, since reproductive success in females is limited by the number of eggs or young they can produce. One suggestion is that females gain by choosing mates of higher 'quality' than their own. In blue tits females mated to attractive, 'high quality' males are less likely to seek extra-pair copulations than females mated to lower quality males (Kempenaers *et al.* 1992). Blue tits are socially monogamous and both sexes help to feed the young. They are also territorial. This means that when a female comes to choose a male, she may find that all the high quality males have already been mated and so has to settle for a less good male than she would ideally like. Kempenaers *et al.* measured 'attractiveness' of males by how many neighbouring females intruded into their territory, and 'quality' by scoring whether the males survived until the next breeding season and how many of their offspring survived. They found that males that had a nest in which there were young fathered by another male were more often abandoned by their own female when she was fertile, received fewer visits from neighbouring females, had fewer surviving young, were smaller and survived less well than males that gained through fathering extra-pair young. By mating with a high quality male who is already mated but having her own mate to help her care for the young, a female blue tit apparently achieves genetic quality as well as paternal care from her mate. In support of this hypothesis is the somewhat surprising finding that females of lekking species rarely copulate with more than one male (Birkhead 1993). This could be because females on a lek have a completely free choice and can choose to mate with high quality males regardless of what other females are doing.

Multiple mating by females is, of course, not in the interests of males and we often find that males take extraordinary steps to ensure their own paternity. Female dragonflies mate with a number of males and store the sperm in a special sac. In some species, the penis of the male is equipped with a elaborate hooks with which the male tries to scoop stored sperm out of the female's sac before he mates with her (Waage 1979). Among birds, female dunnocks (*Prunella modularis*) frequently form bonds and copulate with two (sometimes more) males. Davies (1992) describes how males can be seen pecking at the cloaca of females, attempting to remove the sperm of other males. In a parasitic worm, *Moniliformis dubius*, found in the intestine of rats, the male seals up the female after he has copulated with her. Even when males do not go to quite these lengths, their sperm carry on a subtle form of male–male competition. Simply by making large amounts of sperm, males that copulate with an already mated female help to make sure that theirs is most likely to fertilize the eggs. The aquatic warbler (*Acrocephalus paludicola*) goes one step further in sperm competition. In this species, the females rear the young without any help from the male, and males and females lead largely separate lives, meeting only to copulate. Both sexes have multiple mates and nearly half the broods contain offspring with three or four different fathers (Birkhead 1993). In most small birds, copulation lasts 1–2 seconds, but in the aquatic warbler it lasts 25 minutes. The male and female lie on the ground with the male on top and the male has huge testes from which he repeatedly passes sperm to the female. The most likely explanation of this extraordinary behaviour is that, by copulating with the female for so long and making so many sperm, the male is trying to ensure that he is the winner in the battle between sperm.

Multiple mating by females will, of course, be particularly serious for species where the male contributes paternal care, as by caring for young that are not his own, the male is 'wasting' his time and energy. Some males seem to adjust the amount of care they give depending on their genetic relatedness to the brood. Davies (1992) found that dunnocks fed nestlings more when they were most likely to be the father, but it is not easy for males to establish paternity.

As with so many biological phenomena, to explain the diversity of mating systems we have to consider the combined effects of many different factors. In this case, they range from the spacing of resources and genetic relatedness to the many different factors that can affect the survival chances of the young under different conditions (Reynolds 1996). The mating system that evolves under any given conditions can be regarded as the outcome of a conflict between male and female in which each individual has to cooperate with a member of the opposite sex, at least as far as mating is concerned. Even here there is conflict over which mate to choose as well as over which sperm fertilizes an egg. Where both parents are needed to feed and care for the young, social monogamy will tend to evolve, but where one parent can rear a brood on its own, the other may benefit from deserting and mating again with another individual. In mammals, the retention of young in the female's body and her feeding through lactation means that the male can often do little to improve the survival chances of his young. Consequently it is often in his best reproductive interests not to give parental care but to seek more mates. Hence polygamy (more strictly, polygyny) or promiscuity are the commonest mating systems in mammals. Only where the male can contribute substantially, for example in jackals where males feed the young through regurgitation, or in marmosets where the male carries the young through a long period of dependency, does monogamy appear in mammals. In birds, by contrast, eggs are incubated outside the female's body and the young are often fed by laborious collection of food, tasks that can as well be performed by males as by females and often very much more effectively if both parents are present. This probably explains why over 90% of bird species are socially monogamous while monogamy is the exception in other groups.

As we have seen in the case of the pied flycatcher, females may choose males because of the resources they are defending. Even where a male does not contribute to the rearing of his young by feeding them, he may nevertheless aid their survival by defending the nest against predators or intruders of the same species. This does not, however, necessarily lead to strict monogamy (one male mating with one female) since, as we have seen, a male's ability to defend resources or groups of females will also affect the nature of the mating system. One male may have several females nesting in his territory and provide resources and even paternal care for all their broods. The 'polygyny threshold' model we discussed in the last chapter, as a way of describing the conditions under which it was advantageous for a female to mate polygamously with such a male, has turned out to be useful but something of an over-simplification. Conflict between the sexes means that the model does not always fit the data.

Catchpole *et al.* (1985) show how such conflict may operate in practice. The European great reed warbler is, like the pied flycatcher, a migratory species in which the males arrive on the breeding grounds before the females and defend large territories. Female great reed warblers are attracted by the quality of the males' territories and sometimes several females settle in the particularly good territory of one male. According to the polygyny threshold model, the breeding success of polygynously mated females in good territories should be equal to or greater than the breeding success of monogamously mated females in poorer territories, but it is not: the second- or third-arriving females show reduced breeding success. Catchpole *et al.* argue that the females are deceived into choosing already mated males because the large territories in the reed beds prevent them from seeing the females that the male already has there. Alatalo *et al.* (1981) argue that pied flycatchers go one stage further and defend several different territories at once, deceiving several females into apparently 'monogamous' pairings. The male benefits but the females would seem to be worse off than if they had chosen an unmated male.

Social dominance

Throughout this book, we have emphasized that natural selection often leads animals to compete with one another. Sometimes different individuals will have common interests and will show a high degree of cooperation, but even when they do, competition at the individual or gene level does not disappear. It merely takes on a new form in which individual interests are best served by aiding others (p. 338). At other times, conflicts of interests become more overt, and animals can be highly aggressive toward one another. Male red deer and elephant seals spend much of the breeding season fighting each other in dramatic and often bloody conflict. Huge male elephant seals (Fig. 7.10) dominate stretches of beach where the much smaller females have come ashore to have their young. They fight over access to the females and thus for paternity of next year's young. Only the very strongest can dominate a large area and in some seasons a mere 4% of the males are responsible for 85% of the matings (Le Boeuf & Peterson 1969; Le Boeuf 1974). The reproductive prizes are thus very great but the costs may be considerable too. The fights between males can be very damaging and few males manage to stay successful for more than one or two seasons. The same is true of red deer stags, where the reproductive life of a female may be up to 20 years, but that of a male very much less. In lions, too, a coalition of male lions holds a pride of females for little more than 2–3 years before they are displaced by younger and stronger rivals.

In our discussion of animal communication (Chapter 3), we saw that many animals, including red deer stags, do not always fight even when a group of females is at stake, but have evolved ways of avoiding damaging fights that they are likely to lose. They use signals to 'assess' each other's fighting ability and a challenging stag will retreat if 'out-roared' by a resident stag (Clutton-Brock & Albon 1979). In fact overt fighting occurs when the stags are so

Figure 7.10 Male and female elephant seals showing the great disparity of size which has evolved in these extreme polygynous animals. Mortality among males is commonplace and only a few achieve the size and strength to hold a harem of females on the breeding beaches. (Drawing by Melissa Bateson.)

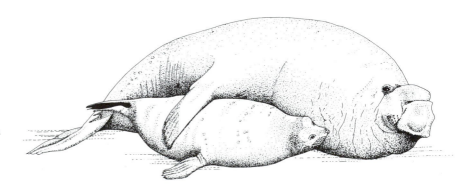

evenly matched in size and physical condition that human observers (and presumably the stags themselves) are unable to predict the outcome of a fight.

Use of signals as substitutes for fighting is one way animals have of avoiding the highest costs of aggression (i.e. death or injury) and we have seen that use of pre-fight 'assessment' signals can be an Evolutionarily Stable Strategy (p. 335), provided, of course, that the signals used are honest and reliable. Otherwise, animals would avoid the costs of fighting but would also miss out on the benefits of the resources they might have had by winning the fight. Many authors, such as Zahavi (1991), have argued that such reliability can only come through the use of signals that are themselves costly or at least difficult for any but the most vigorous and physically strong animals to produce. Very costly signals such as roaring are not, however, the only way animals have of deciding whether or not to fight on a given occasion. They have two other important ways of doing so.

The first method is only possible where the same animals interact frequently enough that they can learn each other's true fighting ability. They can thus avoid fights with animals that are likely to beat them by using their past experience with that same animal. Animals using this method of avoiding overt fighting are often described as having a dominance hierarchy or peck order.

Schjelderup-Ebbe was one of the first to develop the concept of a social dominance hierarchy, using his work on flocks of domestic fowl (1935). He observed that a definite 'peck order' developed amongst a group of hens, one gradually emerging as the dominant in the sense that she could displace all others. Below her was a second-ranking bird who could dominate all except the top bird and so on down the group until at the bottom was a bird displaced by every other in the flock. Figure 7.11 shows one example of a peck order of this type – a linear hierarchy, as it is sometimes called. The hierarchy develops as the birds dispute and often involves a good deal of fighting in the early stages as the birds test each other out. But once it is established, subordinates usually defer without question to the approach of a more dominant bird. Chickens are somewhat unusual in often having clear linear hierarchies. In other species, particularly among primates, the situation may be more complex so that A may dominate B and B may dominate C but A does not necessarily dominate C. In addition, there may be more complex alliances, in which two males cooperate to maintain a joint top position in a hierarchy. De Waal (1996) describes such an alliance that developed between two male chimpanzees at Arnhem Zoo. Neither could have achieved dominance on his own but together they could successfully ward off all challengers.

Not only chickens, but animals of many different species show a marked drop in aggression as they become familiar with each other (Chase 1985). The animals learn that certain other individuals are reliably capable of beating them in a fight and so avoid the risks of a fight they are unlikely to win anyway, by immediately deferring to their superior. They do not even need a costly assessment signal in order to do so and the dominance hierarchy thus enables them to avoid not only the cost of fights they would be likely to lose but also the cost of giving signals. Of course, the stability of such a system (and the advantages of accepting a subordinate position) relies on the fighting ability of the different individuals remaining much the same over long periods of time, and dominance hierarchies can be expected to become unstable when the

Figure 7.11 A perfect linear hierarchy established within a group of 12 domestic hens. Each bird is marked on its legs by colour rings, whose initials identify it. The number of times each bird pecked another flock member is given in the vertical columns (e.g. Y pecked B 22 times and pecked V 8 times) whilst the number of pecks received from another hen is given in the horizontal rows (e.g. VV received 19 pecks from YY and 9 pecks from BB). Note that no bird was ever seen to peck an individual above it in rank. This is an artificial group and hierarchies as perfect as this are probably rare in nature. (From Guhl 1968.)

	Y	B	V	R	G	YY	BB	VV	RR	GG	YB	BR
Y												
B	22											
V	8	29										
R	18	11	6									
G	11	21	11	12								
YY	30	7	6	21	8							
BB	10	12	3	8	15	30						
VV	12	17	27	6	3	19	8					
RR	17	26	12	11	10	17	3	13				
GG	6	16	7	26	8	6	12	26	6			
YB	11	7	2	17	12	13	11	18	8	21		
BR	21	6	16	3	15	8	12	20	12	6	27	

relative fighting abilities of the different members shift. This is exactly what occurs. A dominant hen that becomes sick slips down the peck order and the other individuals may fight to reposition themselves. Similarly, a young male baboon or chimpanzee that grows bigger and stronger than the dominant male may challenge the existing hierarchy and other relationships will change too.

The need for periods in which relative fighting ability is reasonably predictable is probably part of the reason why red deer stags continue to use costly assessment signals throughout the breeding season. A stag that is constantly defending his group of females from the challenges of other stags may have so little time to eat that he loses weight and body condition. His ability to fight is therefore constantly changing and a young challenger who is beaten by him at the beginning of the rut may be more successful later on. Just using individual recognition would therefore be a less successful strategy than using a costly signal like roaring that gives a day-to-day 'read-out' of changing fighting ability of familiar as well as unfamiliar individuals.

When animals encounter too many other animals to learn them all individually, they have a second method of avoiding unncessary fights while keeping signal costs to a minimum. This is to have a dominance hierarchy based not on individual recognition, but on 'badges of status' (Chapter 3) that can be instantly recognized even on unfamiliar individuals. Badges of status, such as the black bibs of male house sparrows or the black head feathers of Harris' sparrows (p. 156) distinguish dominant from subordinate animals. What is striking about them is that they do not appear to be costly and consequently would seem to be vulnerable to cheating. While we can understand why an animal might move away from a piece of food without a fight when a known dominant animal appears (it has always been beaten by this animal in the past), it is less easy to understand why it should defer to an unfamiliar animal simply because it has a dominant status badge. What is to stop a weak animal growing black feathers and so mimicking a dominant? The answer seems to be that there are costs to status badges after all. It may be relatively easy for a weak animal to produce a status badge, but if it is constantly challenged by other dominant individuals (Møller 1987), then it will pay a cost for its deception. The social consequences of cheating may be sufficient to keep cheats under control or at least at a low numbers. In practice, status badges thus have a higher degree of reliability than appears at first sight, although the constant possibility of cheating, albeit at a low level, may be the reason why animals do not always rely on them. They are found particularly in species such as sparrows and great tits, where animals constantly encounter such large numbers of other individuals that hierarchies based on individual recognition are unlikely.

It will be apparent by now that what we refer to as 'social organization' in animals cannot be rigidly categorized as 'territorial' or 'hierarchical', and that mating systems and other social interactions take many different forms and are affected by a number of factors. As we mentioned at the outset, for many vertebrates, the changing seasons involve a change in their social organization. In chaffinches, males tend to dominate females in winter flocks and displace them at feeding sites. But this situation is reversed in spring, when females tend to displace males as the flocks disperse to set up territories. Chickens studied in a truly feral state on an island off Queensland, Australia, by McBride *et al.* (1959), alternate between a territorial system in the breeding season and a more hierarchical flock structure during the winter. The dominant male sets up a territory in the spring with a number of females in his flock. During the winter after the young birds of the year had returned to the flocks, McBride *et al.* found that the alpha male and his harem moved about over a home range with members of subordinate males staying at the periphery of the group, often moving between the home ranges of different alpha males. The alpha male led his flock in every sense. He it was that initiated all movements of the group, particularly across open ground, and his posture was normally more alert than those of his females. He was the first to give alarm calls and even approached predators – feral cats – while the others took cover.

Seasonal changes in behaviour are just as marked in some mammals. The classic study, *A Herd of Red Deer* by Fraser Darling (1935) and the more recent studies on the Hebridean island of Rum by Clutton-Brock and his colleagues (1982) give us a very full picture of the social life of this species. Outside the breeding season, males and females live apart. The males live in loose bachelor herds in which there are consistent linear dominance hierarchies related to body size. High ranking stags are able to displace lower ranked stags from good food patches and often do so by lowering their heads to display their antlers. The females, on the other hand, live in herds which also include young animals of both sexes. Threats and displacing other animals from food are much less common among the hinds, even in winter. In early April, the antlers of the stags are shed and immediately male aggression takes on a new form. The males rear on their hind legs and 'box' with their hooves. New antlers begin growing but remain soft and tender in 'velvet', which is not shed until August. Now the stags become increasingly aggressive. Fighting with the new antlers increases until about mid-September when the male herds break up. The males then go off singly to favoured display areas and begin roaring and gathering hinds who are coming into oestrous, and the 'rut' begins. Oestrus females are attracted to displaying stags, who defend their group of females rather than any particular area. The rutting season lasts only a few weeks and soon both sexes return to their respective herds for the winter.

In many species, of course, social behaviour does not change as dramatically as this from season to season. Animals that cooperate to hunt for prey and to defend themselves often stay together throughout the year. Wolves, for example, hunt in packs and maintain a very stable structure based on an extended family unit. There is usually one dominant male leader, but several other adult males may be included. Many primates also tend to retain a uniform social structure throughout the year. Since their social organization is also characterized by great complexity and subtlety of relationships between the individuals, we will conclude this chapter by considering their social behaviour in more detail.

Primate social organization

Studies on the primates have accelerated very rapidly in recent years. Workers from a number of disciplines – zoologists, psychologists, anthropologists and sociologists – have converged on this group, both for their intrinsic interest and also in the search for information which is relevant to speculations about the origins of human societies. There has been great emphasis in the recent work on studies of natural primate communities in the field. Partly this is a reaction to a sense of urgency as so many primate habitats are under threat from human activities (see Mittermeier & Cheney 1987, in this regard.) In addition, although certain primates, notably the rhesus monkey, have been familiar laboratory animals for some time, it is generally accepted that studies on captive communities are quite inadequate by themselves to reveal the richness of primate social behaviour.

Here we can attempt only a short survey of the primate work. There are two journals dedicated to the group, *Primates* and *Folia Primatologia*, which include many behavioural studies, and a large number of books and symposium volumes. For an introduction to this range the reader is referred to Deag (1980), Dunbar (1988), Hinde (1983), Jolly (1985) and McGrew *et al.* (1996). A particular mention should be made of the volume edited by Smuts *et al.* (1987) because its contributions cover such a diversity of behavioural topics and involve the whole range of living primates.

Primates live in a wide variety of habitats. We tend to think of them as tropical animals, but they were more widely distributed in the recent past and two of the macaques, the Barbary 'ape' (*Macaca sylvanus*) of the Atlas Mountains and the Japanese macaque (*M. fuscata*) live in areas where snow and frost occur every winter. The majority of primates are arboreal: some of them, like the spider monkeys of South America and the colobus monkeys of Africa, exclusively so. However, a number have returned to ground living, such as the baboons and the patas monkey of Africa, whilst chimpanzees and gorillas also spend a lot of time on the ground. In general the more arboreal primates are fruit and leaf eaters; the ground dwellers tend to be more omnivorous, including insects and perhaps small vertebrates in their diet, although a very high proportion of their food remains grass, seeds, bulbs, and so on. Although both mountain and lowland gorillas are almost exclusively vegetarian, chimpanzees and baboons will eat meat if they get the opportunity; certain communities of the former regularly hunt and kill monkeys for food.

The living primates exhibit a wide range of morphological types, from the primitive lemurs which retain a long muzzle, a moist nose and claws on one of their hind toes – little modified from the ancestral primate types – to the monkeys, the great apes and humans (Fig. 7.12). Throughout this series we can observe certain trends: the enlargement of the brain, the development of the grasping hand and, in contrast to many other mammals, the great reliance on full colour vision as a dominant sense for exploration and communication.

It seems certain that from very early in their history the majority of primates were social animals moving around in groups whose organization was

Figure 7.12 Some diverse representatives of the Primates together with a tree shrew (*Tupaia*) (a), which we now recognize is not a primate nor even closely related to them, but which does in its superficial appearance and life as an arboreal, nocturnal insectivore, resemble the stock of early mammals from which the order descended. (b) The tarsier (*Tarsius*) of the Philippines. (c) The ring-tailed lemur (*Lemur catta*) of Madagascar, an island where a number of primates of this general type have survived in isolation. (d) The capuchin monkey (*Cebus*) of South America exhibits the prehensile tail characteristic of a number of highly arboreal New World primates. (e) The pig-tailed macaque (*Macaca nemestrina*) of Southeast Asia has a greatly reduced tail and, in common with a number of other Old World monkeys, is partly terrestrial. (f) The gibbon (*Hylobates*) and (g) the chimpanzee (*Pan troglodytes*) represent the living apes, closest relatives to humans – the latter sharing about 99% of its genetic material with us. (After Le Gros Clark 1965, and not drawn to scale.)

stable. Jolly (1966) makes this point from her study of lemurs; here in these primitive primates we already find small mixed troops (12–20 individuals) which include several adult males and several breeding females – a very typical primate grouping. There is a dominance hierarchy within the troop (and in some lemurs females rank more highly than males) but it remains as a permanent, cohesive unit. Lemurs have group territories within their mixed woodland habitat whose boundaries are often remarkably stable. They are marked by scent in some species, and are defended by calling, which is usually sufficient to cause a neighbouring troop to retreat without further threat or fighting.

Within the troop there are frequent minor disputes but serious fights are rare outside the breeding season. This is very brief in most lemurs – at most 2 weeks – and it is at this time that subordinate males seriously challenge the older dominant ones. Apart from this short period of strife, much of lemur social life is characterized by non-aggressive interactions between individuals – indeed, it is pedantic to avoid using the term 'friendly'. There is always close contact between a mother and her infant, who clings continuously to her at first and is carried around everywhere. As it grows older other adults approach and play with the infant, as they also play with each other. Lemurs have thick, dense fur and groom frequently. Mothers groom their infants and adults frequently groom each other – this being one of the commonest types of friendly contacts between individuals.

In the behaviour of lemurs we can detect most of the elements which characterize all primate societies, although there are many variations on the theme. One of the most obvious types of variation concerns group size, from the almost solitary orang utan, through the single family groups of gibbons, to the commoner multi-male, multi-female groups typical of howler monkeys, vervets, macaques and most baboons up to the veritable herds of gelada baboons and mandrills which may number several hundred. These herds, although they may move together, have social subgroupings within them. Other obvious variations in primate social life concern the social structure within the groups, the type of mating system, the extent of territoriality, and the nature of interchange between groups as animals become sexually mature. A good deal of modern primatological work has been truly socio-biological in its approach and tries to develop a general theory concerning the different roles of the sexes and access to resources in order to explain the wide variation in primate social structures which we observe.

We might begin by citing one primate, and a great ape at that, which uniquely goes against the trend and leads an almost totally solitary life. Orang utans (*Pongo pygmaeus*) associate only for mating and whilst an offspring is dependent on its mother. They are entirely arboreal and both sexes tend to stay in a home range over which they travel throughout the year following the fruiting of trees (Van Schaik & Van Hoof 1996). Moving on we find that, although monogamy is rare amongst primates, it is almost perfectly exemplified by the gibbons, whose life-long pair bond between a male and a female and their strict territoriality, maintained by elaborate 'singing' especially at dawn, show remarkable parallels to some birds (Leighton 1987). The South American marmosets and tamarins also have very small groups with often one, or at the most, three adults of each sex in the group together with their young. Almost all other primates give birth to single offspring, but marmosets always have twins and males help with the care of infants, carrying them for much of the time (Goldizen 1987).

The commonest structure involves groups with a number of males and females living together. Within this general structure we find considerable diversity of social organization even among closely related species such as baboons of the genus *Papio*.* Hamadryas is a dry country species and in their social organization several females are more or less permanently bonded to a single male forming a so-called 'harem group'. A number of such groups band together, moving and foraging as a unit, perhaps 40 or 50 strong. The other baboons also have units of comparable size but here there are no persistent male/female bonds. Adult males form temporary consortships with females as they come into oestrus, but otherwise move generally within the group.

Chimpanzee (*Pan troglodytes*) have even more variable types of grouping. Quite often males associate and travel more together than the females and within a 'community' – a term increasingly used to describe the population of chimps in one area – groups form and disperse for foraging on an hourly basis. Sometimes they may come together to spend the night, but groups may be widely separated and move apart for days on end. Again, there are striking differences both within chimp communities and between them and the closely

* It is worth noting at this point that the baboons of the genus Papio *are now commonly divided into two groups:* P. hamadryas, *the hamadryas baboon, and four other forms which, though generally separated geographically, we know are capable of interbreeding, have very similar types of social organization and are dubiously named as separate species. These are* P. anubis, *the olive baboon;* P. cynocephalus, *the yellow baboon;* P. ursinus, *the chacma baboon; and* P. papio, *sometimes called the common baboon. In fact we know that* P. hamadryas *can also interbreed, at least with* P. cynocephalus, *but it is clearly distinguished from the others on behavioural grounds. In what follows the hamadryas will be compared with all the other forms which we shall simply call 'baboons'.*

related pigmy chimpanzee or bonobo (*P. paniscus*) (see reviews in Wrangham *et al.* 1994, and McGrew *et al.* 1996).

Given such diversity, we have to look widely for factors in the environment and life history of the different species to account for their social organization. The usual approach has been to gather data from as many species as possible and to compare them for a wide range of ecological, morphological and behavioural factors. For example, the nature of the food supply, the degree of pressure from predators, body size, the extent of sexual dimorphism, the number of mature males in the group and group size – all of these and more – may be analysed to look for regular associations. As with most mammals, we have to bear in mind that, in general, the two sexes have different limitations. The reproductive success of females tends to be limited by access to resources, that of males by access to females.

Since food supply is unlikely ever to be permanently abundant, we have to try to explain why almost all the primates live in groups where the food in an area must be shared among several mouths. We have already discussed how living in groups may compensate individuals for having to share food by increasing the efficiency of searching. Groups can also cooperate to defend good food patches. Thus increased feeding efficiency might favour group life. Alternatively, since most primates are fairly small animals and, alone, are very vulnerable to predators such as the big cats, it may be that the extra protection provided in groups is the main factor. Not only is there increased vigilance but several large males combining for defence may be more than a big cat dare take on – this is certainly the case with baboon groups defending themselves against leopards.

The arguments derived from the comparisons of the ecological and behavioural factors outlined above are quite complex and we cannot go into them here. There are challenging reviews by Van Schaik (1983), Wrangham (1987) and Ridley (1986), with no complete agreement on why primate organization is so diverse. Obviously we must expect the balance of factors leading to particular types of social organization to vary and – as with most biological systems – there will never be one solitary factor operating in isolation. Here we are more interested in the remarkable behavioural phenomena exhibited by the primates and move on to discuss some key examples.

Communication and social dominance in primates

One cannot fail to be impressed by the richness and complexity of primate social life and with the high level of intergroup communication that goes on between the members of a group. Each individual is constantly responsive to the posture, movements, gestures and calls of others. We have already referred to some remarkable examples of such responses in Chapter 3. This constant high level of *attention* to others is dramatic, and Chance (1967, 1976) has suggested that the social structure of a monkey group is best regarded as a structure of attention itself, particularly as it relates to social rank, a feature which we shall discuss below.

It is necessary to be somewhat cautious in our emphasis on the complexity of primate groups. We must not automatically assume that primate societies are more organized than those of, say, ungulates or carnivores. Because we ourselves are primates we find it much easier to identify elements in their communication system, in particular the mobility of their faces and the way they watch one another's faces for information on mood and intentions. Nevertheless there are, on the most objective criteria, good grounds for our assumption. Firstly, primates have an extended period of infancy, to which is coupled considerable longevity. The larger primates commonly live for 20 or 30 years and this means that a young primate grows up to take its place in a group where – literally – everybody knows everybody else from long experience in their company.

Secondly, there can be no doubt that primates are amongst the most intelligent of animals with a high learning ability and a correspondingly high flexibility in their behavioural response to a changing social situation (see Chapter 5 for further discussion of their learning). Sustaining this capacity is a large and highly developed brain. We may be dealing here with a close coupling between behaviour and evolution. Jolly (1966) and Humphrey (1976) have suggested that the complex demands of social life itself formed one of the main selective factors for the growth of brain size in primates. This evolving brain, in turn, offered still more flexibility and complexity and so a mutual evolutionary relationship was established. The relationship reaches its most extreme manifestation in the human brain with its capacity to mediate speech and all the richness of social interaction this allows.

We have already discussed the concept of social rank in connection with other species such as chickens. In primates, too, dominance and subordination are important features of their relationships with others. Dominance always involves the threat of physical displacement or attack, even though it may be rarely necessary once rank is established. However, it is most important to remember that relationships within primate groups are as much characterized by positive interactions as negative ones. There are many friendly contacts between animals, as when they move and rest together, invite grooming or offer to groom another. Mutual grooming, which we first mentioned in the lemurs, is very important as a placatory gesture in primates (see Fig. 7.13). Often a dominant animal will 'allow' itself to be groomed by a subordinate following a brief threat to which the subordinate has deferred. Sexual presentation as an appeasement gesture (see p. 245) is very common in baboons and chimpanzees and is made by males or females towards a domi-

Figure 7.13 Friendly grooming in the Barbary macaque. The juvenile male on the right is about 3 years old, he is grooming a subadult male aged about 4½, who ranks above him in the social hierarchy. A baby is being held in the sub-adult's lap. (Photograph by John Deag.)

nant animal who threatens, or even if the subordinate wants to pass close to the dominant one (Fig. 7.14).

The pattern of grooming relationships in a group is often a good measure of its detailed structure. It will show us which animals associate together and is a good index of the cohesive forces which maintain the group

Figure 7.14 Sexual presentation in the chacma baboon. (a) 'Genuine' presentation by an oestrus female; she characteristically looks back over her shoulder towards the male. (b) Appeasement presentation towards a threatening dominant animal; the general stance is the same but the facial expression is one of fear. (After Bolwig 1959.)

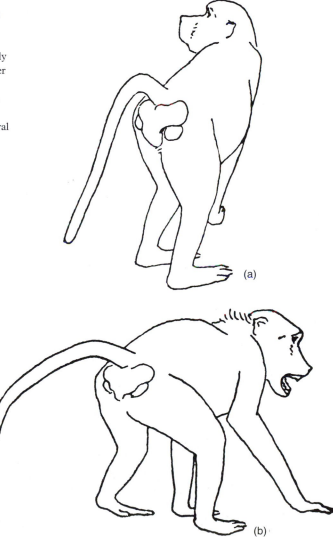

(a)

(b)

as a real social unit. For example, Seyfarth (1984) was able to show that frequent grooming between certain members of a vervet monkey group set up relationships such that one monkey was more likely to call upon another to aid it in a conflict situation, and was more likely to be successful if it did so. It is also very common to find that a mother and her offspring, particularly her daughters, form the most enduring of mutual support groups within the overall social structure.

However, in association with such patterns of cooperation, observations on almost all primate groups reveal the existence of a dominance hierarchy, such as that shown in Fig. 7.15, which is one typical example based upon agonistic interactions. Certainly the role of aggression in primate societies has attracted a great deal of attention. As may be imagined, the non-human primates have been used as evidence by both sides in the dispute about the nature of human aggression. Primates vary greatly in the degree of fighting which is observed both within and between groups. Even within a single species there may be considerable variations; langur monkey populations in northern India are apparently far more peaceful than those in the south. Population densities are much greater in the areas where the aggressive groups live and this is probably one factor involved. Fighting may occur in crowded zoo colonies which rarely give sufficient space for subordinate animals to keep out of the way of more dominant ones. Density and the accompanying stress are obviously crucial factors, but it is unlikely that all primate species will respond to them in the same way; some may be inherently more aggressive than others.

SF1 > JM1

SF3 JF1   IF2

AM1 > AM2 > AM3 > AF1 > SM1 > SM2 > AF2 > JM1 > AF3 > JM2 > AF4 > SF1 > SF2 > AF5 > = > = > JF2 > JF3 > IM1 > = > BF1 > BM2

AF6 JM3   IF3

Figure 7.15 The agonistic hierarchy recorded by Deag in a troop of wild Barbary macaques in the Middle Atlas Mountains of Morocco. It is based upon the distribution of threat and avoidance between pairs of individuals. The hierarchy is reasonably linear, especially at the upper end, where adult males are dominant over all other animals. Amongst the younger animals there is less strict hierarchy, and this reflects changing status as animals grow up. Baby animals are not much involved in the hierarchical system and rarely give or receive threats. There is one clear exception to linearity higher up: subadult female 1 threatened juvenile male 1 and he avoided her. A, Adult; J, juvenile; I, infant; B, baby; M, male; F, female. (From Deag 1977.)

However it is achieved, the stable situation in almost all primate groups can be interpreted in terms of a ranked hierarchy which determines to a greater or lesser extent how the individuals behave. Perhaps the best and most neutral definition of rank is that the behaviour of a high ranking or dominant animal is not limited by other individuals, whilst that of a subordinate is so limited. The limitations imposed by rank are diverse. For instance, having high rank in a group might determine access to food, preferred resting places and females. Thus when they come into oestrus, female baboons may move close to a dominant male and form a temporary 'consort relationship' with him.

A number of primate workers have criticized the way in which the concept of dominance has been used. Rowell (1974) quotes examples of imprecision in the use of the term dominance and points out that the different types of supposed advantage conferred by high rank – access to food, to oestrus females, to resting places, etc. – may not always be correlated. She suggests that hierarchies, where they exist, are as much hierarchies of submission as of aggression. Certainly, it has often been observed that while rank is initially determined by threat and fighting, once established it is maintained as much by the deference of subordinate animals as by any display of threat by the dominants. She concludes that hierarchies, so obvious in some captive colonies, may be less important in the wild.

Rowell's criticisms have some justification. As mentioned above, even with the largest enclosures, primate colonies in zoos *are* overcrowded by natural standards. When obvious dominance hierarchies have been observed in the wild they could sometimes result from human interference. For example, the rigid hierarchies described for Japanese macaque troops are probably influenced by their particular situation. The detailed work on these monkeys has been greatly helped by the success of the Japanese workers in getting them to visit artificial stations where food is regularly provided (see Frisch 1968). Some troops stay close to these stations and providing food in one spot is likely to reinforce dominance structure. Goodall (1968) found just the same thing when she provided caches of bananas for her chimpanzees. The food acts as a focus which all animals are trying to approach, the dominant males sit at the best places and the rest of the group tends to space out according to rank. The hierarchy is less obtrusive when food has to be searched for individually.

Despite these valid criticisms, it remains the case that dominance hierarchies *can* be observed in undisturbed wild primates. Deag's (1977) study of a troop of Barbary macaques revealed the hierarchy illustrated in Fig. 7.15, which is remarkably linear for the adults. He discusses how such a regular system of rank might arise from all individuals behaving to their own best advantage. The cohesion of the troop is vital for everyone's survival and some predictability about social interactions will be advantageous not only to the top animals but to low ranking ones too. Fighting is reduced, and probably stress also, since subordinate animals can keep clear of others to whom they will predictably lose in competitive encounters. It is common to observe that there is very little disputing between animals widely separated in the hierarchy – the subordinate avoids or immediately defers to its superior – but animals of similar rank are much more disputatious. Johnson (1989) found that close ranking juvenile baboons would pick a fight over food items that would never normally lead to a dispute. It may be worth their while because a win will enhance their future prospects when more important resources are at stake. Even if they lose, it may help them to recognize their position in the hierarchy more accurately so as to avoid future punishment. It will always be worth a low ranking animal biding its time because no hierarchy remains stable indefinitely. Subordinates are often observed to take over higher ranking positions as previously dominant animals become become weaker with age.

Some of the most interesting complexities of primate social behaviour are revealed by studies of how ranks are acquired and changed. Linear hierarchies are by no means universal and, particularly at the top end where the adult males of a group are usually competing, we may find shifting triangular relationships. One male may be dominant over either of the next two individually, but these latter may frequently gang up to form an alliance and displace him. Subsequently he, in turn, may try to break up this pair by making conciliatory approaches to one of them, and so on.

Such relationships are nowhere better described than in de Waal's fascinating account of his long-term studies on chimpanzees in the Arnhem Zoo, *Chimpanzee Politics* (1982). There have also been detailed studies of alliances in rhesus monkeys of the free-ranging colony on Cayo Santiago Island (off Puerto Rico) and in African vervet monkeys (Colvin 1983, Cheney 1983a). Alliances which form, break up and re-form produce a shifting, dynamic pattern of social life which greatly modifies the nature and the effects of ranking.

The acquisition of rank is also a complex process, differing between the sexes, which relates to their very different life histories. In many species of monkey, males never breed in the group into which they were born. The stability of primate groups might lead to a potential problem with inbreeding, but this is avoided with the transfer of newly mature males between groups. By contrast, in these same species the females invariably remain in their natal troop and in close association with their mothers and sisters.

We have already mentioned how enduring is the matriarchal grouping of a mother with her daughters, and their rank is determined by that of their mother. With remarkable regularity, in many cases, daughters slot in below their mothers in reverse order of age, i.e. the youngest daughter, once beyond infancy, ranks just below her mother and above her sisters. This is often because the mother supports her the most persistently (Gouzoules & Gouzoules 1987; Datta 1988). It makes good evolutionary sense to act in this way because, if she survives the dangers of infancy, a young daughter has a longer potential reproductive life ahead of her than an older one. It will pay mothers, on average, to apportion their support accordingly (see Gadgil 1982, for a full discussion of this idea).

The rank of young males develops along a different route. As sons grow up, they become independent of their mother and their rank amongst other members of the group is largely determined by their size. In baboons, Johnson (1987) found that juvenile males defeated juvenile females in almost all disputes, no matter what the rank of their mothers. There is a rise in disputing between young males as they become sexually mature; they may even challenge the adult males and because they are rarely successful, this is the commonest age for them to leave the group. They wander briefly, but soon approach other local groups where they are usually accepted quite rapidly (Cheney 1983b).

Most of the examples we have been discussing have been drawn from those primates such as baboons or vervet monkeys (*Cercopithecus aethiops*) whose social organization involves several adult males and adult females living together with no permanent male/female bonds. This is perhaps the commonest type of primate society, but we have already indicated others. As mentioned for hamadryas baboons, there are societies in which male/female bonds are more permanent, based upon 'harem groups' which consist of a single adult male, several females and their offspring. A number of highly arboreal species such as colobus monkeys also have groups with a single breeding male, but in these cases the group moves singly, often within a defended territory. Since males are driven out from the group as they mature, such an organization is bound to leave a surplus of adult males. These usually form all-male bands, members of which occasionally challenge those with harems.

However, the most conspicuous examples of such harem group organization remain the hamadryas (Fig. 7.16) and the gelada (*Theropithecus*

Figure 7.16 Three male hamadryas baboons, whose long capes are familiar from Egyptian art and hieroglyphics, with their harems of females closely herded together. Sexual selection has resulted in divergence between the sexes more marked than in many other primates. (Photograph by Hans Kummer.)

gelada), another baboon-like monkey of the Ethiopian uplands. In these species, a number of harem groups associate together and in the gelada one can observe hundreds of animals moving around together in a loose herd structure (Dunbar 1979). Hamadryas harem groups also move together during the day, although they may spread out and split up to comb an area for food. They always come together as a troop at night to sleep on cliffs which provide protection from predators.

For a concluding example to illustrate the subtlety and flexibility of primate social behaviour, we cannot do better than to stay with these hamadryas baboons and describe some remarkable field experiments by Kummer and his collaborators. Within the one-male units the male closely herds his three or four females; they move with him at all times and are not allowed to stray. Unattached females are rapidly acquired by males, so it is the more striking that there is scarcely any poaching of females between units. This remains the case even if one male is completely dominant over another in terms of access to food or resting places – the subordinate's females remain his own.

Kummer and his colleagues (Kummer *et al.* 1974; Bachmann & Kummer 1980) worked for a number of years with hamadryas in the dry areas of Ethiopia and were able to devise some experiments with these animals in the wild. The team caught some baboons of both sexes and confined them in cages on site. If two males were kept together and an unattached female introduced, they would threaten and sometimes fight each other but eventually the female became attached to one male – usually, of course, the winner of the conflict. However, it was possible to bias the situation by first caging an otherwise subordinate male alone along with the female, although in full sight of and very close to the dominant one. This gave the former two a chance to interact and, depending on his attractiveness, the female would pair, more or less strongly, with the subordinate male. Now the three animals would be put all together. If she was not strongly paired to the subordinate, the dominant would attempt to attract the female, but if she was paired he made no attempt whatsoever. Further, such a male introduced to an established pair appeared 'ill at ease', sitting as far away from the pair as possible and with his head turned away from them. There is complete social inhibition of poaching between males of a troop; dominance is not an absolute character, even where it could easily be expressed in order to gain a resource it is completely constrained by the social context.

Kummer's work also revealed something of the way individuals may acquire behaviour appropriate for the hamadryas social organization. The herding of females into a close unit is accomplished by threat. If, as she forages, a female strays too far from the male he first threatens her and, if she does not return to his side, chases and bites her. Customarily females return to the male at once when he 'stares' at them – a low intensity threat gesture. The yellow baboon (*Papio cynocephalus*) is a close relative and in fact hybridizes with hamadryas. As with other baboons, females in yellow baboon troops do not associate with particular males save at their brief oestrus period when they form a temporary consort relationship with a dominant male. For most of the time they move freely and associate with many different members of the troop.

Kummer (1968) transplanted a few yellow baboon females into a troop of hamadryas. Very quickly males began to herd them into their own units. At first the females wandered away and, if they responded at all to a male's threat, it was to flee. This of course produced the opposite effect to that which they intended. The male instantly pursued them and drove them back to his harem. Kummer found that within a few hours the yellow baboon females had learnt their harsh lesson and stayed close to their male.

Such results demonstrate vividly that some of the variation in social organization which we observe between primate species is of cultural and not genetic origin. There are also many examples of variations within a species which lead us to the same conclusion. The young hamadryas baboon grows up in a harem group, the young yellow baboon in the less restricted social climate of the larger group. They may have similar potential and yet develop quite differently. In Chapter 2 we discussed the cultural transmission of minor items of behaviour such as food-washing habits, but clearly with the extended infancy of primates, their long life and great learning capacities, cultural effects can extend much beyond this.

However, throughout this book we have tried to emphasize the ways in which internal and external factors, inherited and acquired information, interact in the course of an individual's behavioural development. With the baboons also we can see this in operation. There are genetic biases in the behavioural development of these two species. We can detect them because it is possible to study hybrids between the two species; in fact, natural hybridization occurs in some areas of Ethiopia, and the hybrids show intermediate forms of behaviour (Sugawara 1979). This interplay of genetic and cultural influences is certainly to be expected from the primates above all other groups. Cultural factors will play a major part in primate evolution, not just directly but because they will alter the background upon which genetic factors will operate.

Because of their close affinity to ourselves, the primates will always hold a particular fascination. Certainly, quite apart from any light they may throw on the origins of human social life (Hinde 1983, 1987) they offer some of the best opportunities for the study of behavioural development. Experimental approaches, which might include cross-fostering infants and cultural transplantation, have scarcely begun. We must allow all the diversity of primates room to live in the natural habitats to which they have adapted, both for their intrinsic worth and for the important information a combination of captive studies and field work is certain to yield.

References

ABLE, K.P. (1993). Orientation cues used by migratory birds: review of cue–conflict experiments. *Trends in Ecology and Evolution* 8: 367–71.

ALATALO, R.V., CARLSON, A., LUNDBERG, A. & ULFSTRAND, S. (1981). The conflict between male polygamy and female monogamy; the case of the pied flycatcher, *Ficedula hypoleuca*. *American Naturalist* 117: 738–53

ALATALO, R.V., HÖGLAND, J. & LUNDBERG, A. (1991). Lekking in the black grouse – a test of male viability. *Nature* 352: 155–6.

ALATALO, R.V., LUNDBERG, A. & GLYNN, C. (1986). Female pied flycatchers choose territory quality not male characteristics. *Nature* 323: 152–3.

ALCOCK, J. (1984). *Animal Behavior: An Evolutionary Approach*, 3rd edition. Sunderland, Mass.: Sinauer Associates.

ALLEE, W.C. (1938). *The Social Life of Animals*. New York: Norton.

ALTMANN, J. (1980). *Baboon Mothers and Infants*. Cambridge, Mass.: Harvard University Press.

ALTMANN, S.A. (1962). A field study of the sociobiology of rhesus monkeys, *Macaca mulatta*. *Annals of the New York Academy of Sciences* 102: 338–435.

ANDERSSON, M. (1982). Female choice selects for extreme tail length in a widowbird. *Nature* 299: 818–20.

ANDJUS, R.K., KNOPFELMACHER, F., RUSSELL, R.W. & SMITH, A.U. (1955). Effects of hypothermia on behaviour. *Nature* 176: 1015–16.

ANDREW, R.J. (1985). The temporal structure of memory formation. *Perspectives in Ethology* 6: 219–59. New York: Plenum.

ARAK, A. & ENQUIST, M. (1993). Hidden preferences and the evolution of signals. *Philosophical Transactions of the Royal Society of London B* 340: 207–14.

ARCHER, J. (1973). Tests for emotionality in rats and mice: a review. *Animal Behaviour* 21: 205–35.

ARCHER, J. (1979). *Animals Under Stress*. Studies in Biology No. 108. London: Edward Arnold.

AREY, D. (1992). Straw and food as reinforcers for prepartural sows. *Applied Animal Behaviour Science* 33: 217–26.

ASCHOFF, J. (1981). *Biological Rhythms*. Handbook of Behavioural Neurobiology 4. New York: Plenum.

ASCHOFF, J. (1989). Temporal orientation: circadian clocks in animals and humans. *Animal Behaviour.* 37: 881–96

AXELROD, R.D. & HAMILTON, W.D. (1981). The evolution of cooperation. *Science* 211: 1390–6.

BACHMANN, C. & KUMMER, H. (1980). Male assessment of female choice in Hamadryas baboons. *Behavioral Ecolology and Sociobiology* 6: 315–21.

BAERENDS, G.P. (1976). The functional organization of behaviour. *Animal Behaviour* 24: 726–38

BALDA, R.P. & KAMIL, A.C. (1992). Long-term spatial memory in Clark's nutcracker, *Nucifraga columbiana*. *Animal Behaviour* 44: 761–9.

BALDWIN, J.D. & BALDWIN, J.L. (1973). The role of play in social organisation: comparative observations of squirrel monkeys (*Saimiri*). *Primates* 14: 369–81.

BALMFORD, A (1991). Mate choice on leks. *Trends in Ecology and Evolution* 6: 87–92.

BALTHAZART, J. & BALL, G.F. (1995). Sexual differentiation of brain and behavior in birds. *Trends in Endrocrinology and Metabolism* 6(1): 21–9.

BAPTISTA, L.F. & PETRINOVITCH, L. (1984). Social interaction, sensitive phases and the song template hypothesis in the white-crowned sparrow. *Animal Behaviour* 32: 172–81.

BAPTISTA, L.F. & PETRINOVITCH, L. (1986). Song development in the white-crowned sparrow: social factors and sex differences. *Animal Behaviour* 35: 1359–71.

BAPTISTA, L.F. & MORTON, M.L. (1988). Song learning in montane white-crowned sparrows: from whom and when. *Animal Behaviour* 36: 1753–64.

BARFIELD, R.J. (1971). Gonadotrophic hormone secretion in the female ring dove in response to visual and auditory stimulation in the mate. *Journal of Endrocrinology* 49: 305–10.

BARINAGA, M. (1996), Guiding neurons to the cortex. *Science* 274: 1100–1.

BARLOW, H.B. (1972). Single units and sensation: a neuron doctrine for perceptual psychology? *Perception* 1: 371–94.

BARNETT, S.A. (1963). *A Study in Behaviour*. London: Methuen.

BARNETT, S.A. (1964). *The Concept of Stress*. Viewpoints in Biology 3: 170–218. London: Butterworth.

BARTLETT, J. (1987). Filial cannibalism in burying beetles. *Behavioral Ecology and Sociobiology* 21: 179–83.

BASOLO, A. (1990). Female preference predates the evolution of the sword in swordtail fish. *Science* 250: 808–10.

BASOLO, A. (1995). A further examination of pre-existing bias favouring a sword in the genus *Xiphophorus*. *Animal Behaviour* 50: 365–75.

BASTOCK, M. & MANNING, A. (1955). The courtship of *Drosophila melanogaster*. *Behaviour* 8: 85–111.

BATESON, P.P.G. (1976). Specificity and the origins of behaviour. *Advances in the Study of Behavior* 6: 1–20.

BATESON, P.P.G. (1979). How do sensitive periods arise and what are they for? *Animal Behaviour* 27: 470–86.

BATESON, P.P.G. (1980). Optimal outbreeding and the development of sexual preferences in Japanese quail. *Zeitschrift für Tierpsychologie* **53**: 231–44.

BATESON, P.P.G. (1983). Genes environment and the development of behaviour. In *Animal Behaviour*, Vol. 3. *Genes, Development and Learning*, ed. T.R. Halliday and P.J.B. Slater, pp. 52–81. Oxford: Blackwell.

BATESON, P.G.G. & BRADSHAW, E. (1997). Physiological effects of hunting red deer (*Cervus elaphus*). *Proceedings of the Royal Society of London B* **264**; 1707–14.

BATESON P., MENDL, M. & FEAVER, J. (1990). Play in the domestic cat is enhanced by rationing of the mother during lactation. *Animal Behaviour* **40**: 514–25.

BATESON, P.G.G. & REESE, E.P. (1969). Reinforcing properties of conspicuous objects before imprinting has occurred. *Psychonomic Science* **10**: 379–80.

BATESON, P.P.G. & WAINWRIGHT, A.A.P. (1972). The effects of prior exposure to light on chicks' behaviour in the imprinting situation. *Animal Behaviour* **21**: 720–5.

BAYLIS, G.C., ROLLS, E.T. & LEONARD, C.M. (1985). Selectivity between faces in the reponses of a population of neurons in the cortex in the superior temporal sulcus of the monkey. *Brain Research* **342**: 91–102.

BEACH, F.A. (1942). Analysis of the stimuli adequate to elicit mating behavior in the sexually inexperienced male rat. *Journal of Comparative Psychology* **33**: 163–207.

BEAR, M.F. (1997). How do memories leave their mark? *Nature* **385**: 481–2.

BEKOFF, A. & KAUER, J.A. (1984). Neural control of hatching: role of neck position in turning on hatching leg movements in post-hatching chicks. *Journal of Comparative Psychology* **145**: 497–504.

BEKOFF, M. (1995). Play signals as punctuation: the structure of social play in canids. *Behavior* **132**: 419–29.

BENNET-CLARK, H.C. (1970). The mechanism and efficiency of sound production in mole crickets. *Journal of Experimental Biology* **52**: 619–52.

BENNET-CLARK, H.C. (1987). The tuned singing burrow of mole crickets. *Journal of Experimental Biology* **128**: 383–409.

BENNET-CLARK, H.C. & EWING, A. (1969). Pulse interval as a critical parameter in the courtship song of *Drosophila melanogaster*. *Animal Behaviour* **17**: 755–9.

BENNETT, A.T.D. & CUTHILL, I.C. (1994). Ultraviolet vision in birds: what is its function? *Vision Research* **34**: 1471–8.

BENTLEY, D.R. & HOY, R.R. (1970). Postembryonic development of adult motor patterns in crickets: a neural analysis. *Science* **170**: 1409–11.

BENTLEY, D. & KESHISHIAN, H. (1982). Pathfinding by peripheral pioneer neurons in grasshoppers. *Science* **218**: 1082–8.

BERTHOLD, P. (1993). *Bird Migration*. Oxford: Oxford University Press.

BERTHOLD, P. & HELBIG, A.J. (1992). The genetics of bird migration: stimulus, timing, and direction. *Ibis* **134** (Suppl. 1): 35–40.

BERTHOLD, P. & QUERNER, U. (1981). Genetic basis of migratory behaviour in European warblers. *Science* **212**: 77–9.

BINGMAN, V.P., IOLAE, P., CASINI, G. & BAGNOLI, P. (1988). Hippocampal ablated homing pigeons show a persistent impairment in the time taken to return home. *Journal of Comparative Physiology* **163**: 559–63.

BIRKHEAD, T. (1993). Avian mating systems: the aquatic warbler is unique. *Trends in Ecology and Evolution* **8**: 390–1.

BIRKHEAD, T.R., ATKIN, L. & MØLLER, A.P. (1987). Copulation behaviour of birds. *Behaviour* **101**: 101–33.

BIRKHEAD, T.R. & MØLLER, A.P. (1992). *Sperm Competition in Birds: Evolutionary Causes and Consequences*. London: Academic Press.

BIRKHEAD, T.R. & MØLLER, A.P. (1993). Female control of paternity. *Trends in Ecology and Evolution* **8**(3): 100–4.

BLAKEMORE, C. & COOPER, G.F. (1970). Development of the brain depends on the visual environment. *Nature* **228**: 477–8.

BLAKEMORE, R.P. (1975). Magnetotactic bacteria. *Science* **190**: 377–9.

BLANCHARD, R.J. & BLANCHARD, D.C. (1981). The organisation and modelling of animal aggression. In *The Biology of Aggression*, ed. P.F. Brain and D. Benton, pp. 529–61. Aphen aan den Rhijn: Sijthoff & Noordhoff.

BLASS, E.M. & EPSTEIN, A.N. (1971). A lateral preoptic osmosensitive zone for thirst in the rat. *Journal of Comparative Physiology and Psychology* **76**: 378–94.

BLEST, A.D. (1957a). The function of eyespot patterns in the Lepidoptera. *Behaviour* **11**: 209–56.

BLEST, A.D. (1957b). The evolution of protective displays in the Saturnioidea and Sphingidae (Lepidoptera). *Behaviour* **11**: 257–309.

BLÜM, V. & FIEDLER, K. (1965). Hormonal control of reproductive behavior in some cichlid fish. *General and Comparative Endocrinology* **5**: 186–96.

BLURTON-JONES, N. (1959). Experiments on the causation of the threat postures of Canada geese. *Report of the Severn Wildfowl Trust, 1960*, 46–52.

BOAKES (1984). *From Darwin to Behaviourism*. Cambridge University Press.

BOESCH, C. (1994). Cooperative hunting in wild chimpanzees. *Animal Behaviour* **48**: 653–67.

BOGDANY, F.J. (1978). Linking of learning signals in honeybee orientation. *Behavioral Ecology and Sociobiology* **3**: 323–36.

BOLHUIS, J.J. (1995). Development of perceptual mechanisms in birds: predispositions and imprinting. In *Neuroethological Studies of Cognitive and Perceptual Processes*, ed. C.F. Moss and S.J. Shettleworth, pp. 150–76. Boulder, Colorado: Westview Press.

BOLHUIS, J.J. & BATESON, P. (1990). The importance of being first: a primary effect in filial imprinting. *Animal Behaviour* **40**: 472–83.

BOLLES, R.C. (1970). Species-specific defense reactions and avoidance learning. *Psychology Review* **77**: 32–48.

BOLWIG, N. (1959). A study of the behaviour of the chacma baboon *Papio ursinus*. *Behaviour* **14**: 136–63.

BONNER, J.T. (1980). *The Evolution of Culture in Animals*. Princeton, NJ: Princeton University Press.

BOSSEMA, I. & BURGER, R.R. (1980). Communication during monocular and binocular looking in European jays *Garrulus g. glandarius*. *Behaviour* **74**: 274–83.

BOTTJER, S.W. & ARNOLD, A.P. (1984). Hormones and structural plasticity in the adult brain. *Trends in the Neurosciences* **7**(5): 168–71.

BOWLBY, J. (1969). *Attachment and Loss*, Vol. I. *Attachment*. London: Hogarth.

BOWLBY, J. (1973). *Attachment and Loss*, Vol. II. *Separation*. London: Hogarth.

BOYD, H. & FABRICIUS, E. (1965). Observations on the incidence of following of visual and auditory stimuli in naïve mallard ducklings (*Anas platyrhynchos*). *Behaviour* **25**: 1–15.

BRAITHWAITE, V.A. & GUILFORD, T. (1991). Viewing familiar landscapes affects pigeon homing. *Proceedings of the Royal Society of London B* **243**: 183–6.

BREEDLOVE, S.M. (1992). Sexual dimorphism in the vertebrate nervous system. *Journal of Neuroscience* **12**: 4133–42.

BRELAND, K. & BRELAND, M. (1961). The misbehavior of organisms. *American Psychologist* **16**: 681–4.

BRENNER, S. (1974). The genetics of *Caenorhabditis elegans*. *Genetics* **77**: 71–94.

BRINES, M.L. & GOULD, J.L. (1982). Skylight polarisation patterns and animal orientation. *Journal of Experimental Biology* **96**: 64–81.

BROCKMANN, H.J. (1980). The control of nest depth in a digger wasp (*Sphex ichneumoneus* L.). *Animal Behaviour* **28**: 426–45.

BROCKMANN, H.J., GRAFEN, A. & DAWKINS, R. (1979). Evolutionarily stable nesting strategy in a digger wasp. *Journal of Theoretical Biology* **77**: 473–96.

BRONSON, F.H. (1979). The reproductive biology of the house mouse. *Quarterly Review of Biology* **54**: 265–99.

BROOM, D.M. & JOHNSON, K.G. (1993). *Stress and Animal Welfare*. London: Chapman & Hall.

BROWN, C.R. (1986). Cliff swallow colonies as information centres. *Science* **234**: 83–5.

BROWN, J.L. (1969). The buffer effect and productivity in tit populations. *American Naturalist* **103**: 347–54.

BROWN, R.E. & EKLUND, A. (1994). Kin recognition and the major histocompatibility complex: an integrative review. *American Naturalist* **143**: 435–61.

BROWN, R.E. & MACDONALD, D.W. (eds.) (1985). *Social Odours in Mammals*, Vols. 1 and 2. Oxford: Clarendon Press.

BRUNER, J.S. JOLLY, A. & SYLVA, K. (eds.) (1976). *Play: Its Role in Development and Evolution*. London: Penguin.

BULL, H.O. (1957). Conditioned responses. In *The Physiology of Fishes*, 2 vols., ed. M.E. Brown, pp. 211–28. New York: Academic Press.

BULL, J.J. (1980). Sex determination in reptiles. *Quarterly Review of Biology* **55**: 3–20.

BURGHARDT, G. M. (1970). Defining 'communication'. In *Advances in Chemoreception*. Vol. I, ed. J.W. Johnston, D.G. Moulton and A. Turk, pp. 5–18. New York: Appleton Century Crofts.

BURKHARDT, D. (1989). UV vision: a bird's view of feathers. *Journal of Comparative Physiology A* **164**: 787–96.

BUTLIN, R.K. & RITCHIE, M.G. (1989). Genetic coupling in mate recognition systems: what is the evidence? *Biological Journal of the Linnean Society* **37**: 237–46.

BUTLIN, R.K. & RITCHIE, M.G. (1991). Variation in female mate preference across a grasshopper hybrid zone. *Journal of Evolutionary Biology* **4**: 227–40.

BYRNE, R. & WHITEN, A. (eds.) (1988). *Machiavellian Intelligence*. Oxford: Clarendon Press.

CADE, W.H. (1981). Alternative mate strategies: genetic differences in crickets. *Science* **212**: 563–4.

CAMHI, J.M. (1984). *Neuroethology: Nerve Cells and the Natural Behavior of Animals*. Sunderland, Mass.: Sinauer Associates.

CAMHI, J.M., TOM, W. & VOLMAN, S. (1978). The escape behavior of the cockroach *Periplaneta americana*. II. Detection of natural predators by air displacement. *Journal of Comparative Physiology* **128**: 203–12.

CAMPBELL, B. & LACK, E. (eds.) (1985). *A Dictionary of Birds*. Berkhamstead: T. & A.D. Poyser.

CANNON, W.B. (1974). *The Wisdom of the Body*. London: Kegan Paul.

CAPRANICA, R.R. & MOFFAT, A.J.M. (1975). Neurobehavioral correlates of sound communication in anurans. In *Advances in Vertebrate Neuroethology*, ed. J.P. Ewert, R.R. Capranica and D. Ingle, pp. 701–30. New York: Plenum.

CARO, T.M. (1980). The effects of experience on the predatory patterns of cats. *Behavioral Neurobiolology* **29**: 1–28.

CARO, T.M. (1987). Indirect costs of play: cheetah cubs reduce maternal hunting success. *Animal Behaviour* **35**: 295–7.

CARO, T.M. (1994). *Cheetahs of the Serengeti Plains: Group Living in an Asocial Species*. Chicago: University of Chicago Press.

CARO, T.M. (1995). Short-term costs and correlates of play in cheetahs. *Animal Behaviour* **49**: 333–45.

CARTER, C.S. & MARR, J.H. (1970). Olfactory imprinting and age variables in the guinea-pig, *Cavia porcellus*. *Animal Behaviour* **18**: 238–44.

CATCHPOLE, C. (1980). *Vocal Communication in Birds*. Studies in Biology No. 115. London: Edward Arnold.

CATCHPOLE, C., LEISLER, B. & WINKLER, H. (1985). The evolution of polygyny in the Great Reed Warbler, *Acrocephalus arundinaceus*: a possible case of deception. *Behavioral Ecology and Sociobiology* **16**: 285–91.

CATCHPOLE, C.K. & SLATER, P.J.B. (1995). *Bird Song. Biological Themes and Variations*. Cambridge: Cambridge University Press.

CHANCE, M.R.A. (1967). Attention structure as the basis of primate rank order. *Man* **2**: 503–18.

CHANCE, M.R.A. (1976). The organisation of attention in groups. In *Methods of Inference from Animal to Human Behavior*, ed. M. Von Cranach, pp. 213–35. The Hague: Mouton Aldine.

CHASE, I.D. (1985). The sequential analysis of aggressive acts during hierarchy formation: an application of the 'jigsaw puzzle' approach. *Animal Behaviour* **33**: 86–100.

CHENEY, D.L. (1983a). Extrafamiliar alliances among vervet monkeys. In *Primate Social Relationships* ed. R.A. Hinde, pp. 278–89. Oxford: Blackwell Scientific Publications.

CHENEY, D.L. (1983b). Proximate and ultimate factors related to the distributiion of male migration. In *Primate Social Relationships*, ed. R.A. Hinde, pp. 241–9. Oxford: Blackwell Scientific Publications.

CHENEY, D.L. & SEYFARTH, R.M. (1980). Vocal recognition in free-ranging vervet monkeys. *Animal Behaviour* **28**: 362–7.

CHENEY, D.L. & SEYFARTH, R.M. (1982). How vervet monkeys perceive their grunts: field playback experiments. *Animal Behaviour* **30**: 739–51.

CHENEY, D.L. & SEYFARTH, R.M. (1986). The recognition of social alliances among vervet monkeys. *Animal Behaviour* **34**: 1722–31.

CHENEY, D.L. & SEYFARTH, R.M. (1990). *How Monkeys See the World*. Chicago: University of Chicago Press.

CLARK, D.L. & UETZ, G.W. (1992). Morph-independent mate selection in a dimorphic jumping spider: demonstration of movement bias in female choice using video-controlled courtship behaviour. *Animal Behaviour* **43**: 247–54.

CLARK, R.B. (1960a). Habituation of the polychaete *Nereis* to sudden stimuli. I. General properties of the habituation process. *Animal Behaviour* **8**: 82–91.

CLARK, R.B. (1960b). Habituation of the polychaete *Nereis* to sudden stimuli. II. Biological significance of habituation. *Animal Behaviour* **8**: 92–103.

CLARKE, A.M. & CLARKE, A.D.B. (1976). *Early Experience: Myth and Evidence*. London: Open Books.

CLARKE, A.S., WITTWER, D.J., ABBOTT, D.H. & SCHNEIDER, M.L. (1994). Long-term effects of prenatal stress on HPA axis activity in juvenile rhesus monkeys. *Developmental Psychobiology* **27**: 257–69.

CLUTTON-BROCK, T.H. & ALBON, S.D. (1979). The roaring of red deer and the evolution of honest advertisement. *Behaviour.* **69**: 145–70.

CLUTTON-BROCK, T.H., GUINNESS, F.E. & ALBON, S.D. (1982). *Red Deer: Behaviour and Ecology of Two Sexes*. Edinburgh: Edinburgh University Press.

CLUTTON-BROCK, T.H. & HARVEY, P.H. (1984). Comparative approaches to investigating adaptation. In *Behavioural Ecology*, 2nd edition, ed. J.R. Krebs and N.B. Davies, pp. 7–29. Oxford: Blackwell Scientific Publications.

COHEN-SALMON, C., CARLIER, M., ROBERTOUX, M., JOUHANEAU, J., SEMAL, C. & PAILLETTE, M. (1985). Differences in patterns of pup care in mice. V. Pup ultrasonic emissions and pup care behavior. *Physiology and Behavior* **35**: 167–74.

COLGAN, P. (1989). *Animal Motivation*. London: Chapman & Hall.

COLVIN, J. (1983). Description of sitting and peer relationships among immature male rhesus monkeys. In *Primate Social Relationships*, ed. R.A. Hinde, pp. 20–7. Oxford: Blackwell Scientific Publications.

COOPER, K.W. (1957). Biology of eumenine wasps: V. Digital communication in wasps. *Journal of Experimental Zoology* **134**: 469–514.

COWIE, R. J. (1977). Optimal foraging in great tits, *Parus major*. *Nature* **268**: 137–9.

COWIE, R.J., KREBS, J.R. & SHERRY, D.F. (1981). Food storage by marsh tits. *Animal Behaviour* **29**: 1252–9.

COX, C.R. & LE BOEUF, B.J. (1977). Female incitation of mate competition: a mechanism of mate selection. *American Naturalist* **111**: 317–35.

CREEL, S., WILDT, D.E. & MONFORT, S.L. (1993). Aggression, reproduction, and androgens in wild dwarf mongooses: a test of the Challenea hypothesis. *American Naturalist* **141**; 816–25.

CRESSWELL, W. (1994). Song as a pursuit-deterrence signal and its occurrence relative to other anti-predator behaviors of the skylark on attack by merlins. *Behavioral Ecology and Sociobiology* **34**: 217–23.

CRONIN, G.J., WIEPKEMA, P.R. & VAN REE, J.M. (1985). Endogenous opioids are involved in abnormal stereotyped behaviours of tethered sows. *Neuropeptides* **6**: 527–30.

CULLEN, E. (1957). Adaptations in the kittiwake to cliff-nesting. *Ibis* **99**: 275–302.

CURIO, E., ERNST, V. & VIETH, W. (1978). Cultural transmission of enemy recognition. *Science* **202**: 899–901

DAGAN, D. & VOLMAN, S. (1982). Sensory basis for directional wind detection in first instar cockroaches, *Periplaneta americana*. *Journal of Comparative Physiology* **147**: 471–8.

DANE, B., WALCOTT, C. & DRURY, W.H. (1959). The form and duration of the display actions of the 'goldeneye' *Bucephala clangula*. *Behaviour* **14**: 265–81.

DANTZER, R. (1991). Stress, stereotypies and welfare. *Behavioural Processes* **25**: 95–102.

DARWIN, C. (1871). *The Descent of Man and Selection in Relation to Sex*. London: John Murray.

DARWIN, C. (1872). *The Expression of the Emotions in Man and Animals*. London: University of Chicago Press (1965 edition).

DATTA, S. (1988). The acquisition of dominance among free-ranging rhesus monkey siblings. *Animal Behaviour* **36**: 754–72.

DAVIES, N.B. (1989). Sexual conflict and the polygyny threshold. *Animal Behaviour* **38**: 226–34.

DAVIES, N.B. (1991). Mating systems. In *Behavioural Ecology: an Evolutionary Approach*, ed. J.R. Krebs and N.B. Davies, pp. 263–94. Oxford: Backwell.

DAVIES, N.B. (1992). *Dunnock Behaviour and Social Evolution*. Oxford: Oxford University Press.

DAVIES, N.B. & BROOKE, M. DE L. (1989a). An experimental study of co-evolution between the cuckoo *Cuculus canorus* and its hosts. I. Host egg discrimination. *Journal of Animal Ecology* **58**: 207–24.

DAVIES, N.B. AND BROOKE, M. DE L. (1989b). An experimental study of co-evolution between the cuckoo *Cuculus canorus* and its hosts. II. Host egg markings, chick discrimination and general discussion. *Journal of Animal Ecology* **58**: 225–36.

DAVIES, N.B. & HOUSTON, A.I. (1981.). Owners and satellites: the economics of territory defence in the pied wagtail, *Motacilla alba*. *Journal of Animal Ecology* **50**: 157–80.

DAVIS, W.J., MPITSOS, G.J., PINNEO, J.M. & RAN, J.L. (1977). Modification of the behavioural hierarchy of *Pleurobranchaea*. I. Satiation and feeding mechanisms. *Journal of Comparative Physiology* **117**: 99–125.

DAWKINS, M. (1980). *Animal Suffering: The Science of Animal Welfare*. London: Chapman & Hall.

DAWKINS, M.S. (1990). From an animal's point of view: motivation, fitness and animal welfare. *Behavioral and Brain Sciences* **13**: 1–61.

DAWKINS, M.S. (1993). *Through Our Eyes Only?* Oxford: W.H. Freeman.

DAWKINS, M.S. (1995). *Unravelling Animal Behaviour*, 2nd edition. Harlow: Longman.

DAWKINS, R. (1979). Twelve misunderstandings of kin selection. *Zeitschrift für Tierpsychologie* **51**: 184–200.

DAWKINS, R. (1982). *The Extended Phenotype*. London: Longman.

DAWKINS, R. (1986). *The Blind Watchmaker*. Harlow, Essex: Longman.

DAWKINS, R. (1989). *The Selfish Gene*, 2nd edition. Oxford: Oxford University Press.

DAWKINS, R. & DAWKINS, M. (1974). Decisions and the uncertainty of behaviour. *Behaviour* **45**: 83–103.

DAWKINS, R. & KREBS, J.R. (1978). Animal signals: information or manipulation? In *Behavioural Ecology: An Evolutionary Approach*, 1st edition, ed. J.R. Krebs and N.B. Davies, pp. 282–309. Oxford: Blackwell Scientific Publications.

DEAG, J.M. (1977). Aggression and submission in monkey societies. *Animal Behaviour* **25**: 465–74.

DEAG, J.M. (1980). *Social Behaviour of Animals*. Studies in Biology No. 118. London: Edward Arnold.

DEAG, J. & CROOK, J.H. (1971). Social behaviour and agonistic buffering in the wild barbary macaque *Macaca sylvana* L. *Folia Primatologia* **15**: 183–200.

DE COURSEY, P.J. (1960). Phase control of activity in a rodent. *Cold Spring Harbor Symposia in Quantitative Biology* **25**: 49–55.

DELIUS, J.D. (1992). Categorical discrimination of objects and pictures by pigeons. *Animal Learning and Behavior* **20**: 301–11.

DETHIER, V.G. (1953). Summation and inhibition following contra-lateral stimulation of the tarsal chemoreceptors of the blowfly. *Biological Bulletin* **105**: 257–68.

DETHIER, V.G. (1957). Communication by insects: physiology of dancing. *Science* **125**: 331–6.

DE WAAL, F. (1982). *Chimpanzee Politics*. New York: Harper & Row.

DE WAAL, F. (1996). Conflict as negotiation. In *Great Ape Societies*, ed. W.C. McGrew, L.F. Marchant and M.T. Nishida, pp. 159–172. Cambridge: Cambridge University Press.

DEWSBURY, D.A. (1984). A brief history of animal behavior in North America. In *Perspectives in Ethology*, Vol. 8, ed. P.P.G. Bateson and P.H. Klopfer, pp. 85–122. New York: Plenum.

DIAMOND, J.N., KARASOV, W.H., PHAN, D. & CARPENTER, F.L. (1986). Hummingbird digestive physiology: a determinant of foraging bout frequency. *Nature* **320**: 62–3.

DICKINSON, A. (1980). *Contemporary Animal Learning Theory.* Cambridge: Cambridge University Press.

DIJKGRAAF, S. (1962). The functioning and significance of the lateral-line organs. *Biological Reviews* **38**: 51–105.

DILGER, W.C. (1962). The behavior of lovebirds. *Scientific American* **206**(1): 88–98.

DOUGLAS-HAMILTON, I. & DOUGLAS-HAMILTON, O. (1975). *Among the Elephants.* London: Collins and Harvill.

D'SOUZA, F. & MARTIN, R.D. (1974). Maternal behaviour and the effects of stress in tree-shrews. *Nature* **251**: 309–11.

DUDAI, Y. (1989). *The Neurobiology of Memory.* Oxford: Oxford University Press.

D'UDINE, B. & ALLEVA, E. (1983). Early experience and sexual preferences in rodents. In *Mate Choice*, ed. P.P.G. Bateson, pp. 311–27. Cambridge: Cambridge University Press.

DUFFY, S. (1996). Eusociality in a coral-reef shrimp. *Nature* **381**: 512–14.

DUNBAR, R. (1988). *Primate Social Systems.* London: Croom Helm.

DUNBAR, R.I.M. (1979). Structure of gelada baboon reproductive units. I. Stability of social relationships. *Behaviour* **69**: 72–7.

DUNBAR, R.I.M. & DUNBAR, P. (1988). Maternal time budgets of gelada baboons. *Animal Behaviour* **36**: 970–80.

DUNCAN, I.J.H. & KITE, V.G. (1987). Some investigations into motivation in the domestic fowl. *Applied Animal Behaviour Science* **18**: 387–8.

DUNCAN, I.J.H. & WOOD-GUSH, D.G.M. (1972). Thwarting of feeding behaviour in the domestic fowl. *Animal Behaviour* **20**: 444–51.

DUSENBERRY, D.B. (1992). *Sensory Ecology.* New York: W.H. Freeman.

DUVALL, W.D., MULLER-SCHWAZE, D. AND SILVERSTEIN, R.M. (1986). *Chemical Signals in Vertebrates* 4. New York: Plenum.

EHRHARDT, A.A. AND BAKER, S.W. (1974). Fetal androgens, human central nervous system differentiation, and behavior sex differences. In *Sex Differences in Behavior*, ed. R.C. Friedman, R.M. Richart and R.L. Van de Wiele, pp. 33–57. New York: John Wiley.

EHRHARDT, W.G. (1996). *Female Control: Sexual Selection by Cryptic Female Choice.* Princeton, NJ: Princeton University Press.

ELGAR, M. (1986a). The establishment of foraging flocks in house sparrows: risk of predation and daily temperature. *Behavioral Ecology and Sociobiology* **19**: 433–8.

ELGAR, M. (1986b). House sparrows establish foraging flocks by giving chirrup calls if the resources are divisible. *Animal Behaviour* **34**: 169–74.

ELGAR, M.A. (1989). Predator vigilance and group size among mammals: a critical review of the evidence. *Biological Reviews* **64**: 1–34.

ELGAR, M.A., MCKAY, H. & WOON, P. (1986). Flocking and predator surveillance in house sparrows: a test of a hypothesis. *Animal Behaviour* **29**: 868–72.

ELSNER, N. (1981). Developmental aspects of insect neuroethology. In *Behavioural Development*, ed. K. Immelmann, G.W. Barlow, L. Petrinovitch and M. Main, pp. 474–90. Cambridge: Cambridge University Press.

ELWOOD, R.W. (ed.) (1983). *Parental Behaviour of Rodents.* Chichester: John Wiley.

EMLEN, S. (1975). The stellar orientation system of a migratory bird. *Scientific American* **233**(2): 102–11.

EMLEN, S., DEMONG, N.J. & EMLEN, D.J. (1989). Experimental induction of infanticide in female wattled jacanas. *The Auk* **106**: 1–7.

EMLEN, S. AND ORING, L.W. (1977). Ecology, sexual selection and the evolution of mating systems. *Science* **197**: 215–23.

EMLEN, S.T. & WREGE, P.H. (1989). A test of alternate hypotheses for helping behaviour in white-fronted bee-eaters of Kenya. *Behavioral Ecology and Sociobiology* **25**: 303–20.

ENDLER, J. (1980). Natural selection in color patterns in *Poecilia reticulata. Evolution* **34**: 76–91.

ENDLER, J. (1983). Natural selection on color patterns in poeciliid fishes. *Environmental Biology of Fishes* **9**: 173–90.

ENDLER, J. (1993). The color of light in forests and its implications. *Ecological Monographs* **63**: 1–27.

ERICKSON, C.J. (1985). Mrs Harvey's parrot and some problems of socioendocrine response. In *Perspectives in Ethology*, Vol. 6, ed. P.P.G. Bateson and P.H. Klopfer, pp. 261–86. New York: Plenum.

ESCH, H., ESCH, I. & KERR, W.E. (1965). Sound: an element common to communication of stingless bees and to dances of honeybees. *Science* **149**: 320–1.

EVANS, S.M. (1968). *Studies in Invertebrate Behaviour.* London: Heinemann Education.

EWERT, J.P. (1980). *Neuroethology – An Introduction to the Neurophysiological Fundamentals of Behavior.* Berlin: Springer-Verlag.

EWERT, J.P. & TRAUD, R. (1979). Releasing stimuli for anti-predator behaviour in the common toad *Bufo bufo* L. *Behaviour* **68**: 170–80.

FAGEN, R.M. (1981). *Animal Play Behaviour.* Oxford: Oxford University Press.

FERRUS, A. & CANAL, I. (1994). The behaving brain of a fly. *Trends in the Neurosciences* **17**: 479–86.

FICKEN, M.S. & POPP, J.W. (1995). Long-term persistence of a culturally transmitted vocalization of the black-capped chickadee. *Animal Behaviour* **50**: 683–93.

FISHER, R.A. (1958). *The Genetical Theory of Natural Selection.* Oxford: Clarendon Press.

FITZGIBBON, C.D. & FANSHAWE, J.H. (1988). Stotting in Thomson's gazelles: an honest signal of condition. *Behavioral Ecology and Sociobiology* **23**: 69–74.

FITZSIMMONS, J.T. (1972). Thirst. *Physiological Reviews* **52**: 468–561.

FITZSIMMONS, J.T. & LE MAGNEN, J. (1969). Eating as a regulatory control of drinking. *Journal of Comparative and Physiological Psychology* **67**: 273–83.

FLEMING, A.S. & BLASS, E.M. (1994). Psychobiology of the early mother–young relationship. In *Causal Mechanisms of Behavioural Development*, ed. J.A. Hogan and J.J. Bolhuis, pp. 212–41. Cambridge: Cambridge University Press.

FLEISHMAN, L.J. (1992). The influence of the sensory systems and the environment on motion patterns in the visual displays of anoline lizards and other vertebrates. *American Naturalist* **139**: S36–S61.

FORD, N.B. & LOW, J.R. (1984). Sex pheromone source location by garter snakes: a mechanism for detection of direction in nonvolatile trails. *Journal of Chemical Ecology* **10**: 1193–99.

FRANCIS, R.C. (1992). Sexual lability in teleosts: developmental factors. *Quarterly Review of Biology* **67**: 1–18.

FRASER, A.F. AND BROOM, D.M. (1990). *Farm Animal Behaviour and Welfare*. London: Baillère Tindall.

FRASER DARLING, F. (1935). *A Herd of Red Deer*. London: Oxford University Press.

FRISCH, J.E. (1968). Individual behaviour and intertroop variability in Japanese macaques. In *Primates. Studies in Adaptation and Variability*, ed. P. Jay, pp. 243–52. New York: Holt, Rinehart & Winston.

FULLER, J.L. & THOMPSON, W.R. (1978). *Foundations of Behavior Genetics*. St Louis, Missouri: C.V. Mosby.

GADGIL, M. (1982). Changes with age in the strategy of social behavior. *Perspectives in Ethology* **5**: 489–502.

GALEF, B.F. (1976). Social transmission of acquired behavior: a discussion of tradition and social learning in vertebrates. *Advances in the Study of Behavior* **6**: 77–100

GALLISTEL, C.R. (1995). Is long-term potentiation a plausible basis for memory? In *Brain and Memory*, ed. J.L. McGaugh, N.M. Weinberger and G. Lynch, pp. 328–37. New York: Oxford University Press.

GALLUP: G.G., POVINELLI, D.J., SUAREZ, S.D., ANDERSON, J.R., LETHMATE, J. & MENZEL, E.W. (1995). Further reflections on self-recognition in primates. *Animal Behaviour* **50**: 1533–42.

GARCIA, J. & KOELLING, R. (1966). Relation of cue to consequence in avoidance learning. *Psychonomic Science* **4**: 123–4.

GELLERMAN, L.W. (1933). Form discrimination in chimpanzees and two-year-old children: I. form (triangularity) *per se. Journal of Genetic Psychology* **42**: 3–27.

GERHARDT, H.C. (1974). The significance of some spectral features in mating call recognition in the green treefrog *Hyla cinerea. Journal of Experimental Biology* **61**: 229–41.

GIBSON, R.M. & BRADBURY, J.W. (1985). Sexual selection in lekking sage grouse: phenotypic correlates of male mating success. *Behavioral Ecology and Sociobiology* **18**: 117–23.

GILL, F.B. & WOLF, L.L. (1975). Economics of feeding territoriality in the golden-winged sunbird. *Ecology* **56**: 333–45.

GODDARD, G.V. (1986). Learning. A step nearer a neural substrate. *Nature* **319**: 721–2.

GODFRAY, H.C.J. (1995). Evolutionary theory of parent–offspring conflict. *Nature* **376**: 133–8.

GODWIN, J. (1994). Behavioural aspects of protandrous sex change in the anemonefish, *Amphiprion melanopus*, and endocrine correlates. *Animal Behaviour* **48**: 551–67.

GOLDIZEN A.W. (1987). Tamarins and marmosets: communal care of offspring. In *Primate Societies*, ed. B.B. Smuts, D.L. Cheney, R.M. Seyfarth, R.W. Wrangham and T.T. Struhsaker, pp. 34–43. Chicago: University of Chicago Press.

GOLDMAN, S.A. & NOTTEBOHM, F. (1983). Neuronal production, migration, and differentiation in the vocal control nucleus of the adult female canary brain. *Proceedings of the National Academy of Sciences USA* **80**: 2390–4.

GOODALL, J. VAN LAWICK (1968). The behaviour of free-living chimpanzees in the Gombe Stream Reserve. *Animal Behaviour Monographs* **1**: 161–311.

GOODFELLOW, P.N. & LOVELL-BADGE, R. (1993). SRY and sex determination in mammals. *Annual Review of Genetics* **27**: 71–92.

GOTTLIEB, G. (1971). *Development of Species Identification in Birds*. Chicago: University of Chicago Press.

GOTTLIEB, G. (1976). Early development of species-specific auditory perception in birds. In *Neural and Behavioral Specificity*, ed. G. Gottlieb, pp. 237–80. New York: Academic Press.

GOTTLIEB, G. (1983). Development of species identification in ducklings: X. Perceptual specificity in the wood duck embryo requires sib stimulation for maintenance. *Developmental Psychobiology* **16**: 323–33.

References

GOTTLIEB, G. (1993). Social induction of malleability in ducklings: sensory basis and psychological mechanism. *Animal Behaiour* **45**: 707–19.

GÖTTMARK, F. & ANDERSSON, M. (1984). Colonial breeding reduces nest predation in the common gull. *Animal Behaviour* **32**: 485–92.

GÖTTMARK, F., WINKLER, D.W. & ANDERSSON, M. (1986). Flock-feeding on fish schools increases individual success in gulls. *Nature* **319**: 589–91.

GOULD, J.L. (1975). Honey bee recruitment: the dance language controversy. *Science* **189**: 885–93.

GOULD, J.L., DYER, F.C. & TOWNE, W.F. (1985). Recent progress in understanding the honey bee dance language. *Fortschriften der Zoologie* **31**: 141–61.

GOULD, J.L. & GOULD, C.G. (1988). *The Honey Bee*. New York: Scientific American.

GOULD, J.L. & MARLER, P. (1987). Learning by instinct. *Scientific American* **256**(1): 62–73.

GOULD, S.J. & LEWONTIN, R.C. (1979). The spandrels of San Marco and the Panglossian paradigm: a critique of the adaptationist programme. *Proceedings of the Royal Society of London B* **205**: 581–98.

GOUZOULES, S. & GOUZOULES, H. (1987). Kinship. In *Primate Societies*, ed. B.B. Smuts, D.L. Cheney, R.M. Seyfarth, R.W. Wragham and T.T. Struhsaker, pp. 299–305. Chicago: University of Chicago Press.

GRAFEN, A. (1984). Natural selection, kin selection and group selection. In *Behavioral Ecology: An Evolutionary Approach*, 2nd edition, ed. J.R. Krebs and N.B. Davies, pp. 62–84. Oxford: Blackwell Scientific Publications.

GRAFEN, A. (1990). Biological signals as handicaps. *Journal of Theoretical Biology* **144**: 517–5.

GRAUR, D., DUVT, L. & GOUY, M. (1996). Phylogenetic position of the order Lagomorpha (rabbits, hares and allies). *Nature* **379**: 333–5.

GRAVES, H.B., HABLE, C.P. & JENKINS, T.H. (1985). Sexual selection in *Gallus*. Effects of morphology and dominance on female spatial behaviour. *Behavioural Processes* **11**: 189–97.

GREENBERG, L. (1979). Genetic component of kin recognition in primitively social bees. *Science* **206**: 1095–7.

GRIFFN, D.R. (1984). *Animal Thinking*. Cambridge, Mass.: Harvard University Press.

GROOTHUIS, T.T.G. (1993). The ontogeny of social displays: form development, form fixation and change in context. *Advances in the Study of Behavior* **22**: 26–322.

GROSSMAN, C.J. (1985). Interaction between the gonadal steroids and the immune system. *Science* **227**: 257–61.

GUHL, A.M. (1968). Social inertia and social stability in chickens. *Animal Behaviour* **16**: 219–32.

GUITON, P.E. (1959). Socialization and imprinting in brown leghorn chicks. *Animal Behaviour* **7**: 26–34.

GWADZ, R. (1970). Monofactorial inheritance of early sexual receptivity in the mosquito, *Aëdes atropalpus*. *Animal Behaviour* **18**: 358–61.

HAGEDORN, M. & HEILIGENBERG, W. (1985). Court and spark: electric signals in the courtship of mating of gymnotid fish. *Animal Behaviour* **33**: 254–65.

HALLIDAY, T.R. (1975). An observational and experimental study of sexual behaviour in the smooth newt, *Triturus vulgaris*. *Animal Behaviour* **23**: 291–322.

HALLIDAY, T.R. & SWEATMAN, H.P.A. (1976). To breathe or not to breathe; the newt's problem. *Animal Behaviour* **24**: 551–61.

HAMILTON, W.D. (1964). The genetical evolution of social behaviour. I and II. *Journal of Theoretical Biology* **7**: 1–52.

HAMILTON, W.D. (1971). Geometry for the selfish herd. *Journal of Theoretical Biology* **31**: 295–311.

HAMILTON, W.D., AXELROD, R. & TANSES, R. (1990). Sexual reproduction as an adaptation to resist parasites (a review). *Proceedings of the National Academy of Sciences* USA **87**: 3566–73.

HAMILTON, W.D. & ZUK, M. (1984). Heritable true fitness and bright birds: a role for parasites? *Science* **218**: 384–7.

HAMPTON, N.G., BOLHUIS, J.J. & HORN, G. (1995). Induction and development of a filial predisposition in the chick. *Behaviour* **132**: 451–77.

HARCOURT, R. (1991). The development of play in the South American fur seal. *Ethology* **191**–202.

HARLOW, H.F. (1949). The formation of learning sets. *Psychological Review* **56**: 51–65.

HARLOW, H.F. & HARLOW, M.K. (1965). The affectional systems In *Behaviour of Non-Human Primates*, ed. A.M. Schrier, H.F. Harlow and F. Stollnitz, pp. 287–334. New York: Academic Press.

HARRIS, G.W. & MICHAEL, R.P. (1964). The activation of sexual behaviour by hypothalamic implants of oestrogen. *Journal of Physiology, London* **171**: 275–301.

HARTLINE, H.K., WAGNER, H.G. & RATLIFF, F. (1956). Inhibition in the eye of *Limulus*. *Journal of General Physiology* **39**: 651–73.

HAY, D.A. (1985). *Essentials of Behaviour Genetics*. Oxford: Blackwell.

HEBB, D.O. (1958). Alice in Wonderland or psychology among the biological sciences. In *Biological and Biochemical Bases of Behavior*, ed. H.F. Harlow and C.N. Wolsey, pp. 451–67. Madison: University of Wisconsin Press.

HERBERS, J.M. (1981). Time resources and laziness in animals. *Oceologia* **49**: 252–62.

HERMAN, L.M. & ARBEIT, W.R. (1973). Stimulus control and auditory discrimination learning sets in the bottlenose dolphin. *Journal of Experimental Animal Behaviour* **19**: 379–94.

HERRNSTEIN, R.J, . LOVELAND, D.H. & CABLE, C. (1976). Natural concepts in pigeons. *Journal of Experimental Psychology, Animal Behavior Processes*. **2**: 285–302.

HEYES, C.M. (1995). Self-recognition in primates: further reflections create a hall of mirrors. *Animal Behaviour* **50**: 1533–42.

HINDE, R.A. (1954). Factors governing the changes in strength of a partially inborn response, as shown by the mobbing behaviour of the chaffinch (*Fringilla coelebs*), II. The waning of the response. *Proceedings of the Royal Society of London B* **142**: 331–58.

HINDE, R.A. (1960). Factors governing the changes in strength of a partially inborn response, as shown by the mobbing behaviour of the chaffinch (*Fringilla coelebs*), III. The interaction of short-term and long-term incremental and decremental effects. *Proceedings of the Royal Society of London B* **153**: 398–420.

HINDE, R.A. (1974). *Biological Bases of Human Social Behaviour*. New York: McGraw-Hill.

HINDE, R. A. (1977). Mother–infant separation and the nature of inter-individual relationships. Experiments with rhesus monkeys. *Proceedings of the Royal Society of London B* **196**: 29–50.

HINDE, R.A. (1983). The human species. In *Primate Social Relationships*, ed. R. A. Hinde, pp. 334–9. Oxford: Blackwell.

HINDE, R.A. (1987). Can nonhuman primates help us to understand human behavior? In *Primate Societies*, ed. B.B. Smuts, D.L. Cheney, R.M. Seyfarth, R.W. Wrangham and T.T Struhsaker, pp. 413–20. Chicago: University of Chicago Press.

HINDE, R.A. & FISHER, J. (1952). Further observations on the opening of milk bottles by birds. *British Birds* **44**: 393–6.

HINDE, R.A. & ROWELL, T.E. (1962). Communication by postures and facial expressions in the rhesus monkey (*Macaca mulatta*). *Proceedings of the Zoological Society of London*. **138**: 1–21.

HINDE, R.A. & STEEL, E. (1966). Integration of the reproductive behaviour of female canaries. *Symposia of the Society for Experimental Biology* **20**: 401–26.

HINDE, R.A. & STEVENSON-HINDE, J. (ed.) (1973). *Constraints on Learning*. London: Academic Press.

HITCHCOCK, C.L. & SHERRY, D.F. (1990). Long-term memory for cache sites in the black-capped chickadee. *Animal Behaviour* **40**: 701–12.

HODOS, W. & CAMPBELL, C.B.G. (1969). Scala Naturae: why there is no theory in comparative psychology. *Psychological Review* **76**: 337–50.

HOFFMANN, A.A. (1994). Behaviour genetics and evolution. In *Behaviour and Evolution*, ed. P.J.B. Slater and T.R. Halliday, pp. 7–42. Cambridge: Cambridge University Press.

HOFFMANN, K. (1958). Repetition of an experiment on bird orientation. *Nature* **181**: 1435–7.

HOGAN, J.A. (1967). Fighting and reinforcement in Siamese fighting fish *Betta splendens*. *Journal of Comparative and Physiological Psychology*. **64**: 356–9.

HOGAN, J.A. (1989). The interaction of incubation and feeding in broody junglefowl hens. *Animal Behaviour* **38**: 121–38.

HOLLARD, V.D. & DELIUS, J.D. (1982). Rotational invariance in visual pattern recognition by pigeons and humans. *Science* **218**: 804–6.

HÖLLDOBLER, B. (1971). Communication between ants and their guests. *Scientific American* **224**(3): 86–93.

HÖLLDOBLER, B. & WILSON, E.O. (1990). *The Ants*. London: Springer-Verlag.

HOLMES, W.G. & SHERMAN, P.W. (1982). The ontogeny of kin recognition in two species of ground squirrels. *American Zoology* **22**:491–517.

HOOGLAND, W.G. & SHERMAN, P.W. (1976). Advantages and disadvantages of bank swallow coloniality. *Ecological Monographs* **46**: 33–58.

HOPKINS, C.D. & BASS, A.H. (1981). Temporal coding of species recognition signals in an electric fish. *Science* **212**: 85–7.

HORN, G. (1985). *Memory, Imprinting and the Brain*. Oxford: Clarendon Press.

HORN, G. (1990). Neural bases of recognition memory investigated through an analysis of imprinting. *Philosophical Transactions of the Royal Society of London B* **329**: 133–42.

HORN, G. & HINDE, R.A. (1970). *Short Term Changes in Neural Activity and Behaviour*. Cambridge: Cambridge University Press.

HOTTA, Y. & BENZER, S. (1976). Mapping of behaviour in *Drosophila* mosaics. *Nature* **240**: 527–35.

HOUSTON, A.I. & MCNAMARA, J. (1988). A framework for the functional analysis of behavior. *Behavior and Brain Science* **11**: 117–63.

HUBEL, D.H. & WIESEL, T.N. (1959). Receptive fields, binocular interaction and functional architecture in the cat's visual cortex. *Journal of Physiology* **48**: 574–91.

HUGHES, B.O., DUNCAN, I.J.H. & BROWN, M.F. (1989). The performance of nest-building by domestic hens: is it more important than the construction of the nest? *Animal Behaviour* **37**: 210–14.

HUMPHREY, N.K. (1976). The social function of intellect. In *Growing Points in Ethology*, ed. P.P.G. Bateson and R.A. Hinde, pp. 303–17. Cambridge: Cambridge University Press.

HUNSAKER, D. (1962). Ethological isolating mechanisms in the *Sceloporus torquatus* group of lizards. *Evolution* **16**: 62–74.

References

HUNTINGFORD, F.A. (1976). The relationship between anti-predator behaviour and aggression among conspecifics in the three-spined stickleback, *Gasterosteus aculeatus. Animal Behaviour* **24**: 245–60.

HURST, J. (1989). The complex network of olfactory communication in populations of wild house mice *Mus domesticus* Rutty: markings and investigation within family groups. *Animal Behaviour* **37**: 705–25.

HURST, L.D. & PECK, J.R. (1996) Recent advances in understanding of the evolution and maintenance of sex. *Trends in Ecology and Evolution* **11**(2): 47–51.

HUTCHINSON, J.B. (1976). Hypothalamic mechanisms of sexual behaviour with special reference to birds. *Advances in the Study of Behavior* **6**: 159–200.

HUXLEY, J.S. (1914). The courtship habits of the great crested grebe *Podiceps cristatus. Proceedings of the Zoological Society of London* 1914(2): 491–562.

IMMELMANN, K. (1972). Sexual and other long-term aspects of imprinting in birds and other species. *Advances in the Study of Behavior* **4**: 147–74.

IMMELMANN, K. (1976). The evolutionary significance of early experience. In *Function and Evolution of Behavior*, ed. G. Baerends, C. Beer and A. Manning, pp. 243–53. Oxford: Clarendon Press.

IMMELMANN, K., ROVE, R. LASSEK, R. & BISCHOF, H.-J. (1991). Influence of adult courtship experience on the development of sexual preferences in zebra finch males. *Animal Behaviour* **42**: 83–90.

INCE, S.A., SLATER, P.J.B. & WEISMANN, C. (1980). Changes with time in the songs of a population of chaffinches. *Condor* **82**: 285–90.

ISLAM, M.S., ROESSINGH, P., SIMPSON, S.J. & MCCAVERY, A.R. (1994). Effects of population density experienced by parents during mating and oviposition on the phase of hatchling desert locusts. *Proceedings of the Royal Society of London B* **257**: 93–8.

JACOBS, L.F. (1996). Sexual selection and the brain. *Trends in Ecology and Evolution* **11**(2): 82-6.

JARVIS, J.U.M. (1981). Eusociality in a mammal: cooperative breeding in naked mole rat colonies. *Science* **212**: 571–3.

JARVIS, J.U.M., ORIAIN, J.M., BENNETT, N.C. & SHERMAN, P.W. (1994). Mammalian eusociality: a family affair. *Trends in Ecology and Evolution* **9**(2): 47–51.

JENNI, D.A. (1974). Evolution of polyandry in birds. *American Zoologist* **14**: 129–41.

JERISON, H.J. (1985). Animal intelligence and encephalization. *Philosophical Transactions of the Royal Society of London B* **308**: 21–35.

JOHNSON, J.A. (1987). Dominance rank in juvenile olive baboons, *Papio anubis*: the influence of gender, size, maternal rank and orphaning. *Animal Behaviour* **35**: 1694–1708.

JOHNSON, J.A. (1989). Supplanting by olive baboons: dominance rank difference and resource value. *Behavioral Ecology and Sociobiology* **24**: 277–83.

JOLLY, A. (1966). *Lemur Behavior*. Chicago: University of Chicago Press.

JOLLY, A. (1985). *The Evolution of Primate Behavior*, 2nd edition. New York: Macmillan.

JOSEPH, R. (1979). Effects of rearing and sex on maze learning and competitive exploration in rats. *Journal of Psychology* **101**: 37–43.

KALKO, E.K.V. (1995). Insect pursuit, prey capture and echolocation in pipistrelle bats (Microchiroptera). *Animal Behaviour* **50**: 861–80.

KAMMER, A. & RHEUBEN, E. (1976). Adult motor pattern produced by moth pupae during development. *Journal of Experimental Biology* **65**: 65–84.

KANDEL, E.R. & SCHWARTZ, J.H. (1982). Molecular biology of learning: modulation of transmitter release. *Science* **218**: 433–43.

KANIZSA, G. (1979). *Organization in Vision: Essays on Gestalt Perception*. New York: Praeger Publishers.

KATER, S.B. & ROWELL, C.H.F. (1973). Integration of sensory and centrally programmed components in generation of cyclical feeding activity of *Helisoma trivolvis. Journal of Neurophysiology* **36**: 142–55.

KATZ, L.C. & SHATZ, C.J. (1996). Synaptic activity and the construction of cortical circuits. *Science* **274**: 1133–8.

KEMPENAERS, B., VERHEYEN, G.R., VAN DEN BROECK, M., BURKE, T., VAN BROECKHOVEN, C. & DHONDT, A.A. (1992). Extra-pair paternity results from female preference for high-quality males in the blue tit. *Nature* **37**: 494–6.

KENNEDY, J.S. (1965). Coordination of successive activities in an aphid. Reciprocal effects of settling on flight. *Journal of Experimental Biology* **43**: 489–509.

KENNEDY, J.S. (1992). *The New Anthropomorphism*. Cambridge: Cambridge University Press.

KESHISHIAN, H. & CIBA, A. (1993). Neuromuscular development in *Drosophila*: insights from single neurons and single genes. *Trends in the Neurosciences* **16**(7): 279–83.

KETTERSON, E.D. & NOLAN, V., JR (1994). Hormones and life histories: an integrated approach. In *Behavioral Mechanisms in Evolutionary Ecology*, ed. L.A. Real, pp. 327–53. Chicago: University of Chicago Press.

KEVERNE, E.B. (1985). Reproductive behaviour. In *Reproduction in Mammals*, Vol. 4. *Reproductive Fitness*, 2nd edition, ed. C.R. Austin and R.V. Short, pp. 133–75. Cambridge: Cambridge University Press.

KEVERNE, E.B. (1994). Molecular-genetic approaches to understanding brain development and behaviour. *Psychoneuroendocrinology* **19**: 407–14.

KILNER, R. & JOHNSTONE, R.A. (1997). Begging the question: are offspring solicitation behaviours signals of need? *Trends in Ecology and Evolution* **12**(1): 11–15.

KLOPFER, P.H. & GAMBLE, J. (1966). Maternal 'imprinting' in goats; the role of chemical senses. *Zeitschrift für Tierpsychologie* **23**: 588–92.

KNUDSEN, E.I. & KONISHI, M. (1979). Mechanisms of sound location in the barn owl (*Tyto alba*). *Journal of Comparative Physiology* **133**: 13–21.

KÖHLER, W. (1927). *The Mentality of Apes*, 2nd edition. London: Kegan Paul.

KOLTERMANN, R. (1971). 24-Std.-Periodik der Langzeiterinnerung an Duft- und Farbsignale bei der Honigbiene. *Zeitschrift für Vergleichende Physiologie.* **75**: 49–68.

KOMISARUK, B.R. (1967). Effects of local brain implants of progesterone on reproductive behavior in ring doves. *Journal of Comparative and Physiological Psychology* **64**; 219–24.

KOMISARUK, B.R., ADLER, N.T. & HUTCHISON, J. (1972). Genital sensory field: enlargement by oestrogen treatment in female rats. *Science* **178**: 1295–8.

KONISHI, M. (1965). The rôle of auditory feedback on the control of vocalization in the white-crowned sparrow. *Zeitschrift für Tierpsychologie* **22**: 770–83.

KONORSKI, J. (1948). *Conditioned Reflexes and the Nervous System*. Cambridge: Cambridge University Press.

KORTLANDT, A. (1940). Eine Übersicht der angeborren Verhaltensweisen des Mitteleuropäischen Kormorans *Phalacrocorax carbo sinensis*. *Archives neêrlandaises de Zoologie.* **14**: 401–42.

KREBS, J.R., CLAYTON, N.S., HEALY, S.D., CRISTOL, D.A., PATEL, S.N. & JOLLIFFE, A.R. (1996). The ecology of the avian brain: food-storing memory and the hippocampus. *Ibis* **138**: 34–46.

KREBS, J.R. & DAWKINS, R. (1984). Animal signals: mind reading and manipulation. In *Behavioural Ecology: An Evolutionary Approach*, 2nd edition, ed. J.R. Krebs and N.B. Davies, pp. 380–402. Oxford: Blackwell Scientific Publications.

KREBS, J.R., MACROBERTS, M.H. & CULLEN, J.M. (1972). Flocking and feeding in the great tit, *Parus major* – an experimental study. *Ibis* **114**: 507–30.

KREBS, J.R., SHERRY, D.F., HEALY, S.D., PERRY, V.H. & VACCARINO, A.L. (1989). Hippocampal specialization of food-storing birds. *Proceedings of the National Academy of Sciences USA* **86**: 1388–92.

KROODSMA, D.E. (1982). Learning and the ontogeny of sound signals in birds. In *Acoustic Communication in Birds*, ed. D.E. Kroodsma and D.H. Miller, pp. 1–23. New York: Academic Press.

KRUIJT, J.P. & MEEUWISSE, G.B. (1991). Sexual preferences of male zebra finches: effects of early and adult experience. *Animal Behaviour* **42**: 91–102.

KRUUK, H. (1972). *The Spotted Hyaena*. Chicago: University of Chicago Press.

KUMMER, H. (1968). Two variations in the social organization of baboons. In *Primates: Studies in Adaptance and Variability*, ed. P. Jay, pp. 293–312. New York: Holt, Rinehart & Winston.

KUMMER, H., GOTZ, W. & ANGST, W. (1974). Triadic differentiation: an inhibitory process protecting pair bands in baboons. *Behaviour* **49**: 62–87.

LACK, D. (1943). *The Life of the Robin*. London: H.F. & G. Witherby.

LADE, B. & THORPE, W.H. (1964). Dove songs as innately coded patterns of specific behaviour. *Nature* **202**: 66–8.

LAGERSPETZ, K.M.J. & LAGERSPETZ, K.Y.H. (1974). Genetic determination of aggressive behaviour. In *The Genetics of Behaviour*, ed. J.H.F. van Abeelen. New York: Elsevier.

LALAND, K.N. & PLOTKIN, H.C. (1990). Social learning and social transmission of foraging information in Norway rats (*Rattus norvegicus*). *Animal Learning and Behavior* **18**: 246–51.

LALAND, K.N., RICHARDSON, P.J. & LLOYD, J.E. (1993). Animal social learning: towards a new theoretical approach. *Perspectives in Ethology* **10**: 249–77. New York: Plenum.

LALL, A.B., SELIGAR, B.H., BIGGLEY, W. & LLOYD, J.E. (1980). Ecology of colors of firefly bioluminscence. *Science* **210**: 560–2.

LAND, M.F. (1981). Optics and vision in invertebrates. In *Handbook of Sensory Physiology*, ed. H. Autrum, pp. 471–593. New York: Springer-Verlag.

LASHLEY, K.S. (1950). In search of the engram. *Symposia of the Society for Experimental Biology* **4**: 454–82.

LAZARUS, J. (1979). The early warning function of flocking in birds: an experimental study with captive *Quelea*. *Animal Behaviour* **27**: 855–65.

LE BOEUF, B.J. (1974). Male–male competition and reproductive success in elephant seals. *American Zoologist* **14**: 163–76.

LE BOEUF, B.J. & PETERSON, R.S. (1969). Social status and mating activities in elephant seals. *Science* **163**: 91–3.

LEE, P. (1983). Play as a means for developing relationships. In *Primate Social Relationships*, ed. R.A. Hinde, pp. 82–9. Oxford: Blackwell.

LEE, P. (1987). Allomothering among African elephants. *Animal Behaviour* **35**: 278–91.

LE GROS CLARK, W.E. (1960). *The Antecedents of Man*. Chicago: Quadrangle Books.

LEHRMAN, D.S. (1964). The reproductive behavior of ring doves. *Scientific American* **211**(5): 48–54.

LEIGHTON, D.R. (1987). Gibbons: territoriality and monogamy. In *Primate Societies*, ed. B.B. Smuts, D.L. Cheney, R.M. Seyfarth, R.W. Wrangham and T.T. Struhsaker, pp. 135–45. Chicago: University of Chicago Press.

LENINGTON, S. (1994). Of mice, men and the MHC. *Trends in Ecology and Evolution* **9**: 455–6.

LENINGTON, S. & EGID, K. (1989). Environmental influences on the preferences of wild female house mice for males of differing t-complex genotypes. *Behavior Genetics* **19**: 257–166.

LETTVIN, J.Y., MATURANA, H.R., MCCULLOCH, W.S. & PITTS, W.H. (1959). What the frog's eye tells the frog's brain. *Proceedings of the Institute of Radio Engineers* **47**: 1940–51.

LEVINE, R.B. (1986). Reorganization of the insect nervous system during metamorphosis. *Trends in the Neurosciences* **9**: 315–19.

LEVINE, R.B. & TRUMAN, J.W. (1985). Dendritic reorganisation of abdominal motoneurons during metamorphosis of the moth *Manduca sexta*. *Journal of Neuroscience* **5**: 2424–31.

LINDAUER, M. (1961). *Communication among Social Bees*. Cambridge, Mass.: Harvard University Press.

LINDAUER, M. (1976). Evolutionary aspects of orientation and learning. In *Function and Evolution in Behaviour*, ed. G. Baerends, C. Beer and A. Manning, pp. 228–42. Oxford: Clarendon Press.

LLOYD, J.E. (1965). Aggressive mimicry in *Photuris* firefly *femmes fatales*. *Science*. **149**: 653–4.

LLOYD, J.E. (1975). Aggressive mimicry in *Photuris* fireflies: signal repertoires by *femmes fatales*. *Science* **187**: 452–3.

LOFSTEDT, C., VICKERS, N.J., ROELOFS, W.L. & BAKER, T.C. (1989). diet related courtship success in the Oriental fruit moth *Grapholita molesta* (Tortricidae). *Oikos* **55**; 1402–8.

LORENZ, K. (1937). The companion in the bird's world. *Auk* **54**: 245–73.

LORENZ, K. (1941). Vergleichende Bewegungsstudien an Anatinen. *Suppl. J. Ornith.*, **89**: 194–294. (An English translation appeared in several parts, in Vols. 57–59 of *Avicultural Magazine*.)

LORENZ, K. (1952). *King Solomon's Ring*. London: Methuen.

LORENZ, K.Z. (1966a). *Evolution and Modification of Behaviour*. London: Methuen.

LORENZ, K. (1966b). *On Aggression*. London: Methuen.

LURIA, A.R. (1975). *The Mind of a Mnemonist*. London: Penguin.

LYTHGOE, J.N. (1979). *The Ecology of Vision*. Oxford: Clarendon Press.

MCBRIDE, G., PARER, I.P., & FOENANDER, F. (1959). The social organization and behaviour of the feral domestic fowl. *Animal Behaviour Monographs* **2**: 127–81.

MCFARLAND, D.J. (1971). *Feedback Mechanisms in Animal Behaviour*. New York: Academic Press.

MCFARLAND, D.J. (1974). Time-sharing as a behavioural phenomenon. *Advances in the Study of Behaviour* **5**: 201–25.

MCFARLAND, D.J. & LOYD, I. (1973). Time-shared feeding and drinking. *Quarterly Journal of Experimental Psychology* **25**: 48–61.

MCFARLAND, D.J. & SIBLY, R.M. (1975). The behavioural final common path. *Philosophical Transactions of the Royal Society of London B* **270**: 265–93.

MCGREW, W.C. (1992). *Chimpanzee Material Culture*. Cambridge: Cambridge University Press.

MCGREW, W.C., MARCHANT, L.F. & NISHIDA, T. (eds.) (1996). *Great Ape Societies*. Cambridge: Cambridge University Press.

MACDONALD, D.W. (1986). A meerkat volunteers for guard duty so its comrades can live in peace. *Smithsonian* 1986 (April): 55–64.

MACKINTOSH, N.J. (1965). Discrimination learning in the octopus. *Animal Behaviour* **13**(Suppl. 1): 129–134.

MACKINTOSH, N.J. (1983). General principles of learning. In *Animal Behaviour*, Vol. 3. *Genes, Development and Learning*, ed. T.R. Halliday and P.J.B. Slater, pp. 149–77. Oxford: Blackwell Scientific Publications.

MACKINTOSH, N.J., WILSON, B. & BOAKES, R.A. (1985). Differences in mechanisms of intelligence among vertebrates. *Philosophical Transactions of the Royal Society of London B* **308**: 53–65.

MACPHAIL, E. (1985). Vertebrate intelligence: the null hypothesis. *Philosophical Transactions of the Royal Society of London B* **308**: 37–51.

MACPHAIL, E.M. (1987). The comparative psychology of intelligence. *Behavior and Brain Science* **10**: 645–95.

MAGUIRE, E.A., FRACKOWIACK, R.S.J. & FRITH, C.D. (1996). Learning to find your way: a role for the human hippocampal formation. *Proceedings of the Royal Society of London B* **263**: 1745–50.

MAIER, N.R.F. (1932). Cortial destruction of the posterior part of the brain and its effects on reasoning in rats. *Journal of Comparative Neurology* **56**: 179–214.

MAIER, N.R.F. & SCHNEIRLA, T.C. (1935). *Principles of Animal Psychology*. New York: McGraw-Hill.

MAIER, S.F., WATKINS, L.R. & FLESHNER, M. (1994). Psychoneuroimmunology. *American Psychologist* **49**: 1004–17.

References

MANGER, P.R. & PETTIGREW, J.D. (1995). Electroreception and the feeding behaviour of platypus (*Ornithorhynchus anatinus*): Monotremata: Mammalia). *Philosophical Transactions of the Royal Society of London B* **347**: 359–81.

MANNING, A. (1961). The effects of artificial selection for mating speed in *Drosophila melanogaster*. *Animal Behaviour* **9**: 82–92.

MANNING, A. & MCGILL, T.E. (1974). Neonatal androgen and sexual behavior in female house mice. *Hormones and Behavior* **6**: 19–31.

MARCHETTI, K. (1993). Dark habitats and bright birds illustrate the role of environment in species divergence. *Nature* **362**: 149–52.

MARKL, L. (1977). Adaptive radiation of mechanoreception. In *Sensory Ecology*, ed. M.A. Ali, pp. 49–69. New York: Plenum.

MARLER, P. & PETERS, S. (1977). Selective vocal learning in a sparrow. *Science* **198**: 519–21.

MARLER, P. & PICKERT, R. (1984). Species-universal microstructure in a learned birdsong: the swamp sparrow (*Melospiza georgiana*). *Animal Behaviour* **32**: 673–89.

MARLER, P. & TAMURA, M. (1964). Culturally transmitted patterns of vocal behavior in sparrows. *Science* **146**: 1483–6.

MARSHALL, D.A. & MOULTON, D.G. (1981). Olfactory sensitivity to α-ionone in humans and dogs, *Chemical Senses* **6**: 53–61.

MARTIN, G.M. & LETT, B.T. (1985). Formation of associations of colored and flavoured food with induced sickness in five main species. *Behavioral and Neural Biology* **43**: 223–37.

MARTIN, P. & BATESON, P. (1985a). The influence of experimentally manipulating a component of weaning on the development of play in domestic cats. *Animal Behaviour* **33**: 511–18.

MARTIN, P. & BATESON, P. (1985b). The ontogeny of locomotor play behaviour in the domestic cat. *Animal Behaviour* **33**: 502–10.

MARTIN, P. & CARO, T.M. (1985). On the functions of play and its role in behavioral development. *Advances in the Study of Behaviour* **15**: 59–103.

MARTIN, R.D. (1968a). Reproduction and ontogeny in tree-shrews (*Tupaia belangeri*) with reference to their general behaviour and taxonomic relationships. *Zeitschrift für Tierpsychologie* **25**: 409–95.

MARTIN, R.D. (1968b). Towards a new definition of primates. *Man* **3**: 377–401.

MARTIN, R.D. (1990). *Primate Origins and Evolution*. London: Chapman & Hall.

MASON, G.J. (1991). Stereotypies: a critical review. *Animal Behaviour* **41**: 1015–37.

MASON, G.J. & TURNER, M.A. (1993). Mechanisms involved in the development and control of stereotypies. In: *Perspectives in Ethology*, Vol. 10, ed. P.P.G. Bateson, P.H. Klopfer and N.S. Thompson, pp. 53–85. New York: Plenum.

MASSERMAN, J.H. (1950). Experimental neuroses. *Scientific American* **182**(3): 38–43.

MASTERS, W.M. & MARKL, H. (1981). Vibration signal transmission in spider orb web. *Science* **213**: 363–5.

MASTERS, W.M. & MOFFAT, A.J.M. (1986). Transmission of vibration in a spider's web. In *Spiders*, ed. W.A. Shear, pp. 49–69. Stanford, Calif.: Stanford University Press.

MAY, D.J. (1949). Studies on a community of willow warblers. *Ibis* **91**: 24–54.

MAYNARD SMITH, J. (1964). Group selection and kin selection. *Nature* **201**: 1145–7.

MAYNARD SMITH, J. (1978). Optimization theory in evolution. *Annual Review of Ecology and Systematics* **9**: 31–56.

MAYNARD SMITH, J. (1982). *Evolution and the Theory of Games*. Cambridge: Cambridge University Press.

MAYNARD SMITH, J. & PARKER, G.A. (1976). The logic of asymmetrical contests. *Animal Behaviour* **24**: 159–75.

MAYNARD SMITH, J. & PRICE, G.R. (1973). The logic of animal conflict. *Nature* **246**: 15–18.

MAYNARD SMITH, J. & RIECHERT, S.E. (1984). A conflicting-tendency model of spider agonistic behaviour: hybrid–pure population line comparisons. *Animal Behaviour* **32**: 564.

MEDDIS, R. (1983). The evolution of sleep. In *Sleep Mechanisms and Functions*, ed. A. Mayes, pp. 57–106. London: Van Nostrand.

MENZEL, R. & ERBER, J. (1978). Learning and memory in bees. *Scientific American* **239**(1): 80–7.

MENZEL, R., GREGGERS, U. & HAMMER, M. (1993). Functional organization of appetitive learning and memory in a generalist pollinator, the honey bee. In *Insect Learning: Ecological and Evolutionary Perspectives*, ed. D. Papaj and A. Lewis, pp. 79–125. London: Chapman & Hall.

METCALFE, N.B. & FURNESS, R.W. (1984). Changing priorities: the effect of pre-migratory fattening on the trade-off between foraging and vigilance. *Behavioral Ecology and Sociobiology* **15**: 203–6.

MEYER, A., MORRISSEY, J. & SCHARTL, M. (1994). Molecular phylogeny of fishes of the genus *Xiphophorus* suggests repeated evolution of a sexually selected trait. *Nature* **368**: 539–41.

MICHELSON, A. (1989). Ein mechanisches Modell der tanzenden Honigbiene. *Biologie in unserer Zeit* **19**(4): 121–6.

MICHELSON, A., ANDERSEN, B.B., STORM, J., KIRCHNER, W.H. & LINDAUER, M. (1992). How honeybees perceive communication dances, studied by means of a mechanical model. *Behavioral Ecology and Sociobiology* **30**: 143–50.

MILINSKI, M. (1984). A predator's cost of overcoming the confusion effect of swarming prey. *Animal Behaviour* **32**: 1157–62.

MILINSKI, M. & HELLER, R. (1978). Influence of a predator on the optimal foraging behaviour of sticklebacks (*Gasterosteus aculeatus*). *Nature* **275**: 642–4.

MILLER, N.E. (1956). Effects of drugs on motivation: the value of using a variety of measures. *Annals of the New York Academy of Sciences* **65**: 318–33.

MILLER, N.E. (1957). Experiments on motivation. Studies combining psychological, physiological and pharmacological techniques. *Science* **126**: 1271–8.

MILLER, N.E., SAMPLINER, R.I. & WOODROW, P. (1957). Thirst reducing effects of water by stomach fistula versus water by mouth, measured by both a consummatory and an instrumental response. *Journal of Comparative and Physiological Psychology* **50**: 1–5.

MILLS, M.G.L. & GORMAN, M.L. (1987). The scent-marking behaviour of the spotted hyaena *Crocuta crocuta* in the Southern Kalahari. *Journal of Zoology, London* **212**: 483–97.

MISHKIN, M. & APPENZELLER, T. (1987). The anatomy of memory. *Scientific American* **256**(6): 62–71.

MITCHELL, P. & THOMPSON, N. (eds.) (1985). *Deception: Perspectives on Human and Nonhuman Deceit*. New York: State University of New York Press.

MITTERMEIER, R.A. & CHENEY, D.L. (1987). Conservation of primates and their habitats. In *Primate Societies*, ed. B.B. Smuts, D.L. Cheney, R.M. Seyfarth, R.W. Wrangham and T.T. Struhsaker, pp. 477–90. Chicago: University of Chicago Press.

MOCK, D.W., LAMEY, T.C., WILLIAMS, C.F. & PELLETIER, A. (1987). Flexibility in the development of heron sibling aggression: an intraspecific test of the prey-size hypothesis. *Animal Behaviour* **35**: 1386–93.

MOEHLMAN, P.D. (1979). Jackal helpers and pup survival. *Nature* **277**: 382–3.

MØLLER, A. (1987). Social control of deception among status signalling house sparrows. *Behavioral Ecology and Sociobiology* **20**: 307–11.

MØLLER, A. (1988). Female choice selects for male sexual tail ornaments in the monogamous swallow. *Nature* **332**: 640–2.

MØLLER, A. (1990). Effects of a haematophagous mite on the barn swallow *Hirundo rustica*: a test of the Hamilton & Zuk hypothesis. *Evolution* **44**: 771–84.

MONEY, J. & EHRHARDT, A.A. (1973). *Man and Woman, Boy and Girl*. Baltimore, Md.: Johns Hopkins University Press.

MOORE, B.R. (1973). The role of directed Pavlovian reactions in simple instrumental learning in the pigeon. In *Constraints on Learning*, ed. R.A. Hinde and J. Stevenson-Hinde, pp. 159–88. London: Academic Press.

MORGAN, L. (1894). *An Introduction to Comparative Psychology*. London: Scott.

MORRIS, D. (1958). Homosexuality in the ten-spined stickleback (*Pygosteus pungitius* L.). *Behaviour* **4**: 233–61.

MORRIS, D. (1959). The comparative ethology of grassfinches (*Erythrurae*) and mannikins (*Amadinae*). *Proceedings of the Zoological Society of London* **131**: 389–439.

MORRIS, R.G.M. (1983). Modelling amnesia and the study of memory in animals. *Trends in the Neurosciences* **6**: 479–83.

MORRIS, R.G.M. (1989). Synaptic plasticity, neural architecture and forms of memory. In *Brain Organisation and Memory: Cells, Systems and Circuits*, ed. J.L. McGaugh, N.M. Weinberger and G. Lynch. New York: Oxford University Press.

MORRIS, R.G.M., GARRUD, P., RAWLINS, J.N.P. & O'KEEFE, J. (1982). Place navigation impaired in rats with hippocampal lesions. *Nature* **297**: 681–3.

MOSS, C. (1992). *Echo of the Elephants*. London: BBC Books.

MOYNIHAN, M. (1967). Comparative aspects of communication in New World primates. In *Primate Ethology*, ed. D. Morris, pp. 236–66. London: Weidenfeld & Nicholson.

MUNN, C.A. (1986). Birds that 'cry wolf'. *Nature* **319**:143–5.

MUNN, N.L. (1950). *Handbook of Psychological Research on the Rat*. Boston: Houghton Mifflin.

MURPHEY, R.K. (1985). Competition and chemoaffinity in insect sensory systems. *Trends in the Neurosciences* **8**: 120–5.

MURPHEY, R.K. (1986). The myth of the inflexible invertebrate: competition and synaptic remodelling in the development of invertebrate nervous systems. *Journal of Neurobiology* **17**: 585–91.

NARINS, P.M. & CAPRANICA, R.R. (1976). Sexual differences in the auditory system of the tree frog, *Eleutherodactylus coqui*. *Science* **192**: 378–80.

NELSON, B. (1980). *Seabirds, their Biology and Ecology*. London: Hamlyn.

NEWMAN, E.A. & HARTLINE, P.H. (1982). The infrared 'vision' of snakes. *Scientific American* **246**(3): 98–107.

NOBERG, R.A. (1994). Swallow tail streamer is a mechanical device for self-deflection of tail leading edge, enhancing aerodynamic efficiency and flight manoeuvrability. *Proceedings of the Royal Society of London B* **257**: 227–33.

NOTTEBOHM, F. (1989). From bird song to neurogenesis. *Scientific American* **260**(2): 56–61.

OLSON, G.C. & KRASNE, F.B. (1981). The crayfish lateral giants are command neurons for escape behaviour. *Brain Research* **214**: 89–100.

OPPENHEIM, R.W. (1974). The ontogeny of behavior in the chick embryo. *Advances in the Study of Behavior* **5**: 133–72.

OPPENHEIM, R.W. (1981). Ontogenetic adaptations and retroprogressive processes in the development of the nervous system and behaviour. In *Maturation and Development: Biological and Psychological Perspectives*, ed. K.J. Connolly and H.F.R. Prechtl, pp. 73–109. Philadelphia: J.P. Lippincott.

ORIANS, G. (1969). On the evolution of mating systems in birds and mammals. *American Naturalist* **103**: 589–603.

ORIANS, G. (1980). *Some Adaptations of Marsh-Nesting Blackbirds*. Princeton, NJ: Princeton University Press.

PACKER, C. (1977). Reciprocal altruism in *Papio anubis*. *Nature* **265**: 441–3.

PACKER, C. (1986). The ecology of sociality in felids. In: *Ecological Aspects of Social Evolution*, ed. D.I. Rubenstein and R.W. Wrangham, pp 429–51. Princeton, NJ: Princeton University Press.

PAPI, F. (ed.) (1992). *Animal Homing*. London: Chapman & Hall.

PARTRIDGE, B.L. & PITCHER, T.J. (1979). Evidence against a hydrodynamic function for fish schools. *Nature* **279**: 418–19.

PARTRIDGE, B.L. & PITCHER, T.J. (1980). The sensory basis of fish schools: relative roles of lateral line and vision. *Journal of Comparative Physiology* **135**: 315–25.

PARTRIDGE, L. (1983). Genetics and behaviour. In *Animal Behaviour*, Vol. 3. *Genes and Development and Learning*, ed. T.R. Halliday and P.J.B. Slater, pp. 11–51. Oxford: Blackwell Scientific Publications.

PAUL, R.C. AND WALKER, T.J. (1979). Arboreal singing in a burrowing cricket, *Anurogryllus arboreus*. *Journal of Comparative Physiology* **132**: 217–23.

PAVLOV, I.P. (1941). *Lectures on Conditioned Reflexes*. 2 vols. New York: International Publishers.

PAYNE, R. (1971). Acoustic location of prey by barn owls (*Tyto alba*). *Journal of Experimental Biology* **54**: 535–73.

PAYNE, R.S. & MCVAY, S. (1971). Songs of hump-back whales. *Science* **173**: 585–97.

PAYNE, T.L., BIRCH, M.C. & KENNEDY, C.E.J. (eds.) (1986). *Mechanisms in Insect Olfaction*. Oxford: Clarendon Press.

PEARCE, J.M. (1996). *Animal Learning and Cognition: an Introduction*, 2nd edition. Hove, UK: Psychology Press.

PEPPERBERG, I.M. (1990). Some cognitive abilities of an African grey parrot (*Psittacus erithacus*). *Advances in the Study of Behavior* **19**: 357–409.

PEPPERBERG, I.M. (1991). A communicative approach to animal cognition: a study of conceptual abilities of an African grey parrot. In *Cognitive Ethology: the Minds of other Animals*, ed. C.A. Ristau, pp. 153–86. Hillsdale, NJ: Lawrence Erlbaum Associates.

PERRY, E.A. & STENSON, G.B. (1992). Observations of nursing behaviour of hooded seals, *Cystophora cristata*. *Behaviour* **122**: 1–10.

PICMAN, J., LEONARD, M. & HORN, A. (1988). Anti-predation role of clumped nesting by marsh-nesting red-winged blackbirds. *Behavioral Ecology and Sociobiology* **22**: 9–15.

PINXTHEN, R., HANOOK, O., EENS, M., VERHEYEN, R.F., DHONDT, A.A. & BURKE, T. (1993). Extra-pair paternity and intraspecific brood parasitism in the European starlings *Sturnus vulgaris*: evidence from DNA fingerprinting. *Animal Behaviour* **45**: 795–809.

POOLE, J.H., PAYNE, K., LANGBAUER, W.R. & MOSS, C.J. (1988). The social contexts of some very low frequency calls of African elephants. *Behavioral Ecology and Sociobiology* **22**: 385–92.

POVINELLI, D.J., NELSON, K.E. & BOYSEN, S.T. (1992). Comprehension of role reversal in chimpanzees: evidence of empathy? *Animal Behaviour* **43**: 633–40.

POWELL, G.V.N. (1974). Experimental analysis of the social value of flocking by starlings (*Sturnus vulgaris*) in relation to predation and foraging. *Animal Behaviour* **22**: 501–5.

PREMACK, D. & WOODRUFF, G. (1978). Does the chimpanzee have a theory of mind? *Behavior and Brain Science* **1**: 515–26.

PROCTOR, H. (1991). Courtship behaviour in the water mite *Neumania papillator*. *Animal Behaviour* **42**: 589–98.

PROVINE, R.R. (1976). Eclosion and hatching in cockroach first instar larvae: a stereotyped pattern of behaviour. *Journal of Insect Physiology* **22**: 127–31.

PUSEY, A.E. & PACKER, C. (1987). The evolution of sex-biased dispersal in lions. *Behaviour* **101**: 275–310.

QUELLER, D.C., STRASSMAN, J.E. & HUGHES, C.R. (1993). Microsatellites and kinship. *Trends in Ecology and Evolution* **8**(8): 285–00.

RALLS, K. (1971). Mammalian scent marking. *Science* **171**: 443–9.

RAMSAY, A.O. & HESS, E.H. (1954). A laboratory approach to the study of imprinting. *Wilson Bulletin* **66**: 196–206.

RAND, A.S. & RAND, W.M. (1978). Display and dispute settlement in nesting iguanas. In *Behavior and Neurology of Lizards*, ed. N. Greenberg and P.O. MacLean, pp. 245–52. Rockville, Md.: National Institute of Mental Health.

RESCORLA R.A. (1988). Pavlovian conditioning: it's not what you think it is. *American Psychologist* **43**: 151–60.

REYNOLDS, J. (1996). Animal breeding systems. *Trends in Ecology and Evolution* **11**(2): 68–72.

RIDLEY, M. (1986). The number of males in a primate troop. *Animal Behaviour* **34**: 1848–58.

References

RIECHERT, S. (1984). Games spiders play. III Cues underlying context-associated changes in agonistic behaviour. *Animal Behaviour* 32: 1–15.

RIEDMAN, M. (1990). *The Pinnipeds: Seals, Sea Lions and Walruses*. Berkeley: University of California Press.

RISTAU, C.A. (ed.) (1991). *Cognitive Ethology: the Minds of other Animals*. Hillsdale, NJ: Lawrence Erlbaum Associates.

ROEDER, K.D. (1967). *Nerve Cells and Insect Behavior*. Cambridge, Mass.: Harvard University Press.

ROESSINGH, P., SIMPSON, S.J. & JAMES, S. (1993). Analysis of phase-related changes in behaviour of desert locust nymphs. *Proceedings of the Royal Society of London B* 252: 43–9.

ROLLS, B.J. & ROLLS, E.T. (1982). *Thirst*. Cambridge: Cambridge University Press.

ROLLS, E.T. (1994). Brain mechanisms for invariant visual recognition and learning. *Behavioural Processes* 33: 113–38.

ROMER, A.S. (1962). *The Vertebrate Body*. Philadelphia: W.B. Saunders.

ROPER, T.J. (1983). Learning as a biological phenomenon. In *Animal Behavior*, Vol 3. *Genes, Development and Learning*, ed. P.J.B. Slater and T.R. Halliday, pp. 178–212. Oxford: Blackwell Scientific Publications.

ROPER, T.J. (1984). Response of thirsty rats to absence of water: frustration, disinhibition or compensation? *Animal Behaviour* 32: 1225–35.

ROSE, S. (1992), *The Making of Memory*. London: Bantam Books.

ROSENBLATT, J.S. AND SIEGEL, H.I. (1983). Physiological and behavioural changes during pregnancy and parturition underlying the onset of maternal behaviour in rodents. In *Parental Behaviour of Rodents*, ed. R.W. Elwood, pp. 23–66. Chichester: John Wiley.

ROSENZWEIG, M.R. (1984). Experience, memory and the brain. *American Psychologist* 39: 65–76.

ROSSEL, S. & WEHNER, R. (1986). Polarization vision in bees. *Nature* 323: 128–31.

ROTH, L.M. (1948). An experimental laboratory study of the sexual behaviour of *Aëdes aegypti* (L.). *American Midland Naturalist* 40: 265–352.

ROWELL, C.H.F. (1961). Displacement grooming in the chaffinch. *Animal Behaviour* 9: 38–63.

ROWELL, T. E. (1974). The concept of social dominance. *Behavioral Biology* 2: 131–54.

ROWLAND, W.J. (1989). Mate choice and the supernormality effect in female sticklebacks (*Gasterosteus aculeatus*). *Behavioral Ecology and Sociobiology* 24: 433–8.

ROWLEY, I. & CHAPMAN, G. (1986). Cross-fostering, imprinting and learning in two sympatric species of cockatoos. *Behaviour* 96: 1–16.

ROZIN, P. & KALAT, J.W. (1971). Specific hungers and poison avoidance as adaptive specializations of learning. *Psychological Review* 78: 459–86.

RUTTER (1979). Maternal deprivation 1972–1978: new findings, new concepts, new approaches. *Child Development* 50: 283–305.

RYAN, M.J., TUTTLE, M.D. & RAND, A.S. (1982). Bat predation and sexual advertisement in a neotropical anuran. *American Naturalist* 119: 136–9.

RYAN, M.J., TUTTLE, M.D. & TAFT, L.K. (1981). The costs and benefits of frog chorusing behavior. *Behavioral Ecology and Sociobiology* 8: 273–8.

RYAN, M.J. & WILCZYNSKI, W. (1991). Evolution of intra-specific variation in the advertisement call of the cricket frog (*Acris crepitans*, Hylidae). *Biological Journal of the Linnean Society* 44: 249–71.

SACHSER, N. & KAISER, S. (1996). Prenatal social stress masculinizes the females' behavior in guinea-pigs. *Physiological Behaviour* 60: 589–94.

SACKS, O. (1986). *The Man Who Mistook his Wife for a Hat*. London: Picador.

SAINO, N. (1994). Time budget variation in relation to flock size in carrion crows *Corvus corone corone*. *Animal Behaviour* 47: 1189–96.

SALES, G. & PYE, D. (1974). *Ultrasonic Communication by Animals*. London: Chapman & Hall.

SAVAGE-RUMBAUGH, S., RUMBAUGH, D.M. & BOYSEN, S. (1978). Linguistically mediated tool use and exchange by chimpanzees *Pan troglodytes*. *Behavior and Brain Science* 1: 539–54.

SAVAGE-RUMBAUGH, E.S., WILLIAMS, S.L., FURUICHI, T. & KANO, T. (1996). Language perceived: *Paniscus* branches out. In *Great Ape Societies*, ed. W.C. McGrew, L.F. Marchant and T. Nishida, pp. 173–84. Cambridge: Cambridge University Press.

SCHALLER, G. (1972). *The Serengeti Lion: a Study of Predator Prey Relations*. Chicago: University of Chicago Press.

SCHEIH, H., LANGNER, G., TIDEMANN, C., COLES, R.B. & GUPPY, A. (1986). Electroreception and electrolocation in platypus. *Nature* 319: 401–2.

SCHELLER, R.H. & AXEL, R. (1984). How genes control an innate behavior. *Scientific American* 250(3): 44–52.

SCHJELDERUP-EBBE, T. (1935). Social behaviour of birds. In *Handbook of Social Psychology*, ed. C. Murchison, pp. 947–72. Worcester, Mass.: Clark University Press.

SCHLEIDT, W.M., SCHLEIDT, M. & MAGG, M. (1960). Störungen der Mutter-Kind-Beziehung bei Truthühnern durch Gehörverlust. *Behaviour* 16: 254–60.

SCOTT, N.P. (1962). Critical periods in behavioral development. *Science* **138**: 949–58.

SCOTT, J.P. & FULLER, J.L. (1965). *Genetics and the Social Behavior of the Dog*. Chicago: University of Chicago Press.

SEARCY, W.A. & BRENOWITZ, E.A. (1988). Sexual differences in species recognition of avian song. *Nature* **332**: 152–4.

SEARCY, W.A. & MARLER, P. (1981). A test for responsiveness to song structure and programming in female sparrows. *Science* **213**: 926–8.

SEARCY, W.A., MARLER, P. & PETERS, S.S. (1981). Species song discrimination in adult female song and swamp sparrows. *Animal Behaviour* **29**: 997–1003.

SEELEY, T.D. (1992). The tremble dance of the honey bee: message and meanings. *Behavioral Ecology and Sociobiology* **31**(6): 375–83.

SELIGMAN, M.E.P. & HAGER, J.L. (eds.) (1972). *Biological Boundaries of Learning*. Englewood Cliffs, NJ: Prentice-Hall.

SELYE, H. (1973). The evolution of the stress concept. *American Scientist* **61**: 692–9.

SEVENSTER, P. (1961). A causal analysis of a displacement activity (fanning in *Gasterosteus aculeatus* L.). *Behaviour* (*Suppl.*) **9**: 1–170.

SEVENSTER-BOL, A.C.A. (1962). On the causation of drive reduction after a consummatory act (in *Gasterosteus aculeatus* L.). *Archives neérlandaises de Zoologie* **15**: 175–236.

SEYFARTH, R.M. & CHENEY, D.L. (1986). Vocal development in vervet monkeys. *Animal Behaviour* **34**: 1640–58.

SEYFARTH, R.M., CHENEY, D.L. & MARLER, P. (1980). Vervet monkey alarm calls: evidence of predator classification and semantic communication. *Science* **210**: 801–3.

SHAPIRO, D.Y. (1979). Social behavior group structure and the control of sex reversal in hermaphroditic fish. *Advances in the Study of Behavior* **10**: 43–102.

SHERRINGTON, C.S. (1906). *The Integrative Action of the Nervous System*. New York: Scribner's.

SHERRINGTON, C.S. (1917). Reflexes elicitable in the cat from pinna, vibrissae and jaws. *Journal of Physiology* **51**: 404–31.

SHERRY, D. (1985). Food storage by birds and mammals. *Advances in the Study of Behavior* **15**: 153–88.

SHERRY, D. & GALEF, B.G. (1984). Cultural transmission without imitation: milk bottle opening by birds. *Animal Behaviour* **32**: 937–8.

SHERRY, D., JACOBS, L.F. & GAULIN, S.J.C. (1992). Spatial memory and adaptive specialisation of the hippocampus. *Trends in the Neurosciences* **15**: 298–303.

SHERRY, D.F., MROSOVSKY, N. & HOGAN, J.A. (1980). Weight loss and anorexia during incubation in birds. *Journal of Comparative and Physiological Psychology* **94**: 89–98.

SHETTLEWORTH, S.J. (1983). Memory in food-hoarding birds. *Scientific American* **248**(3): 102–10.

SHETTLEWORTH, S.J. (1995). Comparative studies of memory in food-storing birds: from the field to the Skinner box. In *Behavioural Brain Research in Naturalistic and Semi-naturalistic Settings*, ed. E. Alleva, H.P. Lipp and L. Nadel, pp. 159–92. Dordrecht: Kluwer Academic Publishers.

SHORT, R.V. & BALABAN, E. (eds.) (1994). *The Differences Between the Sexes*. Cambridge: Cambridge University Press.

SIMMONS, J.A. & STEIN, R.A. (1980). Acoustic imaging in bat sonar: echolocating signals and the evolution of echolocation. *Journal of Comparative Physiology A* **135**: 61–84.

SIMPSON, S.J. & RAUBENHEIMER, D. (1993). A multi-level analysis of feeding behaviour: the geometry of nutritional decisions. *Philosophical Transactions of the Royal Society of London B* **342**: 381–402.

SKINNER, B.F. (1938). *The Behavior of Organisms*. New York: Appleton-Century-Crofts.

SLATER, P.J.B. & INCE, S.A. (1979). Cultural evolution in chaffinch song. *Behaviour* **71**: 146–66.

SMUTS, B.B., CHENEY, D.L., SEYFARTH, R.M., WRANGHAM, R.W. & STRUHSAKER, T.T. (eds.) (1987). *Primate Societies*. Chicago: University of Chicago Press.

SONNEMANN, P. & SJÖLANDER, S. (1977). Effects of cross fostering on the sexual imprinting of the female zebra finch *Taeniopygia guttata*. *Zeitschrift für Tierpsychologie* **45**: 337–48.

SPALDING, D. (1873). Instinct: with original observations on young animals. *Macmillan's Magazine* **27**:282–93. (Reprinted in *British Journal of Animal Behaviour* **2**: 1–11, 1954.)

SPARKS, J. (1982). *The Discovery of Animal Behaviour*. London: Collins, BBC.

STENT, G.S. (1980). The genetic approach to developmental neurobiology. *Trends in the Neurosciences* **3**: 49–51.

STENT, G. (1981). Genetics and the development of the nervous system. *Annual Review of Neuroscience* **4**: 163–94.

STEPHENS, D.W. & KREBS, J.R. (1986). *Foraging Theory*. Princeton, NJ: Princeton University Press.

STRUHSAKER, T.T. & LELAND, L. (1987). Colobines: infanticide by adult males. In *Primate Societies*, ed. B.B. Smuts, D.L. Cheney, R.M. Seyfarth, R.W. Wrangham and T.T. Struhsaker, pp. 38–97. Chicago: University of Chicago Press.

SUGAWARA, K. (1979). Sociobiological study of a wild group of hybrid baboons between *Papio anubis* and *P. hamadryas* in the Awash Valley, Ethiopia. *Primates* **20**: 21–56.

TEITELBAUM, P. (1955). Sensory control of hypothalamic hyperphagia. *Journal of Comparative and Physiological Psychology* **48**: 156–63.

TEITELBAUM, P. & EPSTEIN, A.N. (1962). The lateral hypothalamic syndrome: recovery of feeding and drinking after lateral hypothalamic lesions. *Psychological Review* **69**: 74–90.

TEN CATE, C. (1982). Behavioural differences between zebra finches and Bengalese finch (foster) parents raising zebra finch offspring. *Behaviour* **81**: 152–72.

TEN CATE, C., LOS, L. & SCHILPEROOD, L. (1984). The influence of differences in social experience on the development of species recognition in zebra finch males. *Animal Behaviour* **32**: 852–60.

THORPE, W.H. (1961). *Bird Song*. Cambridge: Cambridge University Press.

THORPE, W.H. (1963). *Learning and Instinct in Animals*, 2nd edition. London: Methuen.

TIEFER, L. (1978). The context and consequences of contemporary sex research: a feminist perspective. In *Sex and Behavior: Status and Prospectus*, ed. T.E. McGill, D.A. Dewsbury and B.D. Sachs, pp. 363–85. New York: Plenum.

TINBERGEN, N. (1951). *The Study of Instinct*. Oxford: Oxford University Press.

TINBERGEN, N. (1952). 'Derived' activities, their causation, biological significance, origin and emancipation during evolution. *Quarterly Review of Biology* **27**: 1–32.

TINBERGEN, N. (1959). Comparative studies of the behaviour of gulls (Laridae): a progress report. *Behaviour* **15**: 1–70.

TINBERGEN, N. (1963). On aims and methods of ethology. *Zeitschrift für Tierpsychologie* **20**: 410–33.

TINBERGEN, N. & PERDECK, A.C. (1950). On the stimulus situation releasing the begging response in the newly-hatched herring gull chick (*Larus a. argentatus* Pont). *Behaviour* **3**: 1–38.

TINBERGEN, N., BROEKHUYSEN, G.J., FEEKES, F., HOUGHTON, J.C., KRUUK, H. & SZUK, E. (1962). Eggshell removal by the black-headed gull *Larus ridibundus* L.; a behavioural component of camouflage. *Behaviour* **19**: 74–117.

TOATES, F. (1986). *Motivational Systems*. Cambridge: Cambridge University Press.

TOATES, F. (1995). *Stress. Conceptual and Biological Aspects*. Chichester: John Wiley.

TOLMAN, E.C. (1932). *Purposive Behavior in Animals and Men*. New York: The Century Co.

TOVÉE, M. (1995). Ultraviolet photoreceptors in the animal kingdom: their distribution and function. *Trends in Ecology and Evolution* **10**(11): 455–60.

TRIVERS, R.L. (1974). Parent–offspring conflict. *American Zoologist* **14**: 249–64.

TYNDALE-BISCOE, C.H. (1973). *The Life of Marsupials*. London: Edward Arnold.

UNGERLEIDER, L.G. (1995). Functional brain imaging studies of cortical mechanisms for memory. *Science* **270**: 769–75.

VAN RHIJN, J.G. (1980). Communication by agonistic displays: a discussion. *Behaviour* **74**: 284–93.

VAN SHAIK, C.P. (1983). Why are diurnal primates living in groups? *Behaviour* **87**: 120–44.

VAN SCHAIK, C.P. & VAN HOOF, J.A.R.A.M. (1996). Toward an understanding of the orangutan's social system. In *Great Ape Societies*, ed. W.C. McGrew, L.F. Marchant and T. Nishida, pp. 3–15. Cambridge: Cambridge University Press.

VERNER, J. & WILLSON, M.F. (1966). The influence of habitats on mating systems of North American passerine birds. *Ecology* **47**: 143–7.

VESTERGAARD, K. (1980). The regulation of dustbathing and other patterns in the laying hen: a Lorenzian approach. In *The Laying Hen and its Environment*, ed. R. Moss, pp. 101–20. The Hague: Martinus Nijhoff.

VIDAL, J.M. (1980). The relations between filial and sexual imprinting in the domestic fowl: effects of age and social experience. *Animal Behaviour* **28**: 880–91.

VINCE, M.A. (1969). Embryonic communication, respiration and the synchronization of hatching. In *Bird Vocalizations*, ed. R.A. Hinde, pp. 233–60. Oxford: Oxford University Press.

VINCE, M.A. (1993). Newborn lambs and their dams: the interaction that leads to suckling. *Advances in the Study of Behavior* **22**: 239–68.

VINES, G. (1981). Wolves in dog's clothing. *New Scientist* **91**: 648–52.

VOM SAAL, F.S. & BRONSON, F. (1980). Sexual characteristics of adult females correlates with their blood testosterone levels during development in mice. *Science* **208**: 597–9.

VON FRISCH, K. (1967). *The Dance Language and Orientation of Bees*. Cambridge, Mass.: The Belknap Press of Harvard University Press.

VOWLES, D.M. (1965). Maze learning and visual discrimination in the wood ant (*Formica rufa*). *British Journal of Psychology* **56**: 15–31.

WAAGE, J.K. (1979). Dual function of the damselfly penis: sperm removal and transfer. *Science* **203**: 916–18.

WALKER, S. (1983). *Animal Thought*. London: Routledge & Kegan Paul.

WALLRAFF, H.G. (1993). Correct and false olfactory information of homing pigeons as depending on geographical relationships between release site and home site. *Behavioral Ecology and Sociobiology* **32**: 147–55.

WALSH, E.G. (1964). *Physiology of the Nervous System*, 2nd edition. London: Longman.

WARNER, R.R. (1987). Female choice of sites versus mates in a coral reef fish *Thalassoma bifasciatum*. *Animal Behaviour* **41**: 375–82.

WARREN, J.M. (1965). Primate learning in comparative perspective. In *Behaviour of Non-human Primates*, Vol. I, ed. A.M. Schrier, H.F. Harlow and F. Stollnitz, pp. 249–81. New York: Academic Press.

WARREN, J.M. (1973). Learning in vertebrates. In *Comparative Psychology: A Modern Survey*, ed. D.A. Dewsbury and D.A. Rethlingshafer, pp. 471–509. New York: Academic Press.

WATSON, J.B. (1924). *Behaviorism*. Chicago: University of Chicago Press.

WEDEKIND, C., SEEBECK, T., BETTENS, F. & PAEPHE, A.J. (1995). MHC-dependent mate preferences in humans. *Proceedings of the Royal Society of London B* **260**: 245–59.

WELLS, M.J. (1958). Factors affecting reactions to *Mysis* by newly hatched *Sepia*. *Behaviour*. **13**: 96–111.

WELLS, M.J. (1962). Early learning in *Sepia*. *Symposia of the Zoological Society of London* **8**: 149–69.

WEST, M.J. & KING, A.P. (1988). Female visual displays affect the development of male song in the cowbird. *Nature* **334**: 244–6.

WHITEN, A. & HAM, R. (1992). On the nature and evolution of imitation in the animal kingdom: reappraisal of a century of research. *Advances in the Study of Behavior* **21**: 239–83.

WILCOX, R.S. (1979). Sex discrimination in *Gerris remigis*: role of a surface wave signal. *Science* **206**: 1325.

WILEY, R.H. (1973a). The strut display of male sage grouse: a 'fixed' action pattern. *Behaviour* **47**: 129–52.

WILEY, R.H. (1973b). Territoriality and non-random mating in sage grouse, *Centrocercus urophasianus*. *Animal Behaviour Monographs* **6**(2): 87–129.

WILKINSON, G.S. (1984). Reciprocal food sharing in the vampire bat. *Nature* **308**: 181–4.

WILSON, A.C. (1985). The molecular basis of evolution. *Scientific American* **253**(4): 164–73.

WILSON, E.O. (1965). Chemical communication in the social insects. *Science* **149**: 1064–71.

WILSON, E.O. (1971). *The Insect Societies*. Cambridge, Mass: Belknap Press of Harvard University Press.

WILSON, E.O. (1975). *Sociobiology*. Cambridge, Mass.: Belknap Press of Harvard University Press.

WILSON, J. (1992). A re-assessment of the significance of status signalling in populations of wild great tits, *Parus major*. *Animal Behaviour* **43**: 999–1009.

WILTSCHKO, W. & WILTSCHKO, R. (1988). Magnetic versus celestial orientation in migrating birds. *Trends in Ecology and Evolution* **3**: 13–15.

WILZ, K.J. (1970). Causal and functional analysis of dorsal pricking and nest activity in the courtship of the three-spined stickleback *Gasterosteus aculeatus*. *Animal Behaviour* **18**: 115–24.

WINE, J.J. & KRASNE, F.B. (1972). The organization of the escape behavior in the crayfish. *Journal of Experimental Biology* **56**: 1–18.

WINE, J.J. & KRASNE, F.B. (1982). The cellular organization of crayfish escape behavior. In *The Biology of Crustacea*, Vol. IV. *Neural Integration*, ed. D.E. Bliss, H. Atwood and D. Sandeman, pp. 2241–92. New York: Academic Press.

WINGFIELD, J.C., HEGNER, R.E., DUFFY, A.M., JR & BALL, G.F. (1990). The 'Challenge hypothesis': theoretical implications for patterns of testosterone secretion, mating systems, and breeding strategies. *American Naturalist* **136**: 829–46.

WINN, P. (1995). The lateral hypothalamus and motivated behavior: an old syndrome reassessed and a new perspective gained. *Current Directions in Psychological Science* **4**: 1182–7.

WINN, P., TARBUCK, A. & DUNNETT, S.B. (1984). Ibotenic acid lesions of the lateral hypothalamus: comparisons with the electrolytic lesion syndrome. *Neuroscience* **12**: 225–40.

WIRTSHAFTER, D. & DAVIS, J.D. (1977). Set points, settling points and the control of body weight. *Physiology and Behaviour* **19**: 75–8.

WOOLFENDEN, G.E. & FITZPATRICK, J.W. (1984). *The Florida Scrub Jay*. Princeton, NJ: Princeton University Press.

WRANGHAM, R.W. (1987). Evolution of social structure. In *Primate Societies*, ed. B.B. Smuts, D.L. Cheney, R.M. Seyfarth, R.W. Wrangham and T.T. Struhsaker, pp. 282–96. Chicago: University of Chicago Press.

WRANGHAM, R.W., MCGREW, W.C., DE WAAL, F.B.M. & HELTNE, P.G. (1994). *Chimpanzee Cultures*. Cambridge, Mass.: Harvard University Press.

YDENBERG, R. & DILL, L.M. (1986). The economics of fleeing from predators. *Advances in the Study of Behavior* **16**: 229.

YDENBERG, R. & HOUSTON, A.I. (1986). Optimal trade-offs between competing behavioural demands in the great tit. *Animal Behaviour* **34**: 1041–50.

YOUNG, D. (1989). *Nerve Cells and Animal Behaviour*. Cambridge: Cambridge University Press.

ZAHAVI, A. (1975). Mate selection – a selection for a handicap. *Journal of Theoretical Biology* **53**: 205–14.

ZAHAVI, A. (1987). The theory of signal selection and some of its implications. In *International Symposium on Biological Evolution*, ed. V.P. Delfino, pp. 305–27. Adriatica Editrice.

ZAHAVI, A. (1991). On the definition of sexual selection, Fisher's model, and the evolution of waste and of signals in general. *Animal Behaviour* **42**: 501–3.

ZARROW, M.X., DENBERG, V.H. & ANDERSON, C.O. (1965).
Rabbit: frequency of suckling in the pup. *Science* **150**: 1835–6.

ZEKI, S. (1993). *A Vision of the Brain*. Oxford: Blackwell
Scientific Publications.

ZUK, M. (1994). Immunology and the evolution of behavior.
In *Behavioral Mechanisms in Evolutionary Ecology*, ed. L.A.Real,
pp. 354–68. Chicago: University of Chicago Press.

Index

Page references in *italics* refer to figures or their captions; n refers to a marginal note.

DATE